Routledge Revivals

The Vital Science

In this title, first published in 1984, Peter Morton argues that in late Victorian Britain a group of novelists and essayists quite consciously sought and found ideas in post-Darwinian biology that were susceptible to imaginative transformation. The period between 1860 and 1900 was a time of great confusion in biology; the natural selection hypothesis was in retreat before its acute critics, and no extension of evolutionary theory to human affairs was too bizarre to attract its quota of enthusiasts. Writers capitalised on this prevailing uncertainty and used it to their own artistic or polemic ends. A fascinating and interdisciplinary title, this reissue will interest students of late Victorian literature, as well as historians of biological theory between *The Origin of Species* and Mendel.

The Vital Science

Biology and the Literary Imagination 1860-1900

Peter Morton

Routledge
Taylor & Francis Group

First published in 1984
by George Allen & Unwin Ltd

This edition first published in 2014 by Routledge
2 Park Square, Milton Park, Abingdon, Oxon, OX14 4RN
and by Routledge
711 Third Avenue, New York, NY 10017

Routledge is an imprint of the Taylor & Francis Group, an informa business

Publisher's Note
The publisher has gone to great lengths to ensure the quality of this reprint but points out that some imperfections in the original copies may be apparent.

Disclaimer
The publisher has made every effort to trace copyright holders and welcomes correspondence from those they have been unable to contact.

A Library of Congress record exists under LC control number: 84010991

ISBN 13: 978-1-138-79924-0 (hbk)
ISBN 13: 978-1-315-75613-4 (ebk)

The Vital Science

Biology and the Literary Imagination, 1860–1900

PETER MORTON

London
GEORGE ALLEN & UNWIN
Boston Sydney

**George Allen & Unwin (Publishers) Ltd,
40 Museum Street, London WC1A 1LU, UK**

George Allen & Unwin (Publishers) Ltd,
Park Lane, Hemel Hempstead, Herts HP2 4TE, UK

Allen & Unwin, Inc.,
Fifty Cross Street, Winchester, Mass. 01890, USA

George Allen & Unwin Australia Pty Ltd,
8 Napier Street, North Sydney, NSW 2060, Australia

First published in 1984

British Library Cataloguing in Publication Data

Morton, Peter
The vital science.
1. Darwin, Charles—Influence 2. English fiction—19th century—History and criticism 3. Evolution in literature
I. Title
823'.8'09 PR878.E9
ISBN 0-04-800025-6

Library of Congress Cataloging in Publication Data

Morton, Peter.
The vital science.
Bibliography: p.
Includes index.
1. English literature—19th century—History and criticism. 2. Literature and science. 3. Biology—History. I. Title.
PR468.S34M67 1984 820'.9'356 84-10991
ISBN 0-04-800025-6 (alk. paper)

Set in 10 on 11 point Plantin by Grove Graphics, Tring, Hertfordshire
and printed in Great Britain
by Billing and Sons Ltd, London and Worcester

For Jamie

Contents

Acknowledgements

Any book such as this, whose argument has been fashioned over several years and constantly adjusted in the light of further reading and discussion, must partially lose sight of its origins. I readily admit that in some cases I have forgotten how a particular line of thought developed; and much as I would like to I cannot now acknowledge all my intellectual debts. However, I would particularly like to thank Professor Gene LeMire of the School of Humanities, Flinders University, for some stimulating suggestions when the book was in embryo, and Professor Milton Millhauser of the University of Bridgeport for helping me to avoid some silly mistakes later on.

Thanks for help over particular points are due to: Professor Darko Suvin of McGill University and Professor Norman Mackenzie of the University of Sussex for information on Wells's biological interests and his textbooks of biology respectively; to Mr W. B. Hodgson of Sotheby's, London, Professor John Laird of the University of New South Wales and the Curator of the Dorset County Museum for information about Hardy's personal library and manuscripts; and to Mr J. R. Payne of Austin, Texas, for information on the variant editions of Hudson's *A Crystal Age*.

I would also like to thank Professor T. Craik of Durham University for making some of the resources of the English Department available to me, Gareth Reeves and his family for their hospitality, and the university itself for its generosity in making available a very pleasant house for some months. Most of the final draft of this book was written at Durham. The extraordinarily efficient resources of the California State University's library system made the last-revision stages fairly painless.

Like most married writers I owe my wife a considerable debt. Far too busy with her own career even to read the manuscript let alone type it, prepare the index, or whatever, she mainly restricted her help to the sphere of the financial. I am pleased to be able to record my appreciation now.

PETER MORTON

December 1983 — Adelaide, South Australia

The Vital Science

Introduction:
Definitions and Perspectives

> In the book of nature are written . . . the triumphs of survival, the tragedy of death and extinction, the tragi-comedy of degradation and inheritance, the gruesome lesson of parasitism, and the political satire of colonial organisms. Zoology is, indeed, a philosophy and a literature to those who can read its symbols.
>
> H. G. WELLS (1893)

'The history of the Victorian Age will never be written: we know too much about it . . . Ignorance is the first requisite of the historian.' So runs the glittering epigram which opens *Eminent Victorians*, and it supplied Lytton Strachey with all the justification he needed to abandon historiography for iconoclasm – with what successful result, everyone knows. Strachey's method, however, is a distinctly limited one; for our urge to understand that period, so close in time to us yet so remote in mood, cannot be fully assuaged by detached irony. And yet the historian of ideas, and especially the historian interested in the infiltration of scientific ideas into the wider cultural life, soon comes up against that shrewd paradox of Strachey's. There is a real sense in which we *do* know too much for comfort. The Darwinian revolution is an excellent case in point. About its transcendent importance no one today is likely to quarrel: Darwin has speedily taken his place alongside Copernicus and Newton in that small pantheon of scientists whose work has totally disrupted long-established habits of mind. But while the bones of Newton have been resting in Westminster Abbey for two and a half centuries, Darwin's neighbouring remains were placed there barely one long lifetime ago. The profound revelations he supplied are simply too close for us to get any real perspective on them in order to assess their permanent significance.

Let us suppose some historian were reckless enough to try to write a comprehensive account of the cultural impact of Darwinian biology just as far as the beginning of this century. At once Strachey's complaint would obtrude itself. Such a monumental project would have to range freely over the surrounding sciences, but it would also have to take notice of the contemporary theology, social theory, economics and various turn-of-the-century metaphysical systems like theosophy, Christian Science and spiritualism. Our historian would be obliged to investigate Henry Maine on the 'evolution' of the law, Walter Bagehot on sociology, and W. H. Mallock on conservative political theory. He would have to trace the strong influence of evolutionary notions on Victorian literary history and criticism.

For J. A. Symonds found in the central concepts of *The Origin of Species* usable analogies, a handy vocabulary and a typology for his critical essays; Walter Pater found biological relativism could be pleasingly extended to justify cultural and aesthetic relativism, too; Leslie Stephen found that natural selection could also make comprehensible the rise and fall of literary genres. Then, delving down into popular culture, our historian would be obliged to scrabble through the numerous ephemeral squibs and lampoons which were commonplace in the first post-*Origin* decades; he would have to look at dull set-speeches on evolution in forgotten novels of weakly satirical cast; he would assess the accuracy of pedagogical novels like Verne's *Voyage au centre de la terre* (1864) where the professor-hero and his nephew read off the story of life from the layers of rock as they descend into the earth.

Above all, he would not be able to refrain from a prolonged discussion of Social Darwinism. This politico-philosophical movement, a powerful though vague undercurrent in late-Victorian thought, is usually thought to have been a strong prop of the forces of reaction. Certainly on the extreme Right, as this found expression in certain of Nietzsche's dicta or in the violent doctrines of William Sumner (1840–1910), natural selection was indeed used to justify the gross inequalities of early industrial society. And yet at the same time certain other aggressively illiberal Social Darwinists professed themselves committed socialists, as did Karl Pearson (1857–1936). Even more taxing is the problem of accommodating the conservative Liberalism of Walter Bagehot's *Physics and Politics, or Thoughts on the Application of the Principles of 'Natural Selection' and 'Inheritance' to Political Society* (1872); or the attempted merging of Darwinism and utilitarianism in Leslie Stephen's *The Science of Ethics* (1882); or of Darwinism and social democracy in Benjamin Kidd's *Social Evolution* (1894); or of Darwinism and anarchism in Petr Kropotkin's *Mutual Aid: A Factor of Evolution* (1902). As the very titles of these books suggest, Bagehot and Kropotkin, as well as many others of intermediate shades of opinion, all insisted vociferously that their extrapolation of Darwinism was the legitimate one. All of them were anxious to assert that only *their* political prejudices were the ones licensed by the *Origin*. Those on the moderate Left who were interested in biologically justified social theory were slower to assert themselves, but when they did so one result was an invigorated denial of man's right to his dominion over all living things. The Humanitarian League, which was founded by Henry Stephens Salt (1851–1939) and other Fabians and which was active from 1891 until 1920, supported its twin causes of vegetarianism and anti-vivisection with quotations from *The Descent of Man*. Vegetarianism was defended with new anthropological data on the diet of primitive man and new physiological findings about the human digestive tract. In the more emotional utterances of the time the carnivore is practically identified with the cannibal; and in some speculative fiction like H. G. Wells's 'A Story of the Days to Come' (1897) the dietary habits of the remote past (our present) are closely linked to implications of savagery and bestial conflict.

We are told with some gusto that bland jellies and lucent curds have replaced 'the still recognisable fragments of recently killed animals, hideously charred and hacked' – though it is stressed that in more subtle ways the reformed future society is no less Darwinian for all its dainty eating. In a Utopia of 1882, *A Thousand Years Hence* by 'Nunsowe Green', natural agriculture has been replaced by 'laboratorial organic production' whereby human corpses are decomposed and converted directly to food. The question here about what the most 'natural' method might be of responding to population pressure, and hence to the pressures of Darwinian selection (should it be by technologically assisted food-cycles, or by a return to the alleged nut and fruit diet of our pre-human ancestors?) is a trivial example of the fluid interpretation of biological information at the time. How can one judge between these competing Social Darwinisms? Our historian would not be able to evade this difficult question. Indeed, even before he could start work with a clear conscience on his main task he would have to explain how it happened that even the pre-Darwinian biology of the 1850s had already begun to influence distant areas of thought. To affect historical studies, for instance: the first volume of Henry Thomas Buckle's *History of Civilisation in England* (1857) displays its author's fascination with the part played by the forces of inheritance in the evolving pattern of events; and though the general conclusion Buckle reaches is that such speculations are premature and that historians will have to wait on the biologists' uncovering of general laws, he did not doubt that this would be accomplished within a few generations. The *Origin* gave focus to these tentative ambitions and, less than twenty years after Buckle, the naturalist, explorer and novelist William Winwood Reade tried his hand at a fully fledged and widely read Darwinian 'biohistory', *The Martyrdom of Man* (1872).

The present book has a rather limited ambition compared to the grand imaginary *opus magnum* sketched above. Its aim is much more modest: simply to detail the impact of certain selected topics within post-Darwinian biology on the late-Victorian literary imagination. Surprisingly, perhaps, these interrelationships have not yet attracted the detailed attention they deserve. One indicator of this emerged from the centenary celebrations of the publication of the *Origin* in 1959, a year seeing the appearance of work which does go some way towards giving a comprehensive overview of Darwin's cultural impact. We were told much of Darwin's place in philosophy, in Christianity, ecology and genetics; but his effect on the literary imagination received the sparsest treatment. Two years later, a series of lectures with claimed to give 'the range and general tendency of Darwin's intellectual influence' took as its headings Darwinism and the Bible, and natural theology, and social science – not literature.[1] To be sure, it was Darwin himself who set something of a precedent here, for when at the end of the *Origin* he gave a list of those disciplines he thought most likely to be affected by his theory he did not mention the humanities at all. Ever since, critics and historians have been fundamentally divided in their interpretation of Darwinism *vis-à-vis* the literary milieu. Do, for instance, Darwin's own writings have any claim on our attention as literature in their own right? Judgements here have ranged all the way from dismissive to adulatory. Gertrude Himmelfarb, whose much-criticised *Darwin and the Darwinian Revolution* was published in the centennial year, has hammered away at Darwin's sloppy semantics and the ineptitude of his style – which criticisms were indeed commonplace among Darwin's contemporary supporters, notably T. H. Huxley. Yet, only ten years after Himmelfarb, Michael Ghiselin in his *The Triumph of the Darwinian Method* may be found urging that Darwin in his book on the fertilisation of orchids composed nothing less than a 'metaphysical satire' or a 'biological *Candide*' which is both a literary masterpiece and a century ahead of its time in its philosophical handling of the 'semantics of design'. So was Darwin a clumsy stylist hardly capable of a clear sentence or a witty satirical essayist, the Victorian Voltaire? He could not have been both. The difficulty comes in knowing what kind of evidence in proof would be acceptable to both sides.

As for Darwin's broader influence on the literary imagination, there has been no agreement at all – neither about the nature of that influence nor even (in the extreme view) whether it is to be found at all. Some have asserted that Darwin's self-recognised plight of being jammed between his personal liberal optimism and the ferocious logic of Darwinism generated in others a sort of rueful mirth. Dwight Culler has claimed comedy to be the literary form which most vividly reflects the reversals in evolutionary thought, since randomness is as vital a constituent of comedy as determinism is of tragedy. 'Some found this tragic and moaned about "nature red in tooth and claw",' he has written, 'but the younger generation found it funny and breathed a sigh of relief.'[2] For Culler the most Darwinian literature of all is that work by Lear, Wilde and Carroll – and especially the Alice stories of 1865 and 1871 – which dissolves away nearly all formal artistic structure, leaving the crazy disorder of pure farce. It goes without saying that this is totally opposed to the more traditional view that the most relevant literary genre, the one most fitted in its conception of human destiny to transform into art the biologists' new image of man, is tragedy. In this view the malignancies of nature revealed in *King Lear* are just what might be expected of an organic universe which, as James Sully put it in 1881 in a *Nineteenth Century* article, is 'always and everywhere the worst possible – meaning that it contains as much of weakly suffering life as the conditions allow of'. Over the intervening century there has been no lack of proponents of Sully's mournful view. Ifor Evans, in *Literature and Science* (1954), took the position that no honest writer since the *Origin* has been able to read that work aright and find it funny; that a Darwinian blight has spread up to our own day and has indeed become endemic, with the consequence that 'from this sense of fear and depression in relationship to biological science the imaginative artist has never fully recovered'.[3] In recent years Stanley Hyman has gone even farther than this. In studying Darwin's own writings 'as imaginative organisations, as though they were poems', he finds that the *Origin*'s deep structure is itself that of dramatic tragedy, and that it is relevant to apply the Aristotelian definitions of *agon* and *sparagmos, anagnorisis* and *epiphany* to the driving logic of its argument. And the tragedy enacted in the *Origin* is said to be Darwin's own: the Oedipal fantasy or 'compensatory mother worship' of a man who, too civilised to murder his own hated father, by a cathartic act drove the Heavenly Father from the sphere of natural history.[4] (As is usual in this kind of analysis, the fact that Darwin's writings express nothing but admiration tinged with awe for his father is conclusive evidence that he was repressing his loathing of him.) For those who find this psychoanalytic ingenuity forced, there is as an alternative the brisk scepticism of the historian Walter Cannon. He adamantly refuses to see the *Origin*, or even biological prose generally, as having any rhetorical structure worth noticing or using figurative language at all creatively. For him the *Origin* is a dull monograph whose amateurish metaphors (a tree, a bank, a struggle, a chain, a beehive) are superfluous decoration on a structure built out of the massive freestone blocks of documented fact. These figures of speech, he says, are no more than 'available verbalisms' and Darwinism itself profoundly anti-literary because it proves conclusively in its style and its freedom from moralistic assumptions that science has outgrown all the ancient resources of myth, ritual and drama. No single artist, either contemporary or modern, can hope to master its complex vision, even though misguided writers have done their best to deal with the irritant of Darwin's elegant demonstration by coating it with a pearly and spurious optimistic evolutionism.

Important though the endeavour is, it is hard to imagine how such yawning gulfs between interpretations could ever be bridged. It is worth stressing, however, the strongly historical perspective of Cannon and those who likewise offer us Darwin the plain-speaking, mundane scientist. It just will not do to take insufficient notice, or no notice at all, of the powerful mythopoeic capacities of the nineteenth-century biologists themselves, or of the imaginative suggestiveness of issues within biology apart from the central evolutionary debate. Indeed, since Walter Myers's unhappily forgotten *The Later Realism* (1927) was written, it has hardly been recognised that there were any such issues coming within writers' purview. In his book, by striking out along a most original path of argument, Myers tried to explain how shifts in the technique of characterisation in the novel may to some extent be accounted for by the emergence of new problems in biology – and certainly not only in evolutionary biology. He found connections, for instance, between the attitude to eccentric personality in the great Victorian novelists (deviations from the social norm are, in Meredith and Eliot, said to be treated as amusing or bizarre) and the Darwinian insistence that conformity in a static environment has survival value. By the end of the century, Myers recognises, accounting for mutation and variation has become the crucial biological problem – a problem which, he says, 'enlarged the conception of heredity to a roominess which includes and accounts for wide deviations from the type'.[5] The effect on novelists' technique was to make a more diffuse presentation acceptable. In support of this thesis Myers goes on to propose that Meredith's *The Egoist* (1879) and Darwin's *The Expression of the Emotions in Man and Animals* (1872) belong to one distinct phase in the history of ideas, and D. H. Lawrence's *The Lost Girl* (1920) and Hugo de Vries's lectures on *Species and Varieties* (1905) to quite another. While interesting enough, the tracing of such obscure connections places great pressure on the assumption that there is an unavoidable pool of scientific notions into which a novelist is, willy-nilly, plunged; for how else can any systematic parallel be drawn between the severely technical Vries and the avowedly anti-scientific Lawrence? Possibly there is such a shared milieu, but this study does not attempt to explore it: while readily accepting Myers's conviction that quite complex biological ideas were of literary significance in this period, its concern is restricted to those writers who came to grips consciously and deliberately with such ideas. Our argument will be that in late-Victorian Britain a group of novelists and essayists (not poets: the part played by Darwinism in Victorian poetry poses different questions which have been answered in different ways in several extensive studies) thoroughly searched among the data of the life sciences and found there material peculiarly susceptible to imaginative transformation. They were led to make this search, and were demonstrably successful in it, because during the few decades which elapsed between the publication of the *Origin* and the foundation of Mendelian genetics around the turn of the century evolutionary biology was in a state of extraordinary confusion and ambiguity, and a wide range of writers were able to exploit the science for their own aesthetic or polemic ends. We will expand, define and illustrate this claim by looking in successive chapters at certain biological 'unit-ideas' (in A. O. Lovejoy's phrase) which in this span of forty years were of extreme interest to the theoretical scientist and man of letters alike.

Darwin's Literary Impact: the Dominant View

One of the most thoughtful contributions to the celebration of the *Origin*'s centenary was Morse Peckham's tightly knit essay 'Darwinism and Darwinisticism'. The problem Peckham addresses springs from his warning that 'the locus of cultural history is covert behaviour' and that our understanding of the way in which any new idea is assimilated into society is woefully limited. But Peckham certainly intended his essay to sound more than a monitory note. He was calling for action:

> when one tries to tally up the writers affected and to list the books and make an inventory of passages showing Darwinian influence and

> Darwinian assumptions in novels, poems, and essays, a fog seems to arise in one's mind, through which are discernible twinkles of what may or may not be bits of genuine Darwinism.[6]

We can assuredly sympathise with Peckham's difficulties of recognition, but for present purposes it is important to notice the assumption that real Darwinism is, or should be, a clearly discernible entity: something which is 'genuine'; which can have an unequivocal 'influence' and give rise to 'assumptions'. Peckham's argument moves forward from the position that 'Darwinism' is a clear-cut term in the history of biology and was so from the beginning, and that 'Darwinisticism' is related to this in much the same way that phrenology is related to neurology – or a fake science to a real one. In principle it is possible – and ought to be so in practice – to disperse the metaphorical fog and reveal the actual, precise quantity of Darwinism utilised in any given sample of Victorian prose; to dismiss the 'Darwinistic' or quasi-biological as merely spurious; to explain how (to come down to particulars) Thomas Hardy read Darwin properly and is therefore a Darwinian writer while Samuel Butler did not and is not. In this respect Peckham draws together and makes quite explicit the presumption of earlier historians – the studies by Lionel Stevenson (1932), Leo Henkin (1940) and Georg Roppen (1956) being paramount – that there need be little difficulty in making clear the relevant intellectual background. The nature of Darwin's achievement has seemed unambiguous enough, and the range of possible literary reactions which are true to its spirit quite circumscribed. Without, one hopes, doing too much violence to what are very different books, the consensus among literary historians may perhaps be described in the following terms.

There has been a very general recognition that evolution as such was an ancient doctrine, teaching the essential continuity of all life and the hierarchical relation (or, as some pre-Darwinian speculators had even insisted, the transformation) of lowlier forms to higher. Darwin's achievement was to fit out this venerable doctrine with the new motor of natural selection which, when rigorously considered – as it rarely was, in practice – dispensed with the need to hypothesise any kind of supernatural creative Power. The rise of observational astronomy had already brought home painfully the loneliness of the earth and the minuteness of the habitual human sphere of action. But Darwinian biology went much further in lowering man's prestige. Was it not therefore inevitable that its only emotional effect must be to humble his pride almost unbearably? For the *Origin* tore away man's image of himself as a creature of divine fiat, set by God's deliberate choice on a rung of the ladder of organic being – a little below the angels, to be sure, but many rungs above the beasts. But man now stood revealed as the latest and temporary product of nature's assembly-line, a line which ran all the way back to the primordial ooze, and beyond that to a condensing gas-cloud on the rim of an insignificant galaxy. Nowhere, except in his own imperfect and presumably still evolving brain, could post-Darwinian man discover any evidence of mind at work, of goal-directed processes. In the face of this discovery there was, as the nineteenth century ran towards its close, a flight, best demonstrated and documented by literary writers, into an annihilating scepticism based not on the superficial and easily detected untrustworthiness of the senses, but on that of the brain itself. Darwin himself was affected to some degree, for in certain black moods he was struck by the appalling thought that even the thesis of the *Origin* – rock-solid though it seemed to be – was surely undercut by being the dubious product of an upstart monkey's brain. Why should a brain evolved by blind forces for the purpose of gathering fruit more efficiently be able to demonstrate the grandest law of organic nature? Such a degree of cynicism even Diogenes might have flinched from.

Turning now to the impact of Darwinism on the creative imagination, there can in this reading be only one truly informed response. It is that to be found in the novels and stories written after 1859 where those who are pleased to think themselves free from self-delusion strike the appropriate if rather self-conscious pose:

> From that time be began to show symptoms of insanity . . . He dressed always in black, and said that he was in mourning for mankind. The works of Malthus and Darwin, bound in sombre covers, were placed on a table in his room; the first was lettered outside *The Book of Doubt* and the second *The Book of Despair*. He took long solitary walks in the most secluded parts of the estate, and was sometimes seen gesticulating to the heavens, sometimes seated by the wayside plucking grass and casting it from him with a strange, tremulous movement of the hands.

Arthur Elliott, hero of *The Outcast* (1875) by the fervent Darwinian Winwood Reade, hangs himself after his eccentric bibliographical exercise because he is unwilling to live with his nightmare vision of the world as a botched dramatic production by a young and inexperienced playwright, which has been created with a 'want of symmetry in form and of finish in detail, a prodigal waste of raw material, a roughness of style and execution which bear the stamp of inexperience'. Three years later in 1878 one of the members of Robert Louis Stevenson's Suicide Club wants to die as quickly as possible not because of remorse over some vile crime, but just because he had been 'induced to believe in Mr. Darwin'. One may concede that these fictional melancholiacs took an extreme, even a histrionic, line; but, by any honest reading of the inescapable human import of Darwinism, should not their response be considered logical enough – even as being inevitable when the argument is stripped down to its core? For the *Origin*'s message is surely that the greater number of organisms is not destined to survive long and is of no intrinsic value. Suicide or at least the blackest depression might well be thought to be the only rational response, anything less being the futile gesture of the ostrich. And yet, even if the fullest weight is given to the biological pessimism of Hardy and Tennyson and Reade, it can hardly be claimed that the self-proclaimed Darwinians of the later-Victorian literary world were remarkable for their *Origin*-induced despair. On the contrary: those who were accepted most readily as expressing the spirit of the age – Meredith, Swinburne, Browning and Kingsley come to mind at once – certainly believed, each in his own fashion, that he could be an individualist, a strenuous optimist and a Darwinian simultaneously. Were they mistaken? Surely, the argument runs, they were. To revert to Morse Peckham's terminology, the literary employment of biology by this majority should be termed 'Darwinistic' (or, crudely, erroneous) and not 'Darwinian' at all in the strict sense so clearly established, allegedly, by Darwin himself. There is simply no optimism to be sucked out of the *Origin* when read attentively. Any claim to the contrary, no matter how ingenious, is merely individual and cultural self-delusion, and could have been exposed as such in its own day.

Such, stated as baldly as possible, is the argument of the most elaborate accounts of the literary response to biological controversy in the post-Darwinian period. None considers it plausible that any writer of this generation might have been influenced decisively by the minutiae of biological investigation. Where praise is given to a writer for his biological insight, it is likely to be because he anticipated and exploited those lines of thought that have since proven to be conceptually fruitful. Thus Conway Zirkle, a historian of biology, dislikes the work of Butler and Shaw because it is avowedly anti-Darwinian propaganda; on the other hand, he admires Kipling's *Just So Stories* (1902) because he finds them demonstrating with excellent humour that Lamarckian inheritance 'really belongs in the entertaining myths of childhood' and therefore neither on aesthetic nor scientific grounds need be taken seriously by adult sensibilities.[7] He does not for a moment find it relevant to inquire into Kipling's detailed knowledge, if any, of the pre-Mendelian theories of inheritance of his day. The same attitude pervades the structure of two standard works treating of the literary effects of the Darwinian revolution: Stevenson and Henkin share a simple model of the response to Darwinism which is one of martial conflict; that it was a question of a long-fought struggle extending over several decades but ending with the decisive triumph of reason. They take it for granted that what they are doing is exhuming a short-lived and even a rather insignificant episode in literary history in which the dominant movement is that of writers coming to terms with certain uncongenial facts whose status as facts was not to be questioned. And, the battle being won, there was a very rapid erosion of literary interest. Henkin writes:

> by 1910 the great controversy which had raged around the name of

> Darwin had long subsided. Evolution, then, had invaded every branch of science, ethics, philosophy, and sociology. As such it had entered so generally into the warp and woof of modern thought as to be indistinguishable as an independent factor.[8]

This has the same tone of one conducting an autopsy employed by Stevenson eight years earlier:

> by the nineties the period of evolutionary excitement in English poetry was at an end . . . the theory of evolution had ceased to be an intellectual cataclysm and had assumed its place in the general perspective of knowledge, taken for granted as the prevalent hypothesis, and not of sufficient importance to worry the non-scientific mind.[9]

Here the assumption is explicit that there is literature only in confrontation and conflict, in 'worrying' hypotheses. Henkin identifies this scientific battle with 'evolution' *tout court*, and likewise presupposes the Darwinian debate to have been a single entity waxing and then slowly waning in its power to inspire the literary mind. In Henkin's case the martial model then controls the literary judgements (or, more precisely, the selection of texts) because he places his emphasis on what he takes to be the increasing impurity, with the passage of time, of the literary response to the substantive scientific issues. He goes on to describe the concomitant growth of merely distorted, weird and even preposterous uses of 'evolution' in a string of mediocre novels and, eventually, in comic strips, where the last echoes of controversy at last faded into silence. But Darwinism is not of course a synonym for evolution but one explanation (out of several, readily defensible, current at the time) for it; and no notice at all is taken of the fact that, for instance, the concern of the Naturalistic school with heredity and environment all initially stemmed from Darwin's work but that each quickly branched off with a life – a literary life – of its own.

Setting for a moment our width of focus to its maximum, we seem to find here a belief that the Victorian life sciences and the contemporary literature occupied quite separate spheres. Sometimes this leads on to a demonstration of imaginative writers' intellectual inadequacies, and sometimes into the opposite defiant assertion of the primacy of literary values. Constant, however, is the assumption that the two spheres could communicate only in the very crudest terms; by means, in fact, of a set of clichés. What is more, this assertion tends to reinforce itself. Henkin, when he comes to deal with Thomas Hardy, whose imaginative response to certain quite technical problems of inheritance and variation was (as we shall see in a later chapter) really extraordinarily complex, limits himself to a few brief comments on the romances *A Pair of Blue Eyes* (1873) and *Two on a Tower* (1882). These novels are far from being Hardy's best work; as usually happened when he was not on form he laid the scientific interest over a feeble story and failed to integrate the two. Henkin, by choosing to discuss the mediocre, inevitably concludes that the quality of response to biology, even with a master like Hardy, was mediocre too. We must give full weight to the difficulty of Henkin's task. In our own day we have seen the triumph of Darwinism in its most pristine form, thanks to that 'modern synthesis' (Julian Huxley's phrase) which has gradually emerged from a further century of work by geneticists, morphologists and palaeontologists. It is most difficult to recreate a period when the evidence was shifting, uncertain, and – a most crucial point this, one which will be constantly in view in the succeeding pages – the emotional pressure of speculation very much higher than today.

Only a much closer reading into the strong Victorian tradition of *haute vulgarisation* in biology can begin to open up the curious territory of what might be called the cultural history of a science as opposed to its technical history – a territory where the real data are not the ideas constantly under discussion but the unquestioned assumptions and modes of thought. In particular, some interesting and subtle connections begin to be revealed between the creative writer and the biologist. To trace these connections it is necessary to turn away from Peckham's division – how much Darwinism? how much Darwinisticism? – to focus instead on the degree of creative usefulness of various plausible theories as these appeared to the writer and his contemporaries scientific and literary. Thinking in this direction begins to give us some purchase on such apparently intractable questions as the absolute neglect of the sensational claims of post-Darwinian biology by certain thinkers who were in an excellent position to appreciate their significance. We need to keep constantly before us the fact that even the new biology had no inescapable message for every literate Victorian. For very many of the most eminent and most aware there was no message at all. There had in every case to be some predisposition, some preliminary awareness that it was the data of *this* science that could be a critical component in the formation of his personal vision. If that awareness was missing – and it is to be found in relatively few writers of that fertile time – the literary artist was no less likely than any of his semi-literate brethren to be quite satisfied with the vulgar classification of Darwin's work as 'the ape theory' and Huxley's essays as 'Huxley on Atheism'. Unless the writer could see exciting implications cached within his biological reading he would not respond; or perhaps that reading would simply become one prop of many to broad feelings of hope or despair, and appeal be made to it without real understanding and often only for dramatic effect. But, given that all-important predisposition (and it is to be found most typically where there was a close interconnection of literary and scientific interests and often emerged from some training, formal or informal, in biology), then the uses to which biological information might be put were very various indeed, for ambiguities lay at the heart of most of the problems which the biologists were then grappling with. For the predisposed there was available a great jumbled repository of fact and speculation to inspire the artist.

In some cases the liaison was not made directly between the creative writer and the biologist. The flow of information went first through an interpretative middleman, a populariser, who was indeed sometimes one or other of the parties in a slightly different role. The part played by such middlemen has been much neglected in the past, for it requires an investigation of the wide and vague region of expository and polemic prose in the often uncharted wilderness of the Victorian periodicals. As Alvar Ellegård has shown in great detail, it was in the review article that the most important responses to Darwinism were first made explicit, and these articles collectively constitute the best index of the shifts in educated public opinion on this subject. Indeed, Ellegård has expressed his suspicions of more directly literary responses to biological controversy – those of, say, novelists, essayists, reviewers – in no uncertain terms, on the ground that the mirror they hold up to scientific history is bound to be a distorting one. Such responses are not to be trusted, he feels, because 'most of these people have no means of finding out the opinion of more than a small circle, and they are seldom aware of the danger of generalising from their limited experience. In the second place, they are seldom disinterested seekers after truth.'[10] Ellegård is writing as a scientific historiographer, and his strictures are right and proper. A study of the fate of ideas in literary hands must, however, investigate long-exploded error just as reverently as the truth, and take as its texts those writings which lie well on the subjective side of the relatively public knowledge that is science. Huxley's 'lay sermons' and the close-knit articles directed at the concerned layman by Alfred Russel Wallace and H. G. Wells; the 'popular essays and addresses' of a dozen scientific sages; the lesser productions of popularisers now utterly forgotten: all of these helped to establish the human interest of the new biology in a way not markedly different from the fictional writers, and are just as legitimate objects of historical scrutiny. Such publicists were often researchers or novelists (or both) in their own right. They took most seriously, though never heavily, their obligation – and it was in tune with the times that they often perceived it to be their moral obligation – to offer the public at large their interpretation of biological research. To do so effectively, these higher journalists mastered all the skills of literature. At their best they not only impinged upon the literary culture but were also integral with it.

Final Definitions and Limitations

Any study dealing with the cultural impact of biology in the nineteenth century cannot but reckon with the fact that in this period the life sciences were much more loosely organised than they are today. The very word 'biology' went through several shades of meaning after being coined at the beginning of the century by Gottfried Reinhold. In its English form the word first meant the study of *human* life exclusively and its modern meaning did not gain general currency until the 1850s. As for its subject-matter, information which would now be firmly classified as being within the province of the animal psychologist or ethologist tended to find its way into morphological or entomological texts – and much of it would be rejected altogether today as a mishmash of anecdote and sentimental observation. Should, for example, anthropology be regarded as a discipline distinct from biology when it comes to making out its uses in literature? Certainly in some of its sub-disciplines (physical anthropology, anthropometry) Victorian anthropology is hardly to be distinguished from pseudo-biological systems like phrenology, and even at its most lucid it tended to be used as an adjunct to evolutionary speculation. Darwin himself drew on a massive array of evidence supplied by field anthropologists in *The Descent of Man*, and he was largely uncritical of those workers' penchant for manufacturing whole series of normative judgements based on such data as the cranial capacities of different races. Victorian sociology drew heavily on biology and anthropology, too, as John C. Greene has demonstrated, though these three disciplines were rendered at least formally distinct with the foundation of the Anthropological Society of London in 1863.[11] The appeal of anthropology for the lay writer and reader is well marked in the attention given to it by popular novelists. The pioneering studies of alien cultures gave rise almost at once to what has been called the romance of the clash of races – such romances typically offering the reader a protagonist who reverts dramatically to neanderthalism or displays unexpected atavistic tendencies. An example of the genre is Grant Allen's story 'The Reverend John Creery' (*c.*1880). An African savage, brought to England and educated at Oxford, returns to his homeland as a missionary accompanied by his English wife. As time passes she has to endure his horrifying relapse into savagery, the message being that acquired habits of mind must break down before the urgent impulses of the blood. Actually, Grant Allen's handling of the familiar theme is a little more subtle than that, because although he is generally confident in his placement of the lower races he does oscillate somewhat in his convictions about which conditioning is best for a given environment. This very awareness that even the most mundane habits of life *are* cultural conditioning derives from the relativism which anthropology fostered in the 1870s and later. Fortunately, Brian Street's *The Savage in Literature* (1975), which covers the period 1858–1920, is thorough enough for the literary concern with anthropology to be set on one side here. Certainly biology, anthropology and sociology were recognised as entirely distinct disciplines by the chosen terminal date.

Though the opening and closing dates of this study may seem a little arbitrary – is there, after all, any such thing as 'late-Victorian' biology? does progress in a science obey the limits of a reign? – they are not actually so. Since the immediately preceding ground has been well mapped out,[12] the natural starting-point is the publication of *The Origin of Species* on 26

November 1859 – the event which we will assume to have been the most divisive one in 'Victorian' cultural history. And, oddly enough, at the other end of our period the years 1900 and 1901, though not indeed the end of an era, did see biology move into a new phase of growth after a time of stagnation. Two of the foremost and most persuasive of the anti-Darwinians, St George Mivart and George Douglas Campbell, eighth Duke of Argyll, both died in 1900 – which was also the year of the rediscovery of Mendel's crucial work on the laws of inheritance by Vries, Correns and Tschermak von Seysenegg and their rapid confirmation of those laws. Though not himself an evolutionist, Mendel established experimentally the exact nature of the raw material on which natural selection worked, thereby removing at a stroke the overwhelming chief technical objection to orthodox Darwinism. Even with our rather short perspective on these events, we can see clearly enough that Mendelism in the long run closed down the enduring controversy about the purposive versus the non-purposive nature of the evolutionary process. In 1901 the founding of the journal *Biometrika* under the editorship of Karl Pearson and Francis Galton began the empirical study of human variation. This was the very first methodical – and certainly long overdue – attempt to gather a wide range of data from the population at large, and was supposed to bring some rigour to *ex cathedra* pronouncements on eugenics – something which will be considered, in relation to the parallel imaginative products, in the fifth chapter. In fact, the net effect of the long-term investigations of the biometricians eventually spelled out the finish to Victorian eugenics as first established by Galton, by demonstrating just how naîve his attitude had been to the inheritance of variations and the possible human interference in these processes.

Though it is not so clearly marked, the few years around 1900 also form a natural divide as far as the imaginative response is concerned. The latest work to be considered in detail in these pages is Samuel Butler's *The Way of All Flesh*, published in 1903. About this time (it is impossible to date it precisely) the increased complexity of biological research at last dissolved the very strong sentiment that the life sciences were too important to be left to the professionals. Butler, during the years that he wrote *The Way of All Flesh* (1873–84), really believed, and with some substance, that what he had to say could materially affect the direction of biological inquiry. But after 1903 the only literary figure still engaging in controversy with this supposition in mind was Bernard Shaw; and Shaw had nothing to add until he came to write *Back to Methuselah* in 1921 that had not already been expressed more succinctly by Butler before 1885.

Only in one respect has it proven necessary to make the closing date an elastic one. In the case of those biologists and naturalists whose most original contribution to their science was made before 1900 but who survived well into the new century, their later work has been treated in the appropriate detail to give a rounded account of their thought. Thus Wallace's drift towards supernatural evolutionism was completed well before 1900 but his definitive exposition of it, *The World of Life*, did not come until 1910.[13] Galton's eugenic Utopia *Kantsaywhere* (1910) was similarly the product of his extreme old age, but it completes the pattern of his thinking and it would be impossible to omit this fiction without seriously warping an account of the literary response to eugenics. Something similar is also true of W. H. Hudson. After writing his Utopian romance *A Crystal Age* in 1887, Hudson altered his mind radically about the capacity of natural selection to produce organic perfection. In the light of this change much later on in his life, the variety of evolutionism to which Hudson held in the late 1880s stands revealed as much less secure than it would otherwise seem to be.

One final consequence of adhering to the chosen limits fairly closely is worth mentioning, with respect to *Middlemarch* (1871–2). George Eliot there draws many elaborate parallels between her organically structured novel and the society it mirrors on the one hand, and the researches of Tertius Lydgate into the structure of protoplasm on the other. But the action of the novel is in the early 1830s: Eliot very carefully makes Lydgate unaware of the real structure of the cell. For ironic reasons, so that the futility of the direction of his chosen research may be formally related to Casaubon's absurd mythography, she has Lydgate perplexed by difficulties that in real life were not resolved until the years 1838–9 by Schleiden and Schwann. Eliot took infinite pains to tie biological speculation into the great pattern of her novel, but the biology she is using is pre-Darwinian (as J. M. Forrester has shown, much of it was drawn from G. H. Lewes's pre-*Origin* study, *The Physiology of Common Life* of 1859) and therefore *Middlemarch* remains outside the scope of the present study.

It would be improper to end these preliminary remarks without formally recording the debt, obvious on every page, owed to A. O. Lovejoy. The methodology is a simplification of the one perfected by him to examine the concepts of plenitude, continuity and gradation in *The Great Chain of Being* (1936). Lovejoy's theme does not overlap much with the one treated here. The 'chain of being' metaphor, which refers to a hierarchy of living forms, may be found in conjunction with many varied beliefs: for instance, in the transmutation of species by natural agency (by Lamarckian *besoin* or Darwinian natural selection, or by unknown natural causes as in Robert Chambers); with a belief in the sequential creation of distinct species (by divine fiat or as in Genesis); or simply with a belief in a unity of design throughout and beyond nature (as we find in Pope's earlier poetry or the later Alfred Russel Wallace). The chain of being does not have to be a biological hypothesis at all, though in certain hands it tended to become one.

In *The Great Chain of Being* Lovejoy's range is vast: he covers centuries and tirelessly encompasses all kinds of written records from volumes of sermons to gardening manuals. In the rather smaller span of forty years the central chapters of what follows will take up in turn some of the expressly biological 'unit-ideas' which were of use to late-Victorian novelists, *belles-*

lettristes, scientific journalists and others whose writings had, in the broadest sense, some pretensions to literature. Only that writing is included which attempted to rise above the level of mere conveyance of information to cajole, warn, abuse, encourage or inspire the reader. The selection of writers is limited to those who had a firm background in biology, no matter how grotesque their extrapolations from the limited and self-selected facts available to them – and certainly to modern eyes some of those extrapolations appear very grotesque indeed. The test for inclusion applied throughout is that these writers' use of biological concepts should have been deliberate and imaginatively vital. All the writers considered could have explained in some detail where they stood with respect to the key controversies of their day; and it is for this ability, rather than for the absolute extent of their literary talents, that they are treated here. The first two chapters are given up to an interpretation of the history of Darwinism and extra-Darwinian biology up to the year 1900, from the viewpoint that, despite the great enlargement of systematic inquiry, these decades were, paradoxically, a time when the line separating biology from metaphysics became steadily harder to draw. The cultural interaction between biology and English letters over these years will be investigated in terms of the theory inherited from the Romantics which supposedly explains such interactions, and also by surveying the lines of communication open to the biologist and the use he made of them. After these preliminaries the remaining chapters are devoted to four large 'unit-ideas': namely evolutionism, or a theory of progress based on biology; degeneration; eugenics; and heredity. The first two of these central chapters, by setting two possible and quite self-consistent responses to selected biological information in direct opposition, will try to show the inherent difficulties in insisting on either consonance or dissonance with humanity's wishes as being the 'proper' interpretation of Darwinism, or comedy or tragedy as the genre most appropriate to absorb the new lessons. The third of these chapters, on eugenics, will examine popular rationalisations for interfering in the natural evolutionary processes, while the fourth and fifth will take up certain strands of the debate on heredity, elaborate them, and display their great literary fertility. All five chapters interlock conceptually and all are keyed to a discussion of at least two central texts. These texts are merely representative for the first three unit-ideas, but for the last the texts chosen are supreme products of the imaginative interest aroused by inheritance theory.

A supporter of Lovejoy's approach to the history of ideas has mentioned the accusation that anyone following it could well be led to 'read the third-rate works under the impression that they are literary masterpieces'.[14] In the following pages few purely aesthetic judgements will be made. Rather, it is hoped that they will communicate some sense of how the bare threads of biological inquiry might be woven up into a fabric sometimes dull, sometimes bright, but always at least humanly *significant*. The weaving of this new design, a metaphor for the integration into a culture, after much self-deception, grand posturing and anxious meditation, of a science's new paradigm, does not cease on any given date – it continues endlessly. This book traces the vagaries of that extension over a curious forty-year spell.

Notes: Introduction

1 John C. Greene, *Darwin and the Modern World View.*

2 A. Dwight Culler, 'The Darwinian revolution and literary form', in George Lewis Levine and William Anthony Madden (eds), *The Art of Victorian Prose,* pp. 238–9.

3 B. Ifor Evans, *Literature and Science,* p. 75.

4 Stanley Edgar Hyman, *The Tangled Bank: Darwin, Marx, Frazer, Freud as Imaginative Writers.*

5 Walter Lawrence Myers, *The Later Realism: A Study of Characterisation in the British Novel,* p. 29.

6 Morse Peckham, 'Darwinism and Darwinisticism', *Victorian Studies,* vol. 3 (September 1959), p. 19.

7 Conway Zirkle, *Evolution, Marxian Biology, and the Social Scene,* p. 348.

8 Leo Henkin, *Darwinism in the English Novel, 1860–1910: The Impact of Evolution on Victorian Fiction,* p. 9.

9 Lionel Stevenson, *Darwin among the Poets,* pp. 53, 298.

10 Alvar Ellegård, *Darwin and the General Reader: The Reception of Darwin's Theory of Evolution in the British Periodical Press, 1859–1872,* p. 23.

11 John Wyan Burrow studies Darwin's reliance on data from all three disciplines in his 'Evolution and Anthropology in the 1860s: the Anthropological Society of London, 1863–71', *Victorian Studies,* vol. 7 (December 1963), pp. 137–54.

12 By Milton Millhauser in 'The literary impact of *Vestiges of Creation*', *Modern Language Quarterly,* vol. 17 (September 1956), pp. 213–26; and more fully in his *Just before Darwin: Robert Chambers and 'Vestiges'.*

13 Of *The World of Life,* Malcolm J. Kottler has said that 'coming so late in Wallace's life-past his prime – it was not influential': Alfred Russel Wallace, the origin of man, and spiritualism', *Isis,* vol. 65 (1974), p. 184. Wallace was 87 when he wrote this final statement of his evolutionism and, though its views are extreme, they certainly do not run counter to those Wallace had formed as much as forty years earlier. It merely elaborates them and in a sense makes them more self-consistent. And it was influential enought to attract more than one studied rebuttal.

14 Margaret M. Starkey, 'The history of ideas and literary studies', *Modern Language Quarterly,* vol. 13 (September 1952), p. 265.

1

Darwinism on the Deathbed, 1870–1900:

The Failings of Natural Selection

> We are now standing at the death bed of Darwinism, and making ready to send the friends of the patient a little money to insure a decent burial of the remains.
>
> EBERHART DENNERT (1903)

A thorough scientific history of biology in the first few decades after Darwin has yet to be written. Despite the plethora of studies in this field it remains the considered opinion of a recent authority that not one of them has succeeded in being other than 'desultory' and that all their labours have barely reduced the 'area demanding systematic exploration'.[1] Since the present study cannot pretend to offer even a rudimentary history of the biological sciences over the period in question, there is certainly no intention here of mobilising one view of the history of Darwinism* against another, or of discriminating amidst competing claims in any lengthy way. What is necessary, however, is that the particular bias of the following pages towards one reading of the progressive absorption of Darwinism into Victorian intellectual culture should be stated fairly exactly, since on this reading rests some part of the argument to be pursued. One might begin, then, with what can fairly be termed the traditional description of the fortunes of the Darwinian revolution. According to such accounts, which tend to be fond of the imagery of military strategy, the publication of *The*

* To avoid any subsequent misunderstanding, throughout this study the term *Darwinism* should be taken to mean that the following naturalistic processes together supply a *totally necessary and sufficient* explanation for the variety of life on this planet: that, granted a capacity to reproduce among the lowliest organisms, there will be a struggle for existence among those organisms as a consequence of their geometric rate of reproduction and the inelasticity in the supply of food, space and other resources; that this struggle is survived by those organisms which display random, inheritable variations which are advantageous in permitting a relatively higher reproductive rate; that this selective mechanism will generate constant, permanent and adaptive structural modifications in organisms and that these changes, so defined, exhaust the concept of evolution. In the period under review, therefore, *Darwinism* is not equivalent to 'the thought of Charles Darwin' and bears a rather more restricted meaning than even the *Oxford English Dictionary* 'biological theory of Charles Darwin concerning the evolution of species'.

Origin of Species began a campaign of reason versus irrationality in which a small platoon of Darwinians marched from strength to strength, quickly sweeping aside a feeble opposition of ageing High churchmen. It is traditional, too, for the latter to be vilified as the fit progeny of their illustrious forebear who declined to put his eye to Galileo's telescope. This kind of description has been offered times without number, but never more ably than by Julian Huxley.[2] The bare bones of the sequence he offers are as follows: that after 1859, over a span of some thirteen years, Victorian England took Darwin to its heart; from the early 1870s until close to the end of the century the romance was in full bloom; then an estrangement is said to have set in for a couple of decades with the rise of anti-Darwinian evolutionary theory, but by 1925 this had passed by. Neo-Darwinism of a rigour hardly entertained by Darwin himself has held sway ever since. A similar dating appeals to William Coleman, too, for he argues that by the turn of the century 'no serious alternative, other than the ever-dubious environmental (called neo-Lamarckian) hypotheses, could be seriously maintained'.[3] Another historian of science, describing attitudes in this same period of the late 1890s, concurs with this: 'the climate of opinion was by no means wholly anti-Darwinian in character. Among the strongest supporters of natural selection were three of the most influential names in biology at the turn of the century: A. R. Wallace, John George Romanes [sic], and August Weismann.'[4]

It is worthwhile taking up this latter judgement for a moment, for in every detail it is, as a description of the unanimity of views of the inner Darwinian party, a misleading oversimplification which it is vital to clear up at once. Certainly it is unquestionable that Alfred Russel Wallace, the co-presenter of the natural selection hypothesis in 1858, adhered throughout his long career to the very strictest Darwinism as providing a sufficient explanation for the full variety of life on earth. Even as early as 1870, however, he was already insisting vehemently on one most notable exception to the all-sufficiency of Darwinism: *Wallace excluded human evolution.* Wallace's teleological* evolutionism, which he espoused (publicly at least) on the strictest empirical grounds, will be taken up on a later page. Here

* In the following chapters we shall mean by *teleological theories* of evolution any which appeal to those factors inducing evolutionary change which are purposive or goal-directed. Among such factors proposed were use-inheritance, environmental influences (diet, etc.), organismic '*besoin*', entelechy, orthogenesis, supernaturally controlled mutations, Nägeli's *Vervolkommnungsprinzip*, and so on. For present purposes *neo-Lamarckian factors* will serve as reasonable shorthand for all these, even though Lamarck himself specifically rejected certain external influences. Closer distinctions will be made in the appropriate contexts.

it must suffice to point out that right from its inception Darwinism was prone to heresies from within its own order, as here with Wallace's reservation (stated sufficiently bluntly even in the late 1860s to give Darwin some pain) that natural selection is a power completely impotent to explain the prodigiously rapid evolution of our species. G. J. Romanes was no more of a Darwinian in his later years. Though he admired Darwin personally to the point of adulation, he abandoned after Darwin's death in 1881 whatever faith he had had in the universality of Darwinism. In a survey, for example, five chapters long of the second volume of his *Darwin, and after Darwin* (1892–7) – a survey which is surely the most extensive ever made of the evidence relating to teleological and dysteleological evolutionary mechanisms – Romanes, very much closer to the situation than we are, states categorically that at the time of Darwin's death Wallace, even with his advocacy of natural selection restricted to the animal world, 'was absolutely alone in maintaining this opinion: the whole body of scientific thought throughout the world being against him'.[5] Nor is Romanes' own position left in any doubt. At the end of his massive consideration, he ponderously allies himself with the majority with all the force of a man distilling decades of thought and experiment into a few phrases:

> there are certain general considerations, and certain particular facts, which appear to render it probable that [various teleological factors] have played a highly important part in the process of organic evolution as a whole . . . there must be some other, though hitherto undiscovered, principle at work, which co-operates with natural selection.

And we find further that his anti-Darwinian sentiments formed, or began to form, much earlier. On 30 December 1883, in a private letter to Asa Gray, Romanes may be discovered thoughtfully toying with Wallace's by then firm conviction that the human brain must, on the anatomical evidence, have evolved from the simian under the direct guidance of 'a higher and supreme mind'.[6] In the face of these considered and final remarks it is difficult to see how Romanes can properly be maintained in his place as one of the Darwinian triumvirate. As for the third member, August Weismann's uncompromising neo-Darwinism had only a few years in favour in the early 1890s and had certainly fallen out of the centre of disputation before the decade and the century were over. Starting off as a simple and luminous idea – that the germ plasm, the physical principle of heredity, cannot be violated by the environment – Weismannism quickly became ornamented with qualifications and swamped by internal inconsistencies as it tried to deal with the criticisms levelled at it. The formerly indivisible germ plasm was splintered into no less than nine fragments (molecules, biophores, determinants, ids, idants, idioplasm, somatic-idioplasm, morphoplasm and apical-plasm; some of them externally influenced) which Romanes said drily reminded him of the Circles of the *Inferno*, in that from those regions one returned to the firm footing of science with a sense of having indeed been in another world. Even if we try to extend the search for 'hard' Darwinians beyond the names on Allen's list, remarkably few candidates present themselves. The attitude of T. H. Huxley, 'Darwin's bulldog', to the natural selection hypothesis may well be guessed from his reluctance to support it at all, either publicly or privately. Though he sprang to align himself with the evolutionists, and though he spoke of his readiness 'to go to the stake, if requisite', in support of the evolutionary thesis *per se*,[7] he was cool enough about natural selection in a lecture of 1860 for Darwin to complain about it to their mutual friend Hooker; and twenty years later he managed the uncanny feat of delivering an address celebrating the anniversary of the *Origin* without mentioning its central postulate at all! If he never let himself be branded an apostate, Huxley was never a disciple either.

We are obliged to conclude, therefore, that the reaction against Darwinism in the first decades after the *Origin* was a much more profound and firmly reasoned reaction (though indeed highly charged emotionally) than may be apparent at first sight. Scientific history, no less than political, tends to be the victor's version; what is more, as Howard Gruber has reminded us, in the history of ideas it is often the *persecutors themselves*, the vociferous opponents of a new theory, who go down into silence, with the traces they leave behind on that theory's proponents being quickly effaced by time and historians' indifference.[8] And here the matter is further complicated by the undoubted fact, conceded by all parties, that one part of Darwin's work – namely its validation of evolution – did indeed gain recognition and approval, and at a speed which his great peers Marx and Freud might well have envied. But on inspection post-Darwinian biology is found to relate to the insights of its founder in a much more ambiguous way than do Marxian economics and Freudian psychology. The remainder of this chapter will try to expand on this assertion, taking as its text C. D. Darlington's admittedly somewhat heterodox interpretation that in the period under review Darwinism 'began as a theory that evolution could be explained by natural selection. It ended as a theory that evolution could be explained just as you would like it to be explained.'[9]

Darwinism in Decay, 1870–1900

From our perspective it is difficult to avoid seeing the intellectual history of the 1850s as one inexorable wave sweeping forward with gathering momentum, gathering up all the pre-Darwinian evolutionists into one grand synthesis, and finally breaking with a crash in 1859. We smile to read Darwin's denial that the subject of the origin of species was at all in the air when he was collecting his evidence. And yet there is a sense in which the Darwinian revolution was more of a counter-revolution: in an important way it subverted the persisting Romantic impulse towards progressive evolutionism, which had already demonstrated its power by making Robert Chambers's *Vestiges of the Natural History of Creation* (1844) a bestseller on a scale far outstripping the *Origin*'s immediate success a decade later. Despite Chambers's scandalous handling of Genesis, the ideas of his book were assimilated, even by the Established Church, much more easily than Darwinian natural selection. Right up to 1859, with astronomy and geology between them producing more and more cheering evidence for cosmic progressivism on the Spencerian model, the church's energies were well absorbed by the Oxford Movement, leaving little to spare for any direct confrontation with science. Samuel Butler grew to maturity in these years of somnolent calm and, as he tells us in his account of the childhood of his *alter ego* Ernest Pontifex, there was between 1844 and 1859 'not a single book published in England that caused serious commotion within the bosom of the Church'.[10] Even *In Memoriam*, whose biological assumptions as developed through the whole poem seem to be based at least in part on Chambers, does not constitute a clear exception to the view that the century's fifth decade was a period of quietly optimistic quiescence, marked symbolically at its beginning by the death of Wordsworth and Wordsworthian pantheism and at its end by the publication of *The Origin of Species*. There is little evidence, even when we take into account the writings of earlier evolutionists like William Wells (1757–1817), Patrick Matthew (1790–1864), William Lawrence (1783–1867) and Edward Blyth (1810–73), of the coming storm; and there is no sign of it at all in the serene poeticised geology of Martin Tupper's *Proverbial Philosophy* (1838–76) or Kingsley's *Glaucus* (1855). The cracks in pre-Darwinian evolutionism did not begin to gape really wide until Philip Gosse refused both to regard any part of Scripture as allegorical and to reject the evidence of his natural history researches. His *Omphalos* (1857), written, in Gibbon's phrase, 'to reconcile the jarring interests of reason and peity', was ridiculed by all parties precisely because it was too commonsensical in attempting to marry jarring Genesis and geology.

In such a climate of opinion that which was least original in the *Origin* – the proposal of organic evolution as a historical process – gained the speediest assent. Darwin did not invent that theory and did not make it familiar to anyone with even a smattering of the geological knowledge of the first half of the century. By the 1840s the transmutation of species had become a serious and established hypothesis; even earlier, in the 1830s, there was circulating a 'bestial theory' of man's affinity with the higher apes – circulating widely enough for Coleridge to mock it in 1833. In this sense, the *Origin* supplied no more than a shock of recognition. But the addition of natural selection to supply a propulsive force to evolution was truly revolutionary – though not in truth absolutely original, for Darwin was anticipated in some degree by all the evolutionists mentioned above, and by others. It was this addition, this unalloyed Darwinism, which sounded through the literary culture as perplexity, outrage or buoyant inspiration. To some extent the proof of this assertion must be a continuous one; but we will pause here to refer briefly to an invaluable semi-statistical survey by Alvar Ellegård which permits us to bring a little much-needed rigour into this particular area of the history of ideas. On the question of the differing rates of diffusion through Victorian culture of belief in the unit-ideas of 'species-transformation' and 'natural selection' Ellegård is most informative. Ten years after the *Origin* general belief in the evolution (i.e. the progressive transformation) of species ran through every sector of society except, perhaps, a small group of rigidly fundamentalist churchmen. The Darwinian hypothesis, however, in Ellegård's considered opinion, 'clearly met with an incomparably stronger resistance . . . It touched the ideology of the age at a more vital point than did the Evolution theory pure and simple.'[11] Whereas, he contends, the controversy over evolution itself was effectively dead within ten years, Darwin's explanation for it steadily *declined* in popularity. Admittedly, this judgement is offered with the proviso that in the case of human evolution only a small proportion of the authors surveyed were prepared to accept a totally naturalistic explanation for the origin of man, and particularly for the origin of his higher mental faculties. Certainly on what evidence was available this was a good deal to swallow. But the exception Ellegård makes is not a crucial one. The late-Victorian teleologists did not dissent from the evolutionistic orthodoxy of human functions having developed steadily and progressively; they insisted only that the process must have been guided, and so to a large extent the proponents of that highly influential stream of thought still fit into Ellegård's pattern. But he goes farther than this. His careful study of a wide range of periodical articles devoted to these topics leads him to distinguish two fairly distinct phases in the decline of Darwinism. Among the informed and well-disposed, evolution by natural selection made some headway between 1864 and 1869, for the organised technical opposition to it took some time to become fully mobilised. Six years passed before G. M. Humphry's first elaborate critique was placed before the British Association. During this period the hysterical and highly irrational denunciations of Darwin caused a ground swell of support for him among the scientific popularisers and literary men with fewer or no religious convictions to be outraged. However, in the second period up to 1872 the adherents of Darwinism in its 'hard' or classical form fell away sharply. Ellegård's study unfortunately ends in that year; but since the strictly scientific opposition to Darwinism grew steadily stronger after 1872, and was recognised as becoming more and more formidable by all parties, there is every reason to suppose that a statistical inquiry would show this decline continuing at an accelerating pace up to the end of the century and for some time beyond it. By the middle 1890s a small countervailing tendency first becomes detectable in the rise of neo-Darwinism; and it is this that marks the end, effectively, of the curious period of suspension that we are trying to define. In modern biology the significance is well recognised of a phenomenon called *paedomorphosis*, whereby a species can, in order to escape a developmental dead end, regress to an infantile stage in its life-cycle before pushing forward again. The history of neo-Darwinism is just such an example of a *reculer pour mieux sauter* which quite reversed the normal pattern of acceptance, consolidation and a slow decline in explanatory power. It was the genius of the neo-Darwinians, and a little later of the first post-Mendelian geneticists, that they were ready and able to backtrack mentally to 1859 and pick up the one possibility which had been generally neglected: that of eliminating entirely from a species-forming process everything except individual differences, the random bursts of mutation and the consequent modifications induced by selection. This was a bold and unpopular move for some time, because even at that late date it was still violently throwing into reverse gear the main drive of evolutionary thought. The evolution of a science, then, like that of a species, need not proceed smoothly. Evolutionary biology is of course now Darwinian, and it is highly unlikely that it will ever move backwards to embrace modes of explanation now thoroughly discarded. A parallel is to be found in the history of physics: quantum mechanics has there replaced the Newtonian–Galilean model, but the net result has been a shift even farther away, conceptually, from scholastic physics. Such is the trend of a science in the long term; but there can be small fluctuations which run for a while in the opposite direction, and it is inside such a fluctuation that our present attention is directed.

So far our chief emphasis has been placed on the emotional, aesthetic and theological causes for this interregnum in biological explanation. And this is correct. For, despite the title of his great work, Darwin did not believe that species really exist apart from the observer. A philosophical nominalist like Lamarck before him and Vries after, Darwin spoke of species as a mere taxonomic convenience. For in truth it is not species in the abstract which bear the brunt of selection, but unique organisms whose minute differences from each other are so readily overlooked for the benefit of the systematist. Yet it was just these minutiae, and not the intraspecies similarities, that Darwinism in its pre-Mendelian phase stressed as supremely important, for it is these minutiae alone which are the material of evolutionary change. Natural selection therefore cut the ground from under not merely the theological arguments from design, but even more disruptively from under the deep-seated belief in the order, economy and regularity of nature. Out of the Enlightenment, and especially out of its astronomy, had come the apparent discovery that there is a basic rationality in the world order – a rationality finding an encouraging reflection in the *a priori* axioms of mathematics. One could juggle one's indisputably home-made symbols; one could come up with a set of co-ordinates (that is, if one were a Le Verrier or an Adams); a telescope could be directed to a precise point in the sky, and a new planet (Neptune) could sweep into one's ken. Between the inner workings of the human brain and the majestic sweep of the planets a most pleasing concordance seemed to prevail. Darwinism changed all that. The strongly cohesive bonds which held men together in the dignity of *Homo sapiens* lost their grip. Now there were only individual men, more or less wise as the caprice of variation would have it. And how could a natural law whose very hallmark ought to be lucidity, beauty and simplicity work by generating a thousand, a million more waste products than perfected items? Such, at least, is the logic of one popular train of argument. There were others.

This being said, it is important as a corrective to make mention of the very severe difficulties of a strictly technical kind which assailed the Darwinian theory and made it quite possible for anyone who was cold to it to decline to take it seriously. Darwin did his best to make a completely natural science of biology as a whole, to outlaw teleology and to substitute detectable changes happening lawfully over geological ages; but on any analysis regularity and observability were not strong points of the mechanism Darwin posited. Variation followed no law known to man and no one, certainly not Darwin himself, could make any claim to have watched species forming in the wild. Nor did he receive the support on which he had relied from the neighbouring sciences. The enunciation of the basic laws of genetics lay forty years ahead. Palaeontology, despite Lyell's valiant support for his friend and colleague, could not demonstrate the mutability of species in the past but only that the more recent species tend to a greater complexity of function. In embryology and its puzzling datum that ontology repeats phylogeny *in utero* Darwin believed he had an ally; but, though recapitulation was well understood by Baer and Agassiz, these formidable men nevertheless rejected it as evidence of evolution. And biochemistry was at a complete loss to explain the conversion of the same primitive embryonic tissue into quite distinct organs. 'What *directive* agency determines the distribution of the colouring matter,' Wallace was still pertinently asking of embryology and biochemistry a decade after the close of our period as he had asked for the whole span of it, 'so that each feather shall take its exact share in the production of the whole pattern and colouring of the bird, which is immensely varied, yet always symmetrical as a whole, and has always a purpose'?[12] To this and allied questions, which step far beyond the limits of evolutionary biology as it was then studied, Darwinism had in the first four decades of its existence no clear and distinct reply.

Still less could Darwinism cope with a challenge from outside biology altogether. No attack on the foundations of the Darwinian edifice gained greater credence than that by the physicist William Thomson, later Lord Kelvin (1824–1907), who came to biological controversy with the fame of his brilliant formulation of the Second Law of Thermodynamics (1851) securely behind him. Overtly at least, Thomson never had much interest in the details of species modification or indeed in the life sciences at all. He restricted himself to arcane problems of geology, mathematics and astrophysics, and he became so challenging an opponent of the Darwinians because he claimed to undercut their position on physical evidence alone.

In the first edition of *The Origin of Species* Darwin laid very great stress on the nearly unimaginable scale of geologic history; on the passage of hundreds of millions of years during which time, if the Uniformitarians were correct, conditions for life had been much as they are today. Sheer time had for Darwin a creative aspect. Was there anything, no matter how improbable, that could not have recurred time and time again somewhere in that vast bulk of years? Could not even natural selection, running its slow and fumbling course, achieve anything, even man himself, in 300 million to 500 million years? Here Darwin had a convincing argument and one which he deployed with considerable skill. Instead of speaking of aeons of time in the abstract he brings before our mind's eye an actual example of sedimentary formations nearly fourteen miles thick; then, instead of supplying us gratis with an estimate of how long it might have taken this to form, he tells us that the mighty Mississippi lays down about 600 feet of sediment in a hundred thousand years (six feet per thousand years has a far less magnificent thud to it). He leaves the further impressive calculation to us. When we connect up this with the fact that whole generations of simple organisms may succeed each other in days or even in hours, this deftly handled ninth chapter of the *Origin* on the geological record becomes the keystone to the book. Man, too, had his place in this geological record. Charles Lyell (1797–1875), the most famous geologist of the day, had rejected transformism in his *Principles of Geology* (1830–3). The effect, then, was all the greater when in his *Antiquity of Man* (1863) he not only faced about but did Darwin the service of pushing back many thousands of years the emergence of *Homo sapiens* and his predecessors, citing new palaeontological evidence for so doing. His reputation helped – but for a while only – to suggest that geology and Darwinism were in harmony.

But Thomson asked a simple question and, in doing so, cooly sliced away the Darwinians' vast womb of time. If all life depends on the sun's heat, how old is the sun? In forming his estimate it goes without saying that Thomson had no inkling of the Bethe reaction, the thermonuclear conversion process (not described until 1939) which sustains the sun's colossal energy output. He could only imagine a mechanical source of heat, or a chemical one. In a typical early paper he dismisses such sources of energy as meteors falling into the sun or the sun's contracting under its own gravitational attraction. He is led to assume that the sun is one huge glowing coal, a mere 'incandescent liquid mass cooling'.[13] From there, given the known properties of combustible materials, it was a relatively trivial calculation to come up with the judicious statement that 'all geological history showing continuity of life, must be limited within some such period of past time as one hundred million years'.[14] Victorian astrophysics was not a quickly moving science. Nineteen years later Thomson had few additional facts to offer, but those he did have made him slash his allowance to the biologists down to a mere 20 million years[15] – a time now known to have covered the emergence of only the most recent mammals. Thomson continued to insist on this alarmingly small figure right down to the end of the century.[16]

The consternation within the Darwinian circle produced by Thomson's calculations was immense, and it certainly assisted materially in the decline of Darwinism from the early 1870s. The alarm, however, grew slowly. None of the Darwinians gave any public recognition of the deadly significance of Thomson's work before 1868 at the earliest, though they did so privately. The first four editions of *The Origin of Species* (i.e. from 1859 to 1866) do not mention Thomson by name, although as early as the second edition the time mentioned as being available for evolution to work in was scaled down considerably. Thomson could be ignored: he had been cautious in making his foray into alien territory and perhaps had muted the inevitable biological conclusions in the hope that others would make the necessary connections. Meanwhile, the Darwinians' attentions were fully occupied by the immediate threat posed by the most carefully worked-out of the early attacks – that of Fleeming Jenkin. The burden of Jenkin's mathematical analysis of 1867 was that, even if a favourable variation should occur by chance in an individual, the likelihood of its being spread through the population was remote unless both parents displayed it. Otherwise the useful novelty would be diluted (Jenkin wrongly took the blending of characteristics for granted) and lost like a drop of ink in the ocean. Faced with these relatively straightforward puzzles the Darwinians and their public could pass over the unworldly abstractions of Thomson's cosmogonical theories. But, whatever the reason for neglect, Thomson threw down the gauntlet unmistakably during his lecture before the Geological Society of Glasgow in February 1868. The traumatic effect on Darwin personally left many traces in his writings. A little more than a year after this lecture Thomson had become one of Darwin's 'sorest troubles' and two years after this 'an odious spectre' – strong language for the mild Darwin, even to an intimate.[17] In the spring of 1869, while revising the *Origin* for its fifth edition, Darwin privately summarised one important concession that he was about to make rather misleadingly as 'I have been led . . . to think the lapse of time, as measured by years, not quite so great as most geologists have thought',[18] whereas in the forthcoming edition he was in fact going to curtail his estimate of the epochs available for the evolution of life hardly to one 'not quite so great' as before but to a figure less than one-third the size of his original. But this was in a private letter: in public Thomson received more justice than one might have expected. When the fifth edition came out (1869) we hear of his 'formidable objection',[19] and by the sixth (1872) this has been strengthened to 'probably one of the gravest as yet advanced' (XIV:55.1:f) and one detects also a rather desperate quality in the countering arguments. When we set Darwin's final words that 'we do not know at what rate species change as measured by years' (XIV:55.1:f) against his earlier assurance that 'an infinite number of generations' (IX:74) have undergone almost infinitely slow change, we have the most direct evidence of the destructive power of Thomson's papers. A particular irony of the situation was that every increase in the known length of man's prehistory made the problem that much more acute. As man and his pre-human ancestors were traced back more than a million years, how could the whole succession of life from amoeba to man be jammed into an allowance of years only some twenty times longer? Most of the Darwinians eventually broke down at this point: natural selection could not have worked within such a period. Only some teleological mechanism from within, or some transcendent will from without, could have worked speedily enough. Others retreated into sophistries. T. H. Huxley's lecture 'Geological Reform' of 1869 in particular marks a very low point indeed in that skilled debater's powers to muster an irrefragable argument. Despite Darwin's own stress on the capacity of slowly passing time to build on random and irregular variations Huxley does his best to convince us that

> biology takes her time from geology. The only reason we have for believing in the slow rate of the change in living forms is the fact that they persist through a series of deposits which, geology inform us, have taken a long while to make. If the geological clock is wrong, *all the naturalist will have to do* is to modify his notions of the rapidity of change accordingly.[20]

The bland speciousness of this is alarming. It is a characteristic of Huxley's rhetoric that it becomes more facile as the insecurity of his stance grows; and by this measure Thomson had made him very insecure indeed. The rate of change of living structures was no mathematical variable to be stretched or chopped to fit the Procrustean bed Thomson had prepared for it. Either that change had come about thanks to the Darwinian mechanism assisted by time, or there was some other *vera causa* of evolution – if evolution there had been. Thomson's reply a month or so later is a model of amiable butchery. Joint by joint Huxley's feeble defence is dismembered. The Darwinian mechanism will only work where there is a wealth of slow time. Cut down the allowance by a factor of ten or more and the conclusion is inevitable:

> the limitation of geological periods, imposed by physical science, cannot, of course, disprove the hypothesis of transmutation of species; but it does seem sufficient to disprove the doctrine that transmutation has taken place through 'descent with modification by natural selection'.[21]

Evolution is a fact, but something, or someone, has guided its path. If life had had to wait on some 'fortuitous concourse of atoms', then there would even now be nothing but a wash of protein molecules in a dead sea. As far as Thomson, most other geophysicists and an ever-increasing number of laymen were concerned, that was that for the remainder of the century. Thomson happily reprinted his lectures in 1889 and 1894 without altering a word.

In 1903 the discovery of Curie and others that the element radium has a temperature permanently above the ambient level wrecked Thomson's elegant structure, for it was then only necessary to suppose that radioactive elements are present in the sun for Thomson's estimates to be multiplied by any figure one wished. Geological time expanded again so rapidly that no Edwardian physicist dared to offer any calculation as to either the probable or possible age of life on earth, especially after the lead-uranium technique for dating rocks began to give figures in the billion- rather than the hundred-million-year range. The twistings and turnings of the Darwinians proved in the end to have been unnecessary. For several decades, however, the physical sciences seemed linked in a reactionary conspiracy with the fundamentalists to make the creation recorded in Genesis seem less of an outrage on common sense than that proposed by the new biology. It followed that one surrendered no intellectual respectability by refusing to be a Darwinian.

Darwinism reached the nadir of its reputation in the three decades round the turn of the century. It had been abused from all sides, replaced time and again by some version of evolutionism and abandoned by one after another of its advocates including to quite some extent even Darwin himself before his death. It is only the first edition of *The Origin of Species* that has very much argumentative consistency. Later editions became palimpsests of qualification and contradiction as Darwin struggled with his major critics, and the fifth and sixth editions contain more than half of the total number of revisions made by the author in his lifetime. Nor were these revisions well received. Darwin himself mentioned in a letter to Hooker of July 1868 that he had heard 'the belief in Natural Selection is passing away'. Failing to placate his critics, and perhaps knowing full well that some of the criticisms were irrefutable within the prevailing limits of knowledge, Darwin retreated in the last decade of his life to more neutral ground. In treating of the habits of earthworms (1881) he could apply his unmatched gifts for extended and minute observation to the investigation of such matters as the rate of burial of ancient buildings, without having to keep one eye all the time on treacherous definitions which seemed to grow steadily more unruly as the years passed.

Darwin did not retreat, though, without making heroic efforts to cope – efforts which unfortunately compelled him to give full rein to the more credulous side of his nature. Even in the first edition of *The Origin of Species* he neither wished nor was able to dispose entirely of the evidence for the inheritance of acquired characteristics. Despite the taint of teleology Darwin held on to at least two neo-Lamarckian axioms. He allowed from the start that certain variations are not random but must be due either to changes induced by the environment or to the effects of use or disuse on organs or habits, or to both. Nor did he hold out for very long against even Lamarck's notion of a *sentiment intérieur* in each creature; but these constituent parts of the original Lamarckian system tend to become very muddled in Victorian neo-Lamarckism and for present purposes it is not necessary to distinguish them. We need only to notice how under extreme critical pressure several varieties of post-Darwinian speculation allowed purposiveness, design and the concept of a life-force to creep back through the holes in the argument of the *Origin* – an argument in its naked form genuinely proof against all non-naturalism. This is well exemplified in Darwin's own evaluation of Lamarckism, which moves in a full circle from elaborate praise in the early notebooks, through a verdict of 1863 on the *Philosophie zoologique* as being 'a wretched book, and one from which (I well remember my surprise) I gained nothing',[22] on to a renewal of Lamarckian explanation in the *The Variation of Animals and Plants under Domestication* (1868), and at last back to the addition of a whole new chapter describing goal-directed variation in the sixth edition of *The Origin of Species* (1872). In this ambiguous chapter Darwin now appeals in fairly equal proportions to chance (unknown law) and to the conditions of life as the true causes of variation, though to deal effectively with Thomson on his own terms meant that much stress had to be placed on the idea of environmental factors somehow speedily *compelling* a required organ into existence. But, whatever the pressures on Darwin as a theoretician, looking at his new invocation of Lamarckism more broadly obliges us to see in it a reflection of the very pronounced desire among philosophers, social critics and educationalists to believe that acquired characters are transmissible down the generations. In such new guise did the progressivism of the 1840s and 1850s turn up again. Herbert Spencer's adaptive sociology, spanning the pre- and post-Darwinian eras, was rooted in the innate psychology of Gall, but to make his sociology truly evolutionistic after 1859 Spencer had to give a sovereign place to Lamarckian inheritance. Indeed, his reliance on it was so great that to abandon it would have meant abandoning his whole philosophy. Fortunately he was not called upon to do so since the evidence was still ambiguous at the date of his death: it became clear only very much later; and even then, as Zirkle records of the Lamarckians' last gasp in Soviet genetics, 'perhaps no other biological hypothesis has ever been tested so thoroughly and abandoned with such reluctance'.[23] Some of the untenable generalisations in *The Variation of Animals and Plants* sound as if they are cultural reflections, as when Darwin quotes solemnly from Saint-Hilaire that 'the mothers of illegitimate children troubled in their minds and forced to conceal their state, are far more liable to give birth to monsters than women in easy circumstances'.[24] There may be a grain of physiological truth in this, but what Darwin is surely trying to implant is the suspicion that bastards are the unfit, to be weeded out by a selection which is not merely neutral and natural but social too, and in favour of legitimate offspring.

Apart from this kind of transparent response to cultural pressures, the Darwinians made increasing appeals to teleology because it helped to make intelligible the curiously spasmodic irregularities in the frequency of variation. The common supposition of the time was that scientific methodology must rise from the bottom-most rung of explanation if it was to avoid the charge of arguing *ignotum per ignotius*. Certainly Darwin's handling of variation was clumsy; and certainly while Mendelism lay hid in night the ghost of teleology could not be laid. While the lacuna existed there was a constant pressure to fill it up with some directive, and usually divine, process. The American Asa Gray was ready almost as soon as the *Origin* was off the press with the analogy of an irrigation stream. The beds of such streams, he argued, are formed in exactly the same way as the beds of the natural rivers. The water, according to the fixed and impersonal law of gravity, flows to the lowest point and in doing so rolls down stones which wear out a channel by the equally fixed laws of friction. The designer of an irrigation ditch does not try to flout these laws. He limits himself to directing the flow in the required direction. May we not believe, therefore, that, while natural selection orders the details, the evolutionary flow as a whole is the result of the Divine Irrigationist forcing the stream of variation down one channel rather than another? Indeed, Gray insisted, it would be gross scientific irresponsibility 'at least while the physical cause of variation is utterly unknown and mysterious' *not* to believe that 'variation has been led along certain beneficial lines'.[25] Nature should be assumed innocent and supportive until proven uncaring and random.

Darwin's own musings on the broader implications of his theory were rarely more sophisticated than Gray's, but (in private at least) he concurred with Gray that his 'chance' as the cause of variation would not do – it too easily becomes a wastepaper-basket to receive any inconvenient difficulties. When he wrote to Asa Gray that the prime mover in his theory was random variation, he speedily added the rider 'not that this notion *at all* satisfies me. I feel most deeply that the whole subject is too profound for the human intellect. A dog might as well speculate on the mind of Newton.'[26] The odd touch of philosophic idealism in that last sentence (the hint that the organic universe may be more like a great thought than a great roulette-wheel) is indicative of that perennial inclination in Darwinian theory: the tendency to drift into using internal or external purposiveness as a catch-all unless firmly restrained from doing so. In *The Descent of Man* that drift is almost complete. By this time (1871) Darwin would probably have been content, had it not been for the impetus supplied by his public's expectations, to leave the task of elucidating man's own evolution by natural selection to others. We know that he did in fact attempt to pass his notes for the *Descent* on to Wallace, hoping that he would shoulder the enormous burden; but sensibly Wallace declined, almost certainly out of a conviction that it could not be done. It is, then, not very surprising to find that the *Descent* lacks the solid drive of a single sustained argument so characteristic of the early *Origin*. Instead it gives the impression of having grown by accretion, of having fact piled upon fact until it is overweighted with information. And there is a discomfiting toleration of all sources of this information as having equal merit, so that anecdotes from proud pet-owners mingle confusingly with the matter-of-fact expertise of stockbreeders. The refusal to accept variation as random and uncaused is carried to the point where we are told that Europeans in America undergo after a few generations 'a slight but extraordinary [*sic*] rapid change of appearance. Their bodies and limbs become elongated.'[27] Conceptually we are here miles away from that careful experimenter who formerly had weighed duck skeletons and sought for viable seeds in the excrement of seabirds.

Anti-Darwinian Thought Concluded

The purpose of this all-too-rapid excursion through what is still an imperfectly understood sector of scientific history will have been amply fulfilled if it has brought home just how divergent opinion was, even among the most detached theoreticians, of the human significance of evolutionary biology. The opinionative spectrum ran all the way from Spencer's influential belief that progress (moral and spiritual) is *implicit within* the evolutionary progression from spiral nebulae to the refinements of Victorian society and beyond; to the aged Huxley's equally influential belief that cosmic progress is fundamentally amoral and must be vigorously *opposed and rejected* for civilised life to be sustained and improved. No compromise is possible here, and we should not seek to impose one. These divisions emerged out of a particularly static phase in scientific history, and with their relative truth or falsity we are not concerned. When, in the following chapters, there is a charting of the sometimes extreme manipulations of biological concepts it should be borne in mind that there was no reason in the world why even the best-informed novelist or the most cautious speculative essayist, each in his own way concerned to present the latest truths of the life sciences in all their purity, should have adhered to any orthodox line. Natural selection, to mention only the key item, was certainly a bridge which led *from* romantic pantheism and natural theology. But it led *to* no clear destination. Positivism, chiliasm, reworked pantheism, neo-Paleyism, stoical despair, guarded optimism, even mystical exaltation: there were expert defendants for all these and more. None of these conflicting pictures of the world alone represents the supreme contribution of Victorian biology to cultural history.

Our excursion may have given the impression that there was little more for the literary imagination to seize on in Victorian biology than the conflict over randomness versus teleology in variation and inheritance. An important conflict it certainly was, equally for the philosopher concerned for progress, for the reformer intent on social change, and for the moralist worried about the inheritance of virtue and vice. These professional concerns all left a mark on literature. But biology also engendered other equally profound shifts in orientation which were just as amenable to imaginative exploitation. There was the concept of a single primitive substance, protoplasm, as the basic material of all life from bacteria to man. There was the discovery that fundamental metabolic processes work on the same principles in all life-forms, even in those most dissimilar to the human. There was the growing tendency in all the life sciences to reductionism – first to accounting for phenotype in terms of genotype, but later more drastically to the mechanistic assumptions of the contemporary physics. Vitalism was exploded by Wohler's synthesis of urea in 1828, and something which could almost be taken for 'growth' was displayed in inorganic crystals by Schleiden in 1838. The distinction between the organic and the inorganic was thereby shown to be no more than a convenience of terminology. Organisms were of the same substance through and through as the inanimate cosmos, and were subject to all the laws of a universe made out of Daltonian atoms: unchanging, unyielding billiard-balls bounced from one configuration to another by physico-chemical forces. Biology was no sooner thoroughly entangled in this inadequate metaphysical net than developments in the physical sciences began to rip it apart; but while the model lasted it forced biology into a series of antitheses (mind *or* matter, determinism *or* randomness, Darwinism *or* Lamarckism) from which no writer of scientific inclinations could remain aloof. Finally and most crucially there was the common (though not universal) ejection from the new theories of evolution of the ancient doctrine of the chain of being, together with all its inbuilt hierarchical suppositions. To certain mentalities man's image of himself as an intermediate link in this chain subsided into a new picture of a many-branched tree where the branch occupied at present by man might yet fork again, bringing something new, superior and alien. All these conceptions (and they were equally useful whether accepted or denied) were fraught with literary possibilities. Before examining some of these in detail, we need some account of the part played by biology in the extra-scientific culture, and we shall approach this by considering some of the contemporary attitudes to science as a fertilising force on literature.

Notes: Chapter 1

1 William Coleman, *Biology in the Nineteenth Century: Problems of Form, Function and Transformation*, p. 171.

2 Julian Huxley has offered this kind of sequence on several occasions. The dates quoted are as given in his 'The Emergence of Darwinism', *Journal of the Linnean Society of London*, vol. 44 (July 1958), p. 7. Conceptualisations of the early reactions against Darwinism as 'prejudice versus reason' may have a dated air about them, but they

are still current. For instance, the editor of a recent, elaborate reference-work permits the extraordinary judgement that Darwin's opponents were 'mainly members of the privileged upper classes who regarded the theory as a threat to the Establishment': Philip P. Wiener (ed.), *Dictionary of the History of Ideas: Studies of Selected Pivotal Ideas* (New York: Scribner's 1973), 'Evolutionism'.

3 Coleman, *Biology in the Nineteenth Century*, p. 84.

4 Garland E. Allen, 'Hugo de Vries and the reception of the "mutation theory" ', *Journal of the History of Biology*, vol. 2 (Spring 1969), p. 57. Allen denies that sentiments such as Dennert's (which he cites) were of any major significance.

5 George John Romanes, *Darwin, and after Darwin: An Exposition of the Darwinian Theory and a Discussion of Post-Darwinian Questions*, Vol. 2, *Post-Darwinian Questions: Heredity and Utility*, pp. 2, 156–7. Romanes's treatment is so extraordinarily thorough that any further evidence on the point at issue would seem to be redundant. However, Glenn L. Jepsen, 'Selection, "orthogenesis", and the fossil record', *Proceedings of the American Philosophical Society*, vol. 93 (December 1949), pp. 479–500, gives a large choice of quotations from Romanes' contemporaries dating from *c.*1880 onwards, all of them either rejecting natural selection or reducing it to a minor contributory role. So far as I am aware, no case using the documentary evidence can be made against Romanes' assessment.

6 Letter to Gray of 30 December 1883. Ethel Romanes, *Life and Letters of George John Romanes*, p. 154.

7 Letter to Darwin of 23 November 1859. Francis Darwin (ed.), *The Life and Letters of Charles Darwin* (hereafter *Life and Letters*), Vol. 2, p. 231.

8 Howard E. Gruber, *Darwin on Man: A Psychological Study of Scientific Creativity . . .*, *p. 35.*

9 Cyril Dean Darlington, '*The origin of Darwinism*', *Scientific American*, vol. 200 (May 1959), p. 61.

10 Samuel Butler, *Ernest Pontifex, or The Way of All Flesh*, ed. Daniel F. Howard, p. 180.

11 Alvar Ellegård, *Darwin and the General Reader: The Reception of Darwin's Theory of Evolution in the British Periodical Press, 1859–1872*, pp. 65–6.

12 Alfred Russel Wallace, *The World of Life: A Manifestation of Creative Power, Directive Mind and Ultimate Purpose*, p. 295.

13 William Thomson, 'On the age of the sun's heat', *Macmillan's Magazine*, vol. 5 (March 1862), p. 390.

14 William Thomson, 'On geological time', in his *Popular Lectures and addresses,* 2nd edn, Vol. 2, p. 64. This paper was first given as a lecture on 27 February 1868.

15 William Thomson, 'On the sun's heat', in ibid., Vol. 1, p. 397. This paper was first given as a lecture on 21 January 1887.

16 William Thomson, 'The age of the earth as an abode fitted for life', *Science*, vol. 9 (12 May 1899), pp. 665 ff. It has been several times suggested – for instance, by Loren Corey Eiseley in *Darwin's Century: Evolution and the Men Who Discovered It* – that Thomson's somewhat fundamentalist Christian persuasions made him push hard for the lowest figure just compatible with the physical evidence. But no one, not even Huxley, voiced any such suspicion at the time. Surely Huxley at least would have found the temptation to do so irresistible, if he had thought it could be made to stick. One recalls how he dealt with Mivart's citation of Jesuit authorities: by going straight to the original Latin sources and showing how dishonest Mivart's use of them had been. In fact, Thomson's fame made it difficult to question the accuracy of his calculation, though his conclusion could be, and was, denied.

17 These phrases are taken from letters to Alfred Russel Wallace of 14 April 1869 and 12 July 1871. James Marchant, *Alfred Russel Wallace: Letters and Reminiscences*, Vol. 1, pp. 242, 268.

18 Letter to Victor Carus of 4 May 1869: *Life and Letters*, Vol. 3, p. 109.

19 *Origin*, IX:219.1:e. Throughout this book all references to *The Origin of Species* are to the superb and indispensable variorum text edited by Morse Peckham (1959). Roman numerals indicate chapters, arabic numbers the editor's convention for displaying successive

textual changes. The last letter is a code for the edition being cited: a–f represent the six editions published in Darwin's lifetime.

20 T. H. Huxley, 'Geological reform' in his *Collected Essays*, Vol. 8, p. 329 (my italics). This was originally Huxley's presidential address to the Geological Society of London in 1869.

21 William Thomson, 'Of geological dynamics', in his *Popular Lectures and Addresses*, 2nd edn, Vol. 2, pp. 89–90.

22 Letter to Lyell of 12 [March 1863: so dated by Francis Darwin]: *Life and Letters*, Vol. 3, p. 14.

23 Conway Zirkle, *Evolution, Marxian Biology, and the Social Scene*, p. 128.

24 Charles Darwin, *The Variation of Animals and Plants under Domestication*, 2nd edn revised, Vol. 2, p. 318.

25 Asa Gray, 'Natural selection not inconsistent with natural theology', in his *Darwiniana: Essays and Reviews Pertaining to Darwinism*, ed. A. Hunter Dupree, pp. 121–2. This essay first appeared in *Atlantic Monthly* in 1860.

26 Letter of 22 March, dated to 1860 by Francis Darwin: *Life and Letters*, Vol. 2, p. 312.

27 Charles Darwin, *The Descent of Man, and Selection in Relation to Sex*, 2nd edn, p. 304.

2

Victorian Biology and Victorian Letters:

An Overview

> No man reads a book of science from pure inclination. The books that we do read with pleasure are light compositions, which contain a quick succession of events.
>
> SAMUEL JOHNSON (1783)

To establish a way of looking at the relations between literature and science which is true to all the facts has long been, and still remains, a task puzzling to the aesthetician and practical critic alike. Every shade of opinion possible on this issue has been expressed in the past and continues to be restated in the terminology of the present. Advocates for the view that the 'technologico-Benthamite' influence (which is sometimes made to include pure science and sometimes to exclude it) will eventually destroy the literary tradition seem evenly matched against advocates for the mutual fruitfulness of both ways of knowing. Despite the continuation of these longstanding disagreements into our own time, however, there is perceptible over the last two centuries a growing willingness to compromise. There has been a great expansion in our knowledge about human creativity – knowledge which has served to increase our awareness that artistic and scientific creativities cannot be compartmentalised with the only communication between them being certain nebulous 'influences'. In the province of the history of ideas this knowledge has been reflected in moderate accounts of literary and scientific interaction which examine changes in *both* as variant responses to shifts in strata a long way below the ground level of the contemporary cultural consciousness. In such moderate accounts it is not of course supposed that the two cultures will match smoothly at every point. Science progresses by having its hypotheses (which sometimes first attain the dignity of theories or even received facts) absorbed in yet larger and larger syntheses: quantum mechanics swallowed up Newtonian mechanics; the Heisenberg model of the atom swallowed up the Rutherford–Bohr model. Ideally, science hails the truth wherever it is and no matter how unpleasant it may be; art has no such obligations but takes from science only that part of the truth which it finds palatable at any point in time.

Such accounts of their interaction are therefore rather prone to take on a certain fuzziness, since it requires less hard thought to permit science and literature to merge into a single identity – two modes of reaching after the 'truth' at a specific point in history – than it does to insist on the manifest differences between them. Still, it is an attractive theory in the hands of a skilled exponent like Alex Comfort, as when he tells us that 'when a Darwin or a Freud produces a large change in the human self-estimate, the fact that he is able to do so shows that it was already changing, and reflecting the change simultaneously in art'.[1] By setting up a tandem model as a replacement for the tug-of-war model the dignity of both parties is elevated; and, moreover, such a conceptualisation helps any investigation such as the present one with its tricky problems of influence. For instance, as we shall see in a later chapter, the purely literary concern for the manifold problems of heredity certainly antedated the eventual resolution of these problems; and yet that layman's interest was at its peak when Mendel's paper of 1865 was rediscovered by three researchers in three countries simultaneously. In such cases it is good sense to speak of such upsurges of interest as arising from a common substrate than it is to speak of the concern in one part of a culture stimulating attention and investigation in another part – whether the flow is taken to be from science to literature or vice versa.

Such recent treatments of the problem are of course a continuation of a strain of thought which runs back to the Huxley–Arnold debate of a century ago and farther back still to the distrust of science among certain influential Romantic theorists. The later Victorian scientists, as the fairly amicable tone of the Huxley–Arnold debate testifies, believed (not unjustifiably) that their relations with the world of letters was a good healthy one and mutually very supportive. It is, indeed, with the consequences of such a belief that we are at present concerned. We must more or less take for granted the complex pattern of changes which temporarily pushed to one side the Romantics' fears of (in I. A. Richards's happy phrase) the 'neutralisation of nature' by scientific inquiry. Yet we must clarify just what these changes amounted to. To take any particular thinkers as representative of earlier nineteenth-century beliefs about the eventual consequences for the literary sensibility of scientific habits of thought must be a dangerous and arbitrary decision; but of the many who addressed themselves to these subjects Macaulay and John Henry Newman were among the most influential and widely discussed spokesmen. Now, at first sight Macaulay's expression of his conviction, in his famous 'Milton' of 1825, that science must soon become the factor most to be reckoned with in the evolution of poetic language is a good deal more generous than the Newtonian–Benthamite devaluation of poetry as a kind of ingenious nonsense. But as the essay gets under way it begins to emerge that Macaulay agrees about the erosive effect of scientific apprehension on literature. This is the charge of his potent metaphor of poetry as a magic lantern which operates most effectively in the gloom of ignorance and superstition. Such a theory of interaction (if obliteration may be called interaction) is really a prognosis of the gradual dominance of science over art. The contrast between these two ways of knowing, and the superiority of the former, is pointed up with the illustration of the tyro mathematician who 'by resolutely applying himself for a few years' can readily 'learn more than the great Newton knew after half a century of study and meditation'.[2] Macaulay works hard at persuading us, and perhaps himself too, that any notions the artist may form 'respecting the lacrymal glands, or the circulation of the blood' should not cast too much of a shadow over 'the tears of his Niobe, or the blushes of his Aurora'. But only by placing a quite disproportionate weight on *furor poeticus*, the artist's 'certain unsoundness of mind', is he able to do this; and when he comes to speak at last of the truths of poetry being the truths of madness we cannot but feel things have been brought to a desperate pass; that the rainbow is being unwoven once and for all and that literary perception is in headlong retreat. Nor are we much reassured by Macaulay's promise that in the future Scientific State there will be an 'abundance of verses, and even of good ones'. Beside such a prospect Newton's ingenious nonsense sounds quite appealing! Macaulay's unquestioned assumption that the two kinds of creativity are at war – which he took to be an obvious and inevitable consequence of the rise of Baconian empiricism – was restated time and again in the first half of the nineteenth century. As a proposition of stylistics it was still circulating in 1852, the year of Newman's *Idea of a University*. For him the facts of nature turned up by empiricism are anti-poetic, but so intrinsically is the terminology of science. As the terms of any of the exacter sciences approach the symbology of mathematics and thereby retreat from the warm equivocalities of the language of men, it is inevitable that 'however many we use, and however we may perpetuate them by writing, we never could make any kind of literature out of them'.[3] This assertion of the complete antipathy between objective and subjective knowing is as uncompromising as anything earlier. It amounts to the crudity of seeing the province of literature as thought, and that of science as things. Newman's account rejects the possibility of introducing ideas into literature at all until they have been thoroughly transformed into symbolic or even mythic material. And at that stage, should it ever be reached, the utilisation of those ideas will not be according to any rational principle; having been torn from their original context, they will be set in place as the underpropping to some imaginative ideal.

Theory and practice in literature tend to march in step, however, and by the 1850s a movement towards détente may be discerned which within ten years had grown to dominance. The major note for the rest of the century may be sounded by putting alongside each other a higher Victorian, post-Darwinian view of the fertilising powers of science on the literary imagination – Herbert Spencer's – and Wordsworth's of sixty years earlier.

In the 1802 preface to *Lyrical Ballads* the latter had persuasively argued for a new alliance between poet and scientist for the new century. For the young Wordsworth the poet (and as he is the highest form of literary enterpriser, the poet alone) had in his bones that unitive vision which the scientist works so hard to attain. The poet's response to scientific endeavour therefore is, or should be, eager participation and close attention to the challenge of 'carrying sensation into the midst of the objects of the science itself'.[4] In the famous words of the manifesto we are asked to look forward to a time when 'The remotest discoveries of the Chemist, the Botanist, or Mineralogist, will be as proper objects of the Poet's art as any upon which it can be employed'. These words are usually quoted as Wordsworth's plea for an immediate *rapprochement* but, if we give the right stress to the proviso which at once follows, perhaps they should be taken as a boast for the powerful digestion of the poetic apprehension. Poetry can in principle swallow down and assimilate chemistry, botany and the rest, but not yet: discoveries in these fields will become 'proper objects' *only* when they 'shall be manifestly and palpably material to us as enjoying and suffering beings'. That this might be a postponement *sine die* is argued by Wordsworth's reworking of Descartes' dream in the fifth book of *The Prelude*, where Euclid's geometry, the prototype and queen of the sciences, is symbolised as a Stone, whereas an Ode is a resplendent Shell; the two, moreover, are held by the knight in separate hands. Their distinctive roles could hardly be more dramatically expressed.

But if Wordsworth had his reservations, biological theory half a century later certainly became 'palpably material' to the literary mind. (Though it needs to be borne in mind that Darwin's researches into speciation were effectively complete long before Wordsworth died.) Post-Wordsworthian literary aestheticians found no reason for delaying to take the Stone and the Shell into one inclusive grasp. The hesitancies of Wordsworth vanish in the stridencies of Spencer:

> Science opens up realms of poetry where to the unscientific all is a blank. Those engaged in scientific researches constantly show us that they realize not less vividly, but more vividly, than others, the poetry of their subjects. . . . Think you that the rounded rock marked with parallel scratches, calls up as much poetry in an ignorant mind as in the mind of a geologist, who knows that over this rock a glacier slid a million years ago? The truth is, that those who have never entered upon scientific pursuits are blind to most of the poetry by which they are surrounded.[5]

Spencer's claims here for what has since become the cliché of 'the poetry of science' are at once more exaggerated than Wordsworth's and far less rigorous in their logic. He equates poetry and poetical musings in a way which, if they were in fact identical, would make a poet of Everyman. Possibly there are poetic subjects to be found in the geological relics of the Ice Ages (though few good poets to date have found inspiration there) but to realise that, and to realise a poem, are two very different activities. He is failing to observe the distinction between the poet's turning his interests into poetry, and merely meditating upon them poetically. Yet this confusion (shared also by several poets of evolution) is the very core to Spencer's argument, and he persists in the assertion that science – and he did in practice largely mean geology and the sciences of life – had in his day reached the point of tossing out prefabricated poetic notions for anyone to grab. Inevitably, then, the poetic *experience* is demeaned by the new relationship, for it is no longer, as Wordsworth would have it, 'the breath and finer spirit of all knowledge'. For Spencer the very act of hunting fossils serves to open one up to their poetical associations, while to pedantise over a Greek ode is to close oneself off from the grand epic of the organic world's development – and Spencer is in no doubt that it *is* an epic. Now, in an odd reversal of Newman, it is poetry that is lacking, that is suffering etiolation from its unwillingness to engage in scientific discourse.

The significance of Spencer's new way of marking out the respective territories of science and literary art rests, not in his particular formulation gaining assent, but in such a fresh attempt being thought necessary. Naturally enough, his call for a poeticised geology fell on deaf ears; for then as now literary artists concentrated their attention on the themes which have always most interested them – and these as a rule did not include theoretical biology. There is no occasion, and one would need a strong stomach, to spend much time assessing the merits of panegyrics to Darwinism. But if the details of Spencer's proposals may safely be forgotten, we must give ear to his confident tone; to the brashness with which he annexes the poetic impulse and attaches it to humdrum scientific inquiry. These bold claims for the real poetry in science must stem from a new appreciation, even a new reverence, for empiricism. Reverence is an apposite enough term to describe the wide following for what was a true cult, that of scientism – a cult which rounded itself out with dogma and doctrinaire intolerance of other philosophies. In its most organised form, positivism, it moved steadily towards sacerdotalism from 1878 onwards, the year in which Richard Congreve broke with the Continental parent body to found his Church of Humanity. Positivism in England was never more than an extreme fringe movement, but the flood of scientifico-moral essays, of what H. G. Wells called 'pulpit biology' and other uplift literature, testifies to the expectations society had of the Victorian scientific sage. His responsibilities were shouldered willingly enough; sometimes so willingly, indeed, that through pressure of time his speciality had to be abandoned altogether. No harm was seen in sugaring edification with entertainment and, thanks to the new media of the lyceum and soon afterwards the commercial lecture-circuit (an institution which supported Wallace in his improvident old age), public scientific discourse gained a mass appeal it had never had before and was soon to lose again. It was an age when (one observer noted) the middle classes tore from the hands of Mudie's librarians books by Darwin, Huxley and Lyell as though they were novels. Geology and palaeontology, the crazes of the 1840s, were succeeded by crazes for natural history and anthropology. Scientists themselves were the objects of acute public interest. The lady who in the 1890s found Huxley on the lecture platform 'faded but still fascinating' was displaying a last flicker of the response, half reverence and half fan-worship, which was specially reserved for certain star scientists who were also folk-heroes. Familiarity with the age blunts us to it, but there is after all something odd about a man like Huxley, who initially made his name working in the recesses of marine biology, in later years becoming accepted and even nationally famous as a voice of conscience on such issues as racism and female emancipation; becoming the spokesman, in fact, for a reasonable humanism. The specialist's promotion to the rank of social moralist is no doubt something familiar in every age. The promotion of the scientist is, however, a feature of the waxing faith through the seventh and eighth decades that science could tackle all problems – not merely technological problems (that was mostly taken for granted) but social and political problems as well. Even, indeed, epistemological problems: scientism, for its really bold proponents like G. H. Lewes, was a cut above all other philosophical schools in that it could make 'linear progress' and not wander for ever in the labyrinths of subjectivity. Despite the fulminations of Morris, Carlyle and Ruskin among many others, and before disillusionment set in among their own ranks in the late 1890s, it is a fact of social history that the leading British intellectuals in influence and prestige, if not in direct power, were the pure scientists. Beatrice Webb's memories of childhood in that period tell how in a certain circle everything was thought available to scientific scrutiny; for, rightly or wrongly,

> it was they who were routing the theologians, confounding the mystics, imposing their theories on philosophers, their inventions on capitalists, and their discoveries on medical men; whilst they were at the same time snubbing the artists, ignoring the poets and even casting doubts on the capacity of the politicians.[6]

In this extensive programme the biologists were well to the fore – so much so that the vulgar idea of how science itself operated was based on the passionate scepticism of a Huxley essay. It is well worth noticing that Webb's remark above forms part of her appraisal of the writings of Winwood Reade (the explorer, novelist and anthropologist who hoped for a while to supplement Darwin with a treatise on the origin of mind), and her conviction that his writings were vital to her own awakening intelligence in the 1870s. Reade was a belligerent polemicist, and the biologists were typically the most belligerent of all the prophets of scientism. Even when they had their radical disagreements within their own ranks they presented as united a front to the outsider as any social clique. The conspiracy of silence which denied Samuel Butler one jot of satisfaction from his critique of Darwinism was directed more against Butler personally as an ungentlemanly controversialist than against his biological opinions, which read rather modestly when set against, say, Wallace's or even Mivart's.

More reasoned defences of scientism by biologists were rare, and therefore Huxley's address of October 1880, 'Science and culture', is notable among his prolific writings in that it marks one of the very few occasions in the period when a biologist made a critical foray into the literary culture. Arnold's *Culture and Anarchy* (1869) and its central double proposition – that culture means an unending criticism of life, and that literature suffices to construct such a criticism – gives focus to Huxley's claim that science, and especially the life sciences, are essential extra ingredients in the recipe for culture. It is a little remarkable just how much ground Huxley was on that occasion willing to surrender to Arnold. He not only pays tribute to Arnold's 'general catholicity of spirit' and his 'true sympathy with scientific thought',[7] but also concedes that culture must be marked out as quite different in kind from training and technical skills in that culture must 'supply a complete theory of life, based upon a clear knowledge alike of its possibilities and of its limitations' (p. 143). He does not baulk until he comes to Arnold's claim for literature as the truly nourishing cultural medium. At this point scientism breaks through, as with his usual vigour and broad sweep Huxley sketches an unremittingly rationalistic résumé of Western intellectual history. Literature deals in dreams, in sweet illusion. Its dealing with fact has ever been tenuous and even, when under imaginative pressure, quite dishonest. Only with Bacon and the rise of the empirical method did a new way of interacting with the world become important and at last preponderant in human affairs. Only science has the unique merit of being an incontrovertible lawgiver:

> This scientific 'criticism of life' presents itself to us with different credentials from any other. It appeals not to authority, not to what anybody may have thought or said, but to nature. It . . . bids the learner seek for truth not among words but among things. It warns us that the assertion which outstrips evidence is not only a blunder but a crime. (p. 150)

Such is the central statement of 'Science and culture', and whatever our final judgement may be we must concede that sincerity vibrates through it. Huxley indeed built his life on the axiom that sorrow gravitates to sin just as surely as an apple to the earth because both are rooted in observable and quantifiable physical law. By making empirical deduction his ultimate court of appeal Huxley expected that the verdicts of that court would eventually come to express immutable moral and physical laws against which there could be no protest. Despite his habitual and famous scepticism Huxley was in reality at one here – though how outraged he would have been to hear it said so – with theologians like Henry Drummond, who was inspired by his reading of the phenomena of parasitism and devolution to see in the occupants of lunatic asylums and prisons the inevitable by-products of the working of natural selection. And although it is a perfect expression of Victorian scientism at its most high-minded, how inadequate is Huxley's 'criticism of life' both as philosophy of science and as an account of the relations between science and letters! Huxley is unable, or unwilling, to consider that 'evidence' may be labile or may be distorted by the mere pressure of observing it; or that it may be selected to fit a moral predisposition, as with those case-histories of reputable doctors of the period who, to approving nods, claimed in print that they had *observed* insanity to follow from working on the Sabbath. The very decision as to whether an assertion 'outstrips evidence' can be a value judgement; and it verges on arrogance to claim that one never goes beyond the facts while simultaneously reserving the absolute right to determine what the significance of those facts might be. So, when Huxley attended to the questions of women's emancipation or the political and ethical consequences of Darwinism, he did so unmistakably as a moralist and social critic while all the time adopting the lofty pose of the disinterested research biologist. Of his later famous 'Evolution and ethics' it has rightly been said that it is 'a masterpiece of concealed debate'.[8] This is no less true of 'Science and culture' and many another essay besides: in them all Huxley assumes without qualification that his biology is value-free, while demonstrating beyond question that his biology certainly is not. If he ranks high in Victorian polemical prose, it is not essentially because of the irrefutability of his logic, or the clarity and expressiveness of his style, or even the accuracy of his facts (unimpeachable though these usually are); he ranks high because his audience held their scientific intelligentsia in esteem.

These same strictures may be applied to Darwin also, and if in his case the relations between himself, his extra-professional public and his medium of communication are more problematic they are more illuminating, too. Darwin's medium was, with the exception of what are by present standards a minute number of technicalities, very much the language of educated common men. Certainly the *Origin* and the *Descent* comprise (as Darwin said) 'one long argument' for a naturalistic explanation of the organic world, and in so far as they are such an argument they are addressed to Darwin's fellow-workers in evolutionary biology. And yet the *Origin*, and far more so the *Descent*, are continually bumping against matters of theology and ethics for the discussion of which Darwin could not preserve a determinedly judicial tone. On occasion, when particularly moved, he adopts a style of dramatic plainness which has been alleged to be quite unlike the style used in typical scientific monographs of the day. Walter Cannon has spoken of the latter as being 'about as complex and incomprehensible to the layman then as they are now';[9] and Richard Grove, in his attempt to describe the *Origin* as a formal rhetorical structure, has called it 'neither a bold popularisation nor a scientific treatise'.[10] Cannon by implication and Grove expressly argue that Darwin cultivated a style that would reach, in its clarity and convincingness, right into the sphere of interest of the educated layman; Grove further maintains that Darwin adopted this strategy not to widen his audience but to defuse the reaction he expected from his compeers when his theory was formally offered to them. Certainly Darwin sweated over his style; harder, in his own comparison, than any 'nigger with lash over him'. Darwin's work wrestles continually with the inadequacies of language and he is well aware that language is itself an evolutionary product and perhaps cannot fully articulate its own deficiencies. He was closely attentive to the problem of constructing a reasoned and courteous discourse that could adjust itself patiently to the most sceptical and abrasive of implied readers. In that sense Darwin conceded to the demands of rhetoric just as Huxley did. But the further suggestion that Darwin deliberately sought a large lay public cannot be substantiated. He himself declared that he would be satisfied with an attentive audience made up of only three persons: Lyell, Hooker and Huxley. He never set himself the task, as Huxley did, of commanding an audience of 'open-minded and impartial laymen'; on the contrary, on one occasion he expressed scorn at courting such an audience: 'I hope I do not care very much for the approbation of the non-scientific readers.'[11] This is supported by Darwin's concern for his publisher, Murray, that the print order for the first edition of *The Origin of Species* (not indeed a large one) was too great for the small select readership he was anticipating. The evidence really suggests that Darwin thought he was writing in the standard scientific prose of the day, fitted nicely to his subject and containing the kind of extrapolations from the facts that would be found acceptable. It may be that the terminology of the exacter sciences was already serving to push *their* key issues beyond the ken of the curious general reader, but this was assuredly not so in biology. Comparison of the *Origin* or *Descent* with other contemporary texts such as Mivart's *Genesis of Species* (1871), Haeckel's *General Morphology* (1866), or Wallace's *Contributions to the Theory of Natural Selection* (1870) shows Darwin's work is about in the middle rank of technicality. At times his prose is clear and fresh, at times tinged with humorous irony, at times impeccably obscure. Such variations in tone surely do not arise, as one might suspect, from any uncertainty about the audience he is addressing. Darwin is not anxious in the sense that he cannot decide whether to speak to the professional or to the layman, because in biology our present distinction between 'technical' and 'routine educated' prose did not then exist. The homely metaphors, the cool assumption of shared values, the introduction without preamble of personal tastes: all these attest to Darwin's unquestioned assumption of a homogenous audience. The anxious movement of his prose comes rather from his training within the same positivist intellectual tradition as that of Hobbes, Locke and Newton. For the mid-Victorian Darwin and his predecessors shared the supposition that science is a unity, so that investigation into any branch should render up general laws applicable everywhere. If each new formulation did not seem readily available for social, religious or philosophical extension, then there were many interpreters – poets, historians, intellectual middlemen of all calibres – to press it into service. And Darwin, like his fellows, took his obligations here very seriously. He moves, and he expects his reader to move with him, from the strictest cataloguing of experimental results to the purest armchair speculation. He does not have to resist the temptation to pontificate. His only concern is to do it well; and here he fails. As an argumentative edifice, the *Origin* lives up to Darwin's proud boast that for years he read only documents that might refute his hypothesis in order to anticipate every possible objection; but as the expansive meditation of a scientist-sage it is muddled. Darwin's awareness of his failure in the role of sage which his culture demanded of him surely stands behind the erratic oscillation between passages of cautious description and passages of conscientiously spiritual uplift which comprise the last section of the *Origin*. If we take the sad remarks of Darwin in his *Autobiography* (1876) to heart, concerning his having lost in his mature years 'all pleasure from poetry of any kind' as well as the power of becoming deeply attached to anyone,[12] we can reasonably surmise that the struggle to force some kind of reconciliation between the various parts of his being eventually proved too much. 'Anaesthetic' is one label that has been affixed to Darwin to describe his entry into the emotional vacuum; its intent is to suggest that he was rendered senseless to art by the numbing power of his theory, which finally dissociated fact from affect.[13] Another biographer has selected the adjective 'fragmentary' for Darwin's progressive failure to achieve the unity of expression demanded by his culture. The imaginative part of his being was, in these terms, so shattered by his discoveries that in time 'he ceased to have any vital life outside his work at all, was just a nice, a very nice, old man'.[14] These explanations may help us set Darwin's undoubted neuroses into a cultural and psychological context, but they certainly have no validity as a general theory of biologico-literary interaction – any theory, that is to say, which stresses the destructive consequences of taking an empirical, biological or utilitarian (the terms tend to run together in this context) stance to nature. T. H. Huxley, exposed to the same disquieting revelations as his friend, and with much the same sort of basic training, did not react by forming callouses over his emotions. Quite the contrary: he remained alive to the merits of every form of artistic expression, delighted in them all and practised several of them himself. He suffered no dryness of the spirit (though indeed prone to depression) possibly because, unlike Darwin, he was temperamentally fitted to that kind of philosophical analysis which interpreting the biology of his day so required, and which his colleague said he found so uncongenial. Huxley's sensitivity to literary expression and his determination to master it arose directly from his studies in classification, and especially his early reorganisation of the Linnean system. These had made him highly aware that all conceptual frameworks are man-made; that the categories of science are far from immutable; even that those of biology are particularly prone to conceal value judgements within them (but, as we have seen, this awareness was intermittent and tended to vanish when the stakes were high). We recall his own prescription for a good style – simply 'clear and forcible expression' – and his stylistic masters: Hobbes for dignity, Swift for concision and clearness, Goldsmith and Defoe for simplicity. It is not accidental that, with the exception of Goldsmith, Huxley was invoking here the great controversialists, and this reminds us, too, of how important an ability to manipulate rhetoric counted in Victorian science. Every biologist with a hypothesis to be tested or forceful theoretical point to make had to come to some rough working knowledge of the skill. Huxley himself said that he hated oratory, but his distaste should never conceal the fact that he and others – Romanes is the name that most readily comes to mind – argued hard and argued to win. Huxley tried to claim that his technical writing was a strict presentation of the facts and that it was only his public lectures that made use of rhetoric to offer personal convictions which were clearly labelled as being such, but it may be doubted whether this distinction is really worth anything. On examination Huxley's mastery of rhetoric may be found in all his writings to include a very sophisticated grasp of stylistics.[15] The very definition of biology offered in 'On the study of biology' (1876) – that is to say, the study of all phenomena exhibited by living things – was chosen so as to license the biologist's exploration of politics, philosophy or education.

If Huxley did his best to exploit the temporary convergence between the 'two cultures' by trading on his professional reputation, he was followed by others; and it was not only popularisers like Grant Allen who were able and willing, because of the lucrative rewards involved, to stand astride the straits separating biology and letters. The energetic anti-Darwinian St George Jackson Mivart turned out a number of sentimental romances, presumably for diversion. Robert Louis Stevenson wrote not only a full biography of the other major critic of Darwinism, Fleeming Jenkin, but also a fictional sketch based on him, 'Cockshot' in *Talk and Talkers* (1910). Even poetry, most demanding of the literary muses, was a channel of expression for at least three of the dominant names in the life sciences. The emotions unleashed by the funeral of Tennyson flowed, for Huxley, through the unaccustomed medium of verse: braving the competition of, among others, F. W. H. Myers and Francis Palgrave, Huxley added his own workmanlike threnody, 'Westminster Abbey: October 12, 1892', to a tribute published in *Nineteenth Century*. Two more pleasant lyrics appeared posthumously in a collection of his wife's poems.[16] The stern Romanes, in public a relentless and unforgiving controversialist, relieved the tensions of his professional life by producing a good deal of verse from 1878 until his death in 1894. Technically, his posthumous collection *Poems* (1896) is capable enough. Their depth of sensibility, though, may be judged from his wife's innocent appraisal that 'he shared Mr Darwin's tastes for simple, pure, love stories . . . *The Heir of Redclyffe* brought tears to his eyes.'[17] In the opposite camp George Douglas Campbell, eighth Duke of Argyll (1823–1900), did not expend all his energies studying geology and ornithology so as to damn the materialism of Darwin in his *The Reign of Law* (1867) and *The Unity of Nature* (1884). He bolstered the struggle against natural selection with the versified propaganda of his *Burdens of Belief* (1894). The title poem, for instance, an outpouring of faith in divinely directed evolution, refers to the difficult case of the electric organs of some fish as being (to the teleologist) 'the secret weapons of design'. The Duke does not entirely trust in his poem to bring out all the significance of this case: he devotes one of the many elaborate notes appended to the title poem to telling us that Darwinism is powerless here, because selection could not have been applied *until* the organs had, in all their complexity, achieved full functioning – a commonplace but potent objection. Still, as a minor 'In Memoriam' Argyll's poem is even less effective than its famous predecessor in reconciling the demands of reason and emotion. It impresses less as a joyful hymn to evolutionism than as a warning about the perils of using poetry for scientific disputation.

Biology and Letters: Some Specific Linkages

In the preceding discussion of the factors which governed the relations between science and literature in the later Victorian period, much of what has been said was true equally of all the sciences, not just of biology. We have mentioned the prevailing scientism and the literary qualities of the scientific prose of the day; we have seen how the responsibilities freely assumed by the scientific sage controlled his use of language and brought him continually before the public as a polemicist or, occasionally, as a literary artist. But the physicists William Thomson and John Tyndall and the statistician Karl Pearson conformed to these pressures no less than did G. J. Romanes and T. H. Huxley. Is it possible to isolate any specific ways in which biology and letters were drawn together – of more import, say, than the bizarre fact that, early in the year of the *Origin*, the eccentric naturalist Frank Buckland rifled Ben Jonson's tomb of his heelbone as a memento?

There were several ways in which the life sciences were unusually accessible to the literary mind during the late nineteenth century; more accessible than the physical sciences. To begin with, since much of the endeavour in biology was still being channelled into systematics, there was a great reliance on the contributions of people who were very far from being professional, full-time researchers. There were indeed very few paid researchers. Pure physics was a handmaid of engineering and to that extent subsidised, but in pure biology an almost indispensable prerequisite for doing elaborate research was a substantial private income: Darwin, Romanes, Lyell and Hooker were all in that fortunate position. T. H. Huxley, upstart from the class of the landless gentry, was the exception proving the rule and, as he himself recognised, was the first major biologist to make a living wholly from lecturing, writing and research. This unsatisfactory state of affairs had one valuable consequence. Those who moved on the fringes of the science (and this especially applied to those who collected under the banner of 'field naturalists') could, and very frequently did, contribute in a meaningful way to technical debates which were readily comprehensible to them. For even the most complex debates were conducted in a language which was a shared heritage. This was a time when Herbert Spencer, without any formal training whatsoever, without even the rudiments of laboratory practice, could feel competent to express an opinion on the most subtle problems arising from the study of inheritance in the pages of the *Nineteenth Century* and gain a respectful ear though having no new facts whatever at his disposal; when Grant Allen could draw applause from Huxley himself for his little observational essays, and a promise from Wallace to cite one of Allen's essays as a reference in his next book; when even Darwin was not too proud to take a fact or two from any man of letters with some informative anecdote to retail. He engaged in courteous disagreement with the fiery and rather insulting W. H. Hudson; he cited the essayist and political theorist Walter Bagehot approvingly five times in the second edition of *The Descent of Man* and a note of Charles Kingsley's on the mating noises of fishes once.[18] The *Descent* itself turned out to be the last major original work of theoretical science able to be grasped in its entirety by the average educated reader, though for about another three decades the layman needed nothing more than a subscription to one of the great general periodicals like the *Fortnightly Review* to avoid the alienation from current science that is part and parcel of the interaction between science and the humanities in our own day. Though the *Fortnightly* and its great companions were media of popularisation, too, they did open their columns to the serious interchange of new information or the considered rejection of a hypothesis. A glance through the files of one of the very few specialist journals, *Nature*, definitely reinforces the impression that not it, but the more general periodicals, had at this time the more vital role to play in disseminating scientific inquiry. For many years after *Nature* was founded in 1869 the readers were regaled with nothing more solid than topographical descriptions, ornithological observations and suchlike; its format was very much that of a magazine for bug-hunting vicars with time on their hands, like the Rev. Camden Farebrother in *Middlemarch*. Biology's raw facts are to be found there, to be sure, but nothing is digested; there is little or no attempt at articles condensing 'the present state of knowledge' into a dozen or so taut pages – a job performed so admirably in issue after issue of the *Fortnightly* and elsewhere. Further, the readership of *Nature* was very small; not more, according to the best estimates we have, than between two or three thousand up to 1872.[19] The *Fortnightly* alone had this circulation, and for the reader who disliked the *Fortnightly*'s politics there was a choice to be made among the *Athenaeum*, the *Academy*, the *Quarterly*, the *Contemporary* and others. All these carried a good load of biological debate. For a few years the main organ of the younger generation of biologists was the *Natural History Review*. T. H. Huxley, ably assisted by several other 'plastically minded young men', founded this and helped maintain its fierce 'bishop-eating' tone; but he ceased to contribute to it after July 1863 (it folded in 1865), choosing instead to write for a journal of more varied appeal, the *Reader*, for a few years and afterwards for a variety of outlets. *Nature* did not start to come into its own until the late 1890s, when articles of some substance and later of major importance began to appear, by which time the pattern of modern scientific journalism was becoming recognisable. In the meantime, such matters as the attempted hammering-out of the exact role of Lamarckian factors in inheritance were to be found between the pages of the *Contemporary*. It is hard to believe that some, even most, readers were not bored by this apparently endless debate, but the fact remains that a general periodical was thought by all the participants – Spencer, Romanes and Weismann – to be a perfectly appropriate forum for resolving problems which were the lifework of all three.[20]

Possibly even more surprising and significant is that this editorial assumption of a high level of tolerance of and interest in biology among the public at large was by no means restricted to the highbrow organs. The same assumption is to be found in many burlesque pamphlets and other fugitive productions for unsophisticated readers. Very little of this extensive sub-literature has been described. One amusing and quite well-informed example must suffice here: Charles William Grant's *Our Blood Relations, or The Darwinian Theory* (1872). Published anonymously, this verse tract offers in its first part a highly critical summary of the *Descent*, making many shrewd hits. In the second part Grant sets out his fundamentalist faith with a special emphasis on the regulative aspect of creationism. He thereby makes it plain that his fear is of the onset of civil and moral chaos, which he suspects will follow hard on any acceptance of Darwin's theory of human evolution. The verse form throughout is rough rhyming couplets which occasionally achieve a memorable vigour, as when Grant lambasts Kingsley's Broad Church liberalism on biological issues (not an easy target to miss, admittedly). Another anonymous squib of 1863 in *Punch* raises the issue, via a Welsh correspondent, of blending inheritance. He supposes innocently that the immediate progenitors of humanity must have been ape infant prodigies, and so back down the evolutionary scale with 'every new species of at least all the higher animals involved in the nearly coincident birth of two infant prodigies which were not hybrids'.[21] The correspondent goes on to ask dubiously whether even all the resources of geological time could have served to bring together two such unlikely variations; for how could such variations have perpetuated themselves otherwise? Behind the vulgarity of dubbing a Welshman a 'semi-ape' for a cheap laugh there is active that very important and serious 'swamping' objection to Darwinism which was not given full expression by Fleeming Jenkin for another four years.

This popular interest and involvement in the technicalities of biological debate was reinforced by the fact that, considered historically and socially, biology at this time was in a somewhat anomalous position with respect to the exact sciences. The physicists and chemists, thanks to their superficially more amenable subject-matters, had confronted and overwhelmed their heterodoxies years, or even centuries, earlier. No serious supporters of phlogiston, or of the music of the spheres, or of angels in the driving-seats, lingered on even as stimulating and thought-provoking eccentrics. The contrast with Victorian biology could hardly be more acute: there, to the community at large (and even to some members of their own ranks) scientists' pronouncements were still in serious contention with the systems of Moses, of the Greeks, of eighteenth-century divines. What one chemist maintained would, after experimental checking, be assented to by his fellow-workers or else be rejected on clearly articulated grounds; but the biologists were riven by schism even over fundamentals. Now, the argument here is that the literary utility of a science is at its highest when its central axioms retain sufficient ambiguity to make their exact bearing on the human condition uncertain but interesting. In the period in question basic research in the exacter sciences led in due course to the submarine telegraph cable, electric light and electromagnetic communication. But the workers who stood behind all this astonishing technology – men of the stamp of Clark Maxwell, Thomson, Crookes – were never more than dull figures in the public imagination. It is hard to imagine anyone bothering to dub Maxwell 'Faraday's bulldog'. By contrast Darwinian biology, though at first sight seeming to promise the same kind of lawful regularity that Mendeleyev brought to chemistry in 1869, soon subsided into a soft-textured subject where imagination could run riot. We will be tracing some of the consequences of this in the following five chapters.

Notes: Chapter 2

1 Alex Comfort, *Darwin and the Naked Lady: Discursive Essays on Biology and Art*, p. 3.
2 *The Life and Works of Lord Macaulay*, Vol. V, pp. 5, 6. 'Milton' was first published in 1825.
3 John Henry Newman, *The Idea of a University, Defined and Illustrated in Nine Discourses . . .*, ed. Martin J. Svaglic, p. 206.
4 Preface to *Lyrical Ballads* (1802), in *The Poetical Works of William Wordsworth*, new edn revised by Ernest de Selincourt, p. 738.

5 Herbert Spencer, *Education: Intellectual, Moral, and Physical* [1861] (London, 1893), pp. 40–1.
6 Beatrice Webb, *My Apprenticeship*, p. 113. A subjective view, of course, but there is hard evidence in that the number of people attending the annual meetings of the British Association remained constant at about 1,200 for the three decades before 1860. But by the year to which Webb is referring (1872) that number had more than doubled to 2,500. Much time was spent at these meetings on the moral responsibilities of science.
7 Thomas Henry Huxley, 'Science and culture', in his *Collected Essays*, Vol. 3, p. 142.
8 By Michael S. Helfand, 'T. H. Huxley's "Evolution and Ethics": the politics of evolution and the evolution of politics', *Victorian Studies*, vol. 20 (Winter 1977), p. 160. Helfand strongly attacks Gertrude Himmelfarb's contention in 'Varieties of social Darwinism' (in her *Victorian Minds*) that Huxley kept Darwinism depoliticised; i.e. that he succeeds in his expressed wish never to give his sanction to the application of Darwinism to any social theory or practice.
9 Walter F. Cannon, 'Darwin's vision in *On the Origin of Species*', in George Lewis Levine and William Anthony Madden (eds), *The Art of Victorian Prose*, pp. 154–5.
10 Richard Stanley Grove, 'A re-examination of Darwin's argument in *On the Origin of Species*', unpublished PhD thesis, University of Missouri, 1969, fo. 3.
11 Letter to W. B. Carpenter of 6 April: Francis Darwin (ed.), *The Life and Letters of Charles Darwin*, Vol. 2, p. 299. Dated by the editor 1860.
12 Charles Darwin, *Autobiography of Charles Darwin, 1809–1882*, ed. Nora Barlow, pp. 44, 115.
13 Donald Fleming, 'Charles Darwin, the anaesthetic man', *Victorian Studies*, vol. 4 (March 1961), pp. 219–36.
14 'Geoffrey West' [Geoffrey Harry Wells], *Charles Darwin: The Fragmentary Man*, pp. 328–9.
15 Considering the influence he wielded over scientific and educational policy, Huxley's prose has, surprisingly, been subjected to little penetrating stylistic analysis. Apart from Helfand above, D. J. Foskett has looked at the technical prose in 'T. H. Huxley and his presentation of scientific material', unpublished MA thesis, University of London, 1955, and Charles Blinderman at the language of the essays in 'Semantic aspects of T. H. Huxley's literary style', *Journal of Communication*, vol. 12 (1962), pp. 171–8.

Walter Houghton has exposed Huxley's frequent rhetorical flourishes in his 'The rhetoric of T. H. Huxley', *University of Toronto Quarterly*, vol. 18 (1948–9), pp. 159–75. The substance of Houghton's important attack is taken up by Foskett, fos 91 ff., in more moderate vein.
16 'Westminster Abbey' was printed in *Nineteenth Century*, vol. 32 (November 1892), p. 831. There are some minor differences in the version published together with 'From Shanklin' and 'Altr'Arno, Florence' in *Poems of Henrietta A. Huxley with Three of Thomas Henry Huxley*. Entirely incorrect is William Irvine's statement in *Apes, Angels and Victorians: A Joint Biography of Darwin and Huxley* that this first poem was the only one 'of his discreeter years' (p. 266). The other two poems are dated only five and seven years earlier, when Huxley was in his early sixties.
17 Ethel Romanes, *The Life and Letters of George John Romanes*, new edn, p. 155.
18 Walter Bagehot's *Physics and Politics* (1872) first appeared as a series of essays in the *Fortnightly Review* from 1867, where Darwin presumably read them. The citation from Charles Kingsley in *The Descent of Man* is to 'Strange noises heard at sea off Grey Town', *Nature*, vol. 2 (19 May 1870), p. 46. This latter title accurately captures the flavour of *Nature* at this point in its history.
19 There is no adequate bibliographical–critical guide to the scientific press in this period. The estimate above is taken from Alvar Ellegård, *The Readership of the Periodical Press in Mid-Victorian Britain*, p. 24. This invaluable study, being a predecessor of his *Darwin and the General Reader*, unfortunately covers only the same dates as the latter (1859–72). There was a very great expansion of the periodical press in these decades: between 1860 and 1870 a new periodical began publication once every three weeks, on average. Though

the percentage of these covering scientific topics must have been small, doubtless there is a good deal of writing by well-known scientists stored in the files of such transient productions.

20 It is worthwhile summarising the progress of this protracted debate, because it demonstrates as nothing else can the almost unlimited appetite of Victorian *lay* readers for the most intricate analysis of issues in biological theory. Herbert Spencer started the ball rolling with his 'The inadequacy of "natural selection" ', vol. 63 (February–March 1893), pp. 153–66 and 439–56. This is a typical piece of anti-Darwinism of the time, rehearsing all the usual objections. In the same volume (April 1893, pp. 499–517), George John Romanes reacted to 'Mr Herbert Spencer on "natural selection" ' with a long-drawn-out demonstration 'that the whole of this line of evidence is practically worthless' (p. 499). Spencer answered Romanes in May (1893, pp. 743–60); and, as if that wasn't enough, the whole debate was reviewed in detail as 'The Spencer–Weismann controversy' (vol. 64, July 1893, pp. 50–9) by Romanes and Marcus Hartog in tandem. Later on in the same year August Weismann himself entered the fray with a two-part article insisting on 'The all-sufficiency of natural selection' (vol. 64, September–October 1893, pp. 309–38 and 596–610). This did not silence the indefatigable Spencer, who offered 'A rejoinder to Professor Weismann' (vol. 64 December 1893, pp. 893–913) and finally, not having been able to draw Weismann again, concluded almost a year later with 'Weismannism once more' (vol. 66, October 1894, pp. 592–608). And this sequence is not atypical.

21 'Notes on development (no. 1)', *Punch*, vol. 44, 21 March 1863, p. 122.

3

Better, Wiser, and More Beautiful Beings:

The Cheerful Doctrine of Evolutionism

> For, judging by analogy, better, wiser, and more beautiful beings will inhabit this planet in the ages to come, according to the law of evolution, than we can now have any conception of.
>
> MATHILDE BLIND (1886)

> To us, the whole purpose, the only *raison d'être* of the world . . . was the development of the human spirit in association with the human body. From the fact that the spirit of man – the man himself – *is* so developed, we may well believe that this is the only, or at least the best, way for its development; and we may even see in what is usually termed 'evil' on the earth, one of the most efficient means of its growth . . . Beings thus trained and strengthened by their surroundings, and possessing latent faculties capable of such noble development, are surely destined for a higher and more permanent existence.

The source of this emotional defence of teleology in evolutionary biology might well be judged, in the lack of other evidence, to be some eminent divine's *cri de cœur* against agnosticism in a number of the *Quarterly Review* of the early 1860s, or perhaps part of an attempt to paraphrase Pope's *Essay on Man*. It is in fact taken from the concluding pages of Alfred Russel Wallace's *Darwinism* (1889).[1] By what was after the events of 1858 a characteristic expression of magnanimity, Wallace was one of the first to use the term 'Darwinism', not simply as a synonym for the work of Darwin, but in its modern, restricted sense. Wallace limited the meaning of the word as he did because by the 1880s he had become convinced that no subsidiary hypotheses whatever, including those offered by Darwin himself, were necessary to an inclusive theory of evolution. He did not believe in the inheritance of acquired characters. Sexual selection he held to be a superfluity in biological explanation to be shorn away by applying Occam's razor. 'Natural Selection is supreme,' he insisted, 'to an extent which even Darwin himself hesitated to claim for it' (p. 441). Supreme: but supreme only in the animal kingdom below the level of *Homo sapiens*. For, as the leading quotation insists so forcibly, Wallace exempted man's past and future history from the natural conflict. Natural selection had not produced man. Though a hard Darwinian in every other respect, Wallace was, for his own species, an uncompromising teleological evolutionist.

That Wallace deserted the inner Darwinian circle over the issue of the origin of man is a fact; and it has been substantiated that that desertion occurred soon after Wallace had become convinced, late in 1866, that his observations in the seance-room were not due to hallucination or trickery.[2] The cause of Wallace's desertion is, however, less important than the need to understand, before we go on to consider some of the cultural and specifically literary manifestations of evolutionism, the reasons for this post-Darwinian resurgence of biological entelechy and the wide range of options this made available to men inclined to appeal to that science for human guidance. Wallace was far from being alone in constructing a complete anti-Darwinian philosophical system out of the new anthropological and palaeontological facts as he saw them. St George Jackson Mivart (1827–1900), a Catholic who in the 1860s was a Darwinian and an intimate of Huxley, trod much the same path as Wallace and to the consternation of the Darwinian circle had also reneged completely by 1869. His *Genesis of Species* (1871) is a formidable work indeed, collecting and deploying all the most stringent arguments ever to be levelled against Victorian Darwinism. Mivart did not bother to conceal the religious prejudices which chiefly motivated him, but his trained legal faculty probed and probed away at the weak points in the *Origin*'s argument to deadly effect. His criticism had a constructive side, too, in that it tried to remedy Darwinism's failings; but it cannot be said that his remedy was one palatable to Darwin's faction. Mivart argues in the *Genesis* for a Christian version of Wallace's interventionism; for there being a continuous manipulation of the stream of inheritable variations by various agencies both terrestrial and what he called 'cosmical'. The high status in the ranks of Victorian science of Wallace and Mivart should therefore make us cast a more sympathetic eye on other, less august exponents of evolutionism; for if such men as these could be persuaded that evolution is really a metaphor for design, then there must have been compelling reasons to make them believe so. The case of Wallace is the more fully documented one.

When we examine Wallace's papers of the 1860s in their original form (for several were later revised to accord with his later views), his last full endorsement of natural selection as the universal mechanism of evolutionary change – even, by implication, reaching as far as man – is found to come in 1867, after Wallace had been 'convinced' by the evidence for the paranormal he had encountered at seances.[3] Less than two years after this date we find Wallace reviewing two new editions of works by Lyell and making for the first time a tentative statement that, in his opinion, certain specifically human faculties just could not have arisen through any natural process at all (not even use-inheritance, which Wallace in any case rejected as unproven) but were instead imposed upon man's immediate ancestors by incorporeal intelligences benevolently involved in his destiny.[4] This was indeed an extraordinary line to take in that since Darwin had yet to publish his *Descent of Man* Wallace could not know in detail how forcible a completely naturalistic explanation might be; nevertheless Wallace was convinced that only his astonishing hypothesis was adequate to explain certain crucial facts. Among these, and the ones that Wallace found totally inexplicable without positing supernatural intervention, were: that the size of the brain-pan is very similar in all known true men of the past and present, yet huge compared with that of even the highest apes; that the complex structure of the hand and foot which permits man's unique use of tools is fully perfected in even the most primitive races; that the larynx is physiologically far more complex than could ever be demanded by the language and song repertoires of savages; that the loss of body hair confers no breeding advantage and therefore could not have been brought about by natural selection; that, finally and most dauntingly, inside the skull of a cave-dweller exists a brain potentially capable of differential calculus. This last point of the 'surplus cortex', though very telling (and indeed puzzling enough to have been revived periodically ever since), is to our present understanding a fairly shaky one. It makes large assumptions about the limited degree of intelligence demanded at the hunting-and-gathering stage of human evolution, and about the correlation between abstract reasoning power and the brain's physical size. Since substantive information was lacking on both these points a century ago, the argument could sway this way and that between authorities without moving forward an inch. Huxley defied Wallace by saying (in his turn, without any strong evidence) that 'the intellectual labour of a "good hunter or warrior" considerably exceeds that of the ordinary Englishman'[5] but, despite him, Wallace – and Argyll and Mivart, too – continued to press hard on their point. How overwhelmingly, they argued, does the intelligence of a contemporary primitive living under palaeolithic conditions exceed his daily opportunities! How could natural selection possibly pick out unerringly *on the test of fitness to survive alone* individuals of superior abstract reasoning capacities, when the most complex mental challenge the environment could set them might be how to trap a small mammal? Wallace found this quite inconceivable. He entirely rejected Darwin's plea for the qualitative similarity of the intelligence of beasts and men. As he put it mockingly in his review of 1869, 'natural selection could only have evolved the savage with a brain a little superior to that of an ape, whereas he actually possesses one but very little inferior to that of the average members of our learned societies'.[6] He felt that the same strictures applied to the evolution of language. The average British rustic, someone had calculated, could live out his life on a total vocabulary of three hundred words without being incommoded to any degree. This probably apocryphal calculation crops up in several contemporary discussions on the genesis of language. The witty Grant Allen once debunked it by acting out in pantomime the activities of Hodge's day, thereby indicating all the objects, relationships and concepts he would have to name. But the precise size of the basic English vocabulary is less important than the puzzling superfluity Wallace mentions: *why* have the blind evolutionary forces acting solely by selectionist pressure gratuitously given that labourer a cortex able to accumulate the word-store of a Shakespeare? These questions struck Wallace with enormous force and his response was effectively to reinstate Paley's evidences and indeed the whole argumentative structure of natural theology. In order for the physiological apparatus of speech to have been evolved in advance of the requirements of its owners, he thought there must be some external power guiding the flow of variations in the desired direction. If variations *are* subtly directed by our mentors, insisted Wallace, we would never come to know of it directly. The products of human artificial selection – the bigger and juicier peach, the seedless banana – are indistinguishable botanically from the naturally occurring inedible varieties. An alien observing the earth miraculously stripped of mankind could only, by indirect inference from the presence in nature of very unlikely variations, reason that a conscious adaptive force had been at work. In the same way man's most distinctively human qualities – complete luxuries from the Darwinian standpoint – must be the net result of deliberate selection, just as much so as any cultivated vegetable which left to itself would never survive, let alone have evolved, under the conditions of the wild. Particularly in the case of the higher aesthetic capabilities Wallace thought he could catch tantalising glimpses of an imperial order of supernatural manipulation by which, as a disciple of his put it in memorable phrasing,

> some higher intelligence has exercised over the world at large the same kind of control which man displays in his farm or in his poultry-yard. This superior intelligence has forced the great life-agencies on the earth out of their natural course for the sake of producing a choice and eminent creature, just as the florist manipulates his roses to produce a Lamarque or a Maréchal Niel.[7]

So, for the energetic but random fecundity of the 'tangled bank' – a metaphor carefully chosen by Darwin to oppose the instances Paley cited of divine organic engineering – Wallace once more brought back the ingenious regulation of nursery and breeding-pen. It might not be very soothing to human pride to be obliged to accept the status of some cosmic gardener's hot-house bloom, but there is after all more dignity in it than the *Descent* allowed. And these appealing and well-argued views were no temporary quirks of Wallace's. Having quite melted away all the conceptual core of Darwinism, Wallace went on to accumulate by the century's end a huge range of material to support the simple but staggering thesis of the last strange mystical book of his extreme old age: that all evolution has laboured towards 'the development of mankind for an enduring spiritual existence'.[8] The phanerogams (the flowering plants, i.e. those of decorative appeal) have, he suggests in *The World of Life*, evolved – or, more correctly, have *been* evolved – for the express and foreseen purpose of improving the charm, delicacy and refinement of life. And not only that. By this means our aesthetic sense was stimulated and sharpened, our wonder and curiosity enlarged – necessary preludes to the rise of the empirical investigation of nature. Appreciation of the rose helped prepare the way for a Newton. The horse, too, was brought to its highest perfection at just the right moment when man could best make use of it as a beast of transport. Everywhere his naturalist's eye roams, Wallace spots examples of some omniscient, non-human mind ordering these valuable collocations in time. The reptiles of the Mesozoic era were granted no more than a small ration of 'intelligent vitality' in order that they might not impede the spreading abroad of higher forms; likewise, small and primitive mammals served the 'temporary purpose' of keeping down the vegetation and of serving as food for the carnivora of that epoch, freeing by their sacrifice the unspecialised and unprotected orders to begin their wonderful diverging lines of advance which led inexorably to man. The endless varieties of species and their range of products so useful to man prove in themselves, to Wallace's eyes, that they were developed expressly for his sustenance and comfort. On the question of the mosquito and housefly he is silent.

Though it is quite true that Wallace's later strange theories are infused with his possibly naîve belief in spiritualism and the occult generally, he did not dismiss all naturalistic accounts of human evolution *because* he was a spiritualist or because, as one historian has said, he was 'unable to endure the tension caused by the emergence of certain problems as evolutionary theory developed'.[9] Rather, it makes more sense to see it the other way round. Wallace detected certain technical inadequacies in Darwinism's handling of man (and who was better qualified to do so than the joint discoverer of natural selection?) and concluded that so radical were those inadequacies that the minimum hypothesis commensurate with the facts required the introduction of the supernatural to remove them. In the unseen world, impalpable to physical instruments, he supposed there to exist superior forces capable and desirous of helping us to ascend the ranks of conscious existence. Wallace remained vague about the precise nature of these forces: they are just 'higher intelligences or . . . one Supreme Intelligence' in 1870; but by 1891 he was insisting that there are supernatural beings between man and God – beings he personifies as 'angels and archangels, spirits and demons'[10] – and it is these beings which have their hands on the evolutionary switches. Within such a cosmology Wallace was readily able to incorporate all the velleities of the biological teleologists back to Aristotle. His ultimate view of evolution was that it is not an impersonal process working on species alone, but something which affects each individual and to which everyone continues subject after death. Further, for Wallace spiritual evolution is progressive in the moral sense, since there exists in some higher region of nature an external value system against which all evolutionary attainment might be measured. Evolution, then, is not only guided; it is guided along the lines which (contrary to most appearances) best support the personal aspirations of human beings alive now. Wallace's creed, in short, is a pure variety of *evolutionism*: a metaphysical position supported, with variations, by his fellow-theorists Spencer, Mivart, Romanes and the Duke of Argyll; and by other writers as diverse as W. H. Hudson, Richard Jefferies, Charles Kingsley, Winwood Reade and Henry Drummond. This chapter will trace out some of these variations with a view to demonstrating that the notion of biologically guaranteed progress certainly survived Darwinism and certainly did not lose its force (or did not do so universally) with that 'fading out of the religious assumptions of purposiveness and benevolence'[11] which a hundred times has been asserted to be the natural sequel to the acceptance of Darwinism. It will try to show that the influence of Comtean positivism from the 1830s, and behind it that of German romanticism, was still strong enough in the 1860s to ride right over the negativism of the new biology; and further, that by the 1860s it was open to the anti-Darwinians not only to insist on the readmission of teleology to remedy the failings of Darwinism, but also for them to advance much farther than that into frankly theistic evolutionism while preserving a front of objectivity.[12] By this means evolutionism and Darwinism itself could be forcibly yoked together into an intellectually respectable unity and used as the basis for at least warily optimistic speculative fictions like W. H. Hudson's *A Crystal Age* (1887).

Evolutionism as Credo: Spencer, Reade and Drummond

The philosophy of evolutionism may be defined in various ways, for it is at best no more than a loose conglomeration of beliefs. An additional confusion arises from the fact that in the common usage of the day – T. H. Huxley's, for instance – the word often meant nothing more than any theory of evolutionary development. But in the most characteristic form which evolutionism took among the post-Darwinian generation there are to be found the following dominant elements: that change is the most fundamental law of organic life and that change is always in the direction of greater complexity; that biological evolution is but a sub-division, and a small one, of a cosmic evolutionary process which, beginning with an amorphous gas-cloud, ends in the creation of planetary systems; that nature is in some deep and vaguely comprehended way in alliance with nineteenth-century Western man's deepest needs and most noble goals; that, despite all appearances to the contrary, inevitable progress is built into the fabric of the universe and by all clear minds can be recognised as such. For the proponent of evolutionism, recognising the sequence from simplicity to complexity, whether this be called progress, advancement, development or what you will, is an intuition logically prior to grasping any specific law. For this sequence, it was claimed, lies at the very root of nature and is a phenomenon undercutting all others. From the macroscopic world to the microscopic, from the spiral nebula down to the bacterium, everything was seen to display a frantic and gorgeous urge to build, ramify, release every iota of its potential. Such an Urge to Become even the most rigorously temporalistic theist could without strain identify with the divine; and in our period this identification was indeed commonplace. Evolutionism was a persuasive doctrine for those unwilling to look too deeply into the raw facts alleged to support it. There was above all its emotional appeal, and the way it dovetailed so neatly with those other strands of Victorian thought which were prone to see all temporal events as coalescing into a single progressive stream. Further, it had the advantage over mere Utopian speculation in its insistence that history does have a meaning, and a meaning to be discovered within the phenomena of biology and hence empirically verifiable.

It is beyond our present scope to offer a detailed account of the ways in which the transition was made from evolutionary theory to evolutionism, for that is something which took place at a conceptual level where (as a philosopher writing on the subject has concluded despairingly) 'full attention is not paid to questions of method, theory-construction, and the validity of appeals to analogy'.[13] Evolutionism was one of those doctrines that mesh so closely with widespread assumptions of the day that they are felt on the pulses and are rarely subjected to any fierce critical analysis. And it could also exist quite separately from any biological considerations at all; for it pre-dates not only Darwinism itself but systematic research into the life sciences altogether. It has formed a coherent body of opinion as long as there have been theories of progress at all, which in modern European terms means from the time of Condorcet's (d.1794) division of all human history into ten stages reaching from barbarism to perfection. Even after 1860 evolutionism may be found being preached with only the most cursory side-glance at Darwinian biology. 'Progressive improvement inevitable': this, the title of the last chapter of William Ellis's *Thoughts on the Future of the Human Race* (1866), by itself expresses the simple and confident faith that the cornucopia of the future must pour forth the benison which some Victorians believed they had the right to expect of it. Not only economists and social theorists like Ellis (he was a disciple of Mill) but also the more radical theologians pandered to this mood. A contributor to *Essays and Reviews* (1860) spoke up for the *Origin* as 'a masterly volume' (Baden Powell, in 'On the study of the Evidences of Christianity'), and several of these famous essays show traces of what might be termed sub-biological evolutionary suppositions. Frederick Temple, for instance, in 'The education of the world' divided the cultural history of mankind into arbitrary phases corresponding to those in the life-cycle of a single man – each successive culture contributing certain skills and items of knowledge to an expanding collective mentality. The purpose of this extended analogy is to convey the thought that each phase advances the race towards a quasi-biological maturity. But Temple's cultural life-cycle is arrested in its prime, which is Temple's own age. He does not allow it to turn onwards into senility and death. His analogy, superficially evolutionistic, is thwarted by the implications of the biological frame of reference from which he has extracted it. Other philosophical evolutionists with fuller training were better able to choose their data and to construct metaphorical systems buttressing their determination to see an idealised version of their own age as the evolutional peak of organic history; though in the following chapter we shall see that the cast of mind which declined to accept the millennial anticipations of evolutionary theory was never in danger of being eclipsed. Prime among these optimistic philosophers was Herbert Spencer. His Synthetic Philosophy is, in one of its aspects, the most coherent expression of an evolutionism developed in full cognisance of Darwinism.

In an essay written in 1899 at the close of his career and unpublished in his lifetime Spencer, in course of reflecting on the chain of thought which led him to evolutionism, incidentally confirms the essential maturity of evolutionary thought before 1859. To him it seemed that the tentative dismissal of Mosaic creationism by Chambers and Lyell left a vacuum to be filled with, among other things, inductive systems 'derived from the general aspects of organic nature',[14] rather than from deduction from wholesale collections of facts – which was the image of Darwin's methodology sedulously promoted by Darwin himself. Though this may give rise to the reasonable suspicion that the aged Spencer, in allegedly giving a synoptic history of a group of connected unit-ideas, is actually only allowing us his own narrow perspective on that history, direct reference to Spencer's earliest papers does reveal the acumen with which he converted the hypothesis of evolution into a metaphysic of unbridled optimism. His evolutionism may have originated in the cosmological theory expressed in 'The development hypothesis' of 1852, and not in biology at all; but when it came to the test he had no trouble in absorbing the immediately post-Darwinian *Principles of Biology* (1864–7) into the plan of the Synthetic Philosophy, even though the outline of this had been laid down in the prospectus which he issued in the 1850s. Another article of those years, 'Progress: its law and cause' (1857), foreshadows the reliance Spencer was later to place on data taken somewhat uncritically from many branches of the life sciences, and it also summarises the trend of thought to which he held, despite others' pessimistic readings of Darwinism, with fierce tenacity for nearly another forty years.[15] It contains, in small, nearly all of his key evolutionistic beliefs; and it has as its largest aim the revelation of the natural transition of all natural processes towards greater heterogeneity, whether the processes belong to sociology or to cosmology. Now, whether this transition is goal-orientated, governed by one increasing purpose, is not an issue which touches Spencer deeply. He is almost uninterested in it. He is content to rest in the simple conviction 'that the increase of heterogeneity so brought about is still going on, and must continue to go on; and that thus Progress is not an accident, not a thing within human control, but a beneficent necessity' (p. 484) – a simple conviction that all, in the last analysis, must be for the best in this best of evolutional worlds. Though he does not address himself to the final end of evolution, it is noteworthy that in order to explore the inner nature of his benevolent juggernaut Spencer draws heavily on just those disciplines whose very pith is increasing complexity over time: palaeontology, anthropology, and especially embryology; for the embryologist, his attention fixed on the biological events lying between conception and cradle, ought to find it easier than most biologists to believe in nature's one increasing purpose. In the process of trying to explain why organic development happens at all, Spencer toys for a while with a selectionist theory which is accidentally but strikingly similar to Darwin's. While speaking of geological upheavals as being the cause of the extinction of some species and the proliferation of others, Spencer describes clearly the likelihood of change coming out of the struggle as one species battles another for survival. But he fails to follow up this insight with the further crucial point of the reproductive advantage that would then be given to certain favourable variations; instead, he soon surrenders to Lamarckism and begins to stress the modifying effects on species of diet, habit and climate. Much later on in the perplexed 1890s Spencer on many occasions defended Lamarckism with some skill and with all the assiduity of one who sees a fundamental, optimistic principle of education and moral training under challenge. The biological foundation of this form of evolutionism, therefore, is a little different from Wallace's. For Spencer, purposiveness is an emerging quality, and it ought to be investigated down among the minutiae of cellular division, not looked for beyond nature entirely. He sees life *as* purposive, not as having purpose imposed upon it. Otherwise there is little to distinguish the biological philosophies of Spencer and Wallace. The point is that the evolutionism of both was resilient enough to absorb Darwinism and yet remain unashamedly teleological.

So far we have considered evolutionism as it appeared among anti-Darwinian biologists; for, as we have seen, Wallace's neo-Paleyism was the outcome of carefully articulated reservations about natural selection, and Spencer's Lamarckism a product of inductive reasoning. Neither had properly to confront the difficulties of holding to an optimistic evolutional metaphysic in the very teeth of Darwinism. Yet others did just this, insisting that they were both optimistic about evolution and true Darwinians to boot. The contradictions which are apparently inherent in such an extravagant claim have been too much for a number of those who have studied the use of Darwinian concepts by important late-Victorian poets. Their discomfiture arises from the fact that, say, Swinburne's 'Hertha' and Hardy's 'The Mother Mourns' both owe their inspiration in a pivotal way to one and the same biological theory – at least, their creators thought so. Puzzlement and disbelief, however, are not unreasonable critical reactions to readings of Darwinism so diametrically opposed that they could inspire simultaneously the ebullience and the gloom of, respectively, Swinburne and Hardy. The question at once presents itself whether any account of nature rooted in empirical observation could really permit this range of interpretation and retain its integrity. Is it not much more likely that, quite simply, we have here an inaccurate, a *bad*, reading of Darwin? That either Swinburne or Hardy read him inattentively? Morse Peckham has argued unequivocally that 'Hertha' is the consequence of

> a misunderstanding of the *Origin*. For the biologic world that Darwin revealed, if you do not read him with the assumptions of metaphysical evolutionism as instruments for understanding the book, is a world totally lacking in the organized and teleological process characteristic of evolutionary metaphysics.[16]

The nub of Peckham's careful argument is that Darwinism may not legitimately be viewed as potentially generating an open-ended series of cultural responses. Whatever Swinburne himself may have believed he was about, he did not discover anywhere in the *Origin* the conceptual universe of *Songs before Sunrise*. The two are just incommensurable: Swinburne's pantheistic, inspirational verse may be 'Darwinistic' but it cannot possibly be called 'Darwinian'. It is not an authentic response to the Darwinian revolution at all.

Peckham's analysis is bold and has the virture of straightforwardness. Merely by tidily classifying the frequently erratic and excessive claims made by uninformed critics on behalf of their subjects he brings order and the prospect of objective evaluation. For him the work of Hardy, Meredith and Nietzsche is imbued with Darwinism because it rejects teleology; whereas Tennyson's and Browning's poetry is only Darwinistic. And so on. Yet when applied to specific texts Peckham's simple bipartite division gives rise to some disappointingly insubstantial literary judgements. He praises, for instance, *The Return of the Native* as a profoundly Darwinian novel; one that relates so truly to the insights of the *Origin* that it even contains 'the implication that we can never fully understand the world in which we live, and our relation to it' (p. 38). One would question whether Hardy's perceptions, Darwinian or not, are ever quite that bathetic. If that is the deepest insight into the nature of the world that Darwin can suply the writer with, is there any literary advantage in being a Darwinian at all? Is it not, rather, a liability? And 'Darwinisticism', too, tends to deteriorate in Peckham's usage into mere derogation: in such a terminology a good proportion of Darwin's own work (including his entire theory of pangenesis) would have to be condemned as Darwinistic rather than Darwinian. This is likely to prove more confusing than helpful. As was suggested earlier, the source of these difficulties lies in giving way to the temptation to read intellectual history backwards. To the disinterested eye which can avert its gaze from the further developments over a century of post-Darwinism it is evident that throughout our period every scrap of evidence which could be forced to yield up the implication that evolution does indeed bear an optimistic meaning for humanity (which is one definition of evolutionism) *was* so forced; and not only by literate laymen but by philosophically inclined biologists, too. It must be conceded at once that some of the biologists' support of evolutionism arose not so much from utter commitment to the facts but from what Charles Gillispie has called the 'semantic starvation' of Victorian science generally.[17] Lacking a truly technical, unambiguous lexicon Darwin was forced into annexing a vocabulary much of which had already a vivid connotative aura: *selection, favoured races, fitness* and *fittest, progress.* We can see with hindsight that these were barely adequate counters for biology to work with, and they were all too easily mistaken for the real currency of thought. It is true also that it was easy enough to produce a damning indictment against any Darwinian evolutionism which relied on natural selection as its *modus operandi*, as both Darwin and Wallace were made painfully aware. Nevertheless, it is worth resisting the simplification that Darwinism, in so far as it supported evolutionism, gave a culturally tolerable face to the biology of the day and, in so far as it promoted natural selection, gave an intolerable one. That seems explicit in Peckham's concept of 'covert culture', which he uses to pin down the perversion of Darwinism by self-deluding writers who (Peckham assumes) refused to accept what he takes to be the unavoidable concomitant of being a Darwinian: 'a willingness to endure considerable loss of energy' – by which he means creative energy. But do the intellectual histories of all affected writers prove this 'loss of energy' to have been inevitable or – more – ineluctable? For the ambivalence of nature was recognised and, further, recognised in biological terms well before Darwin, though the question was at that early date typically posed in the conventional theological form of whether the 'Devil's meaning' or 'God's meaning' predominates in the world. In the era of *Vestiges of Creation*, as in the era of the *Origin*, cosmic optimism was a quite deliberate choice. Charles Kingsley, in his account of a tour of the Devon coast in 1849, almost instinctively chose a biological analogy to express his evolutionism. Nature may present a chaotic front but 'the great organic world is still living, and growing, and feeding, unseen by us, all the black night through'. In the Spencerian manner, Kingsley finds this pullulating activity reassuring precisely because it is going on independently of man. Though we may be ignorant of all the living energies fermenting about us, that microscopic world is still there and is (comforting thought!) 'multiplying as fast as the thoughts of man'.[18] Since Kingsley became an enthusiastic Darwinian some ten years later, it is undeniable that for at least this philosopher-novelist the alienating concept of natural selection could well be absorbed into evolutionism. Far from weakening the creative drive of the artist there could be something exhilarating in the very impersonality of the process. For a few years after 1859, no doubt, the publicists of Darwinian evolutionism were quiescent, with a few of their number reduced to scepticism or even despair. But when conditions became more propitious they became as vigorous as ever, sometimes reshaping their creed into a form Darwin himself would not have blushed to own. For there is the fact (which W. H. Hudson seized upon) that evolutionism is far from extinct in Darwin's own major works – even though Darwin used the word 'evolution' most cautiously in the *Origin* lest his whole argument be contaminated by so dubious a term. But his nicety was in vain, and not really wholehearted. For the layman as for the informed essayist, evolutionism proved for the remainder of the century to be a source of spiritual uplift comforting equally to the most hedonistic and most puritan of temperaments. And it may be demonstrated that this was true even of rigidly Darwinian evolutionism. No failure of energy is perceptible in the writings of William Winwood Reade (1838–75). No feebleness of spirit was induced by the impossibility of remaining spiritually buoyant and an avid reader of Darwin too. Rather, there is the fierce exultation (reflected in the idiosyncratic style) of one rejoicing at the harshness of a natural law which guarantees that only the very best are not found wanting.

The Darwinian evolutionism of Winwood Reade finds a specially pure expression in his *The Martyrdom of Man* (1872), which is an impressionistic survey of 'universal' history by a freethinker contemptuous of all politically pigmented readings of history but respectful almost to excess of the biological vision of the past – and promise for the future. The *Martyrdom* has an extremely simple structure. Three long chapters, entitled successively 'War', 'Religion' and 'Liberty', paint the history of civilisation from the first Egyptian dynasties in slapdash but vivid colours. The fourth and last chapter, 'Intellect', is something of a mishmash. 'Finding that my outline of Universal History was almost complete,' Reade tells us in his preface, 'I determined in the last chapter to give a brief summary of the whole . . . adding to it the materials of another work suggested several years ago by the *Origin of Species*.'[19] The work to which he refers, to have been called 'The Origin of Mind', was to have been based on material Reade accumulated during the second of the two expeditions into central Africa (1868–70), on the subject of what he calls 'the phenomena of savage life'. As the publication of *The Descent of Man* rendered this project superfluous, Reade added the material to the 'Intellect' chapter of the *Martyrdom* and in so doing packed into a small space the outline of a cosmology, a geological history, an etiology of morals, and a vision of the future. The whole chapter is saturated through and through with evolutionism, presenting in essence every last one of the optimistic deductions which could be made out of the Darwinism contemporary with Darwin himself.

The *Martyrdom* is one of those histories in the line which runs from Vico to Toynbee: the line which claims to display some pattern in the inscrutable passage of events. The pattern Reade reveals is that the story of man's social existence has unfolded according to the same natural law which governed his ascent from the inorganic world. The law he has in mind is undoubtedly natural selection, though in his interpretation of it he is closer to the mood of Spencer's *Principles of Biology* than of most of the *Origin*. Nature does not recognise the needs of individuals, he tells us, any more than we recognise each cell of our bodies. But when he considers the destiny of all mankind as one creature in the Spencerian manner he is left in no doubt that the One is 'becoming more and more noble, more and more divine, slowly ripening towards perfection' (p. 429).[20] Such expansive but imprecise quotations could be multiplied, without much profit accruing, almost indefinitely from the *Martyrdom*. Within thirteen years of the *Origin*'s appearance, and without traversing any of its key axiom, Reade's evolutionism stands revealed as an important buttress of Victorian illiberalism, fluidly adaptable to establishing the natural role of women, the necessity of empire, the biological superiority of the European races, and to a worship of technology so extravagant that he can forecast aeroplanes, new energy sources and even interstellar migration. Buoyed up by a very strict Darwinism which intrudes into every corner of life, Reade can without strain denounce the 'sickly school of politicians who declare that all countries belong to their inhabitants, and that to take them is a crime' (p. 415); he can term the conquest of Asia as being 'in reality Emancipation' (p. 415); war as the means by which 'men were first brought into amicable relations with one another' (p. 413); and Christianity, though a lie, a convenient brainwashing technique with which to 'Europeanise the barbarous nations' (p. 415). On even more dubious grounds, where other naturalists and anthropologists contemporary with Reade were loath to tread, he feels able to surmise wildly that all animals 'deficient in conjugal desire or parental love . . . are blotted out of the book of Nature' (p. 365); or that such traits as the human females' 'refined sentiments, their native modesty, their sublime unselfishness' (p. 373) – and even their flowing tresses – are the consequence of long-continued sexual selection. Darwinian evolutionism offered Reade the most transparent window on the world; but more interesting than his crude optimism and cosmic Whiggery is his free and almost joyful acquiescence in the indefinitely prolonged suffering which his version of a progressive philosophy necessarily entailed. As a Darwinian moralist Reade is prepared to point without flinching to pain and waste as the ultimate driving forces in history, not excepting the organic history of the last ten thousand years that we call civilisation:

> when we open the Book of Nature, that book inscribed in blood and tears . . . we see plainly how illusive is this theory that God is Love. In all things there is cruel, profligate, and abandoned waste. Of all the animals that are born a few only can survive; and it is owing to this law that development takes place. The law of Murder is the law of Growth. Life is one long tragedy; creation is one great crime. (p. 427)

Reason enough here, one would think, for the utter decay of the spirit; for those fearsome dreams of Arther Elliott, the central character of Reade's last novel *The Outcast* (1875), who smells from far out in space the carrion stench of the earth. But the fact is that Elliott is a hopeless neurotic and Reade condemns him as such. For Reade himself the observable criminality of nature certainly does not license the crimes of men. When we examine Reade's evolutionism from other directions he is found to be triply protected. First, by his deification of humanity and by his exclusion of a supernatural drive in evolution, whereby he is able to dispense with the problem of evil and is freed from the strain of finding a solution acceptable to biology and theology together. This was a considerable achievement when one considers that even Darwin, despite his studied pose of indifference to philosophical and theological puzzles, was subject to some strain in this matter. His letters to his intimate friends Hooker and Lyell are full of remarks on the inexplicable evil in nature. It was to Hooker, on 13 July 1856, that he expressed the much-quoted sentiment 'what a book a devil's chaplain might write on the clumsy, wasteful, blundering, low, and horribly cruel works of nature!'[21] Confronting the same problem, Wallace, who of course went much farther than Darwin by imagining a benevolent power using natural selection as its agent, devoted an entire chapter of *The World of Life* to the topic 'Is nature cruel? The purpose and limitations of pain', where his method is to advance various *ad hoc* arguments for supposing that as the less fit are eliminated the victims, far from suffering, may actually enjoy being devoured in a way 'not only absolutely painless but slightly pleasurable' (p. 375). Reade's solution got him out of Darwin's impasse, and makes rather more sense than that of Wallace, who was of course caught up in the same difficulty as any Christian theologian in his assumption that the benevolence of the 'angels' who are flicking open and closed the tap of desirable variation is absolute.

Second, Reade is protected by his strong sentiment that to resist nature would be quite wrong. He will tolerate no discord between evolution and ethics. Evolution *is*, must be, ethical in itself; for it is evolution that applies the most telling sanction of all – the only one that counts in the long run – that of progress or eclipse. Against this great judgement, for Reade as for Spencer, no merely human system of ethics could hope to prevail.

Third, and finally, Reade shelters behind the grand organising metaphor of his book: that the men of all generations must in the light of biology be ready to bleed to redeem the future. Our own prosperity has been founded on the agonies of the past; on the brutalities of class and war and religion. 'Is it therefore unjust,' he demands on his last page, 'that we also should suffer for the benefit of those who are to come?' (p. 447) We will of course be martyred to the future, like it or not, for such is nature's logic; Reade only asks us to bear it with conscious joy. But in thus appealing to the masochism of his readers Reade is careful not to demand too much. What he asks his contemporaries to contribute to the misery of the ages is not dreadfully painful; it is not in fact a corporal pain at all, but a spiritual one: the abandonment, in the face of science, of the sweet delusions of religion. So much is asked, but no more. By comparison with the past, the debt due to nature by the people of the nineteenth century seems, happily, to be a modest one.

Reade gauged his audience's mood so well that the *Martyrdom* ran to eight large editions in twelve years, despite attracting reviews that were uniformly hostile and often virulent. Some of his popularity, no doubt, was due to the new demand for potted histories from the self-educated and to his infamously anti-Christian sentiments, but there are also his purely literary qualities to be reckoned with. His fluent and hypnotic style filled with Swinburnian rhythms is adjusted more to the creation of a national epic than to the writing of a global history. And the *Martyrdom* does constantly strive towards the status of epic, with Reade's unquestioning faith in Darwinian evolutionism supplying the underswell of emotional conviction. His role as prophet is a little weakened, we may feel, by his failure to cry doom to the unbelievers as actively as he might have done later. Dying in 1875, he was unable to benefit from the influence of the retrogressionists of the 1880s who, with their warnings about laxity and surrender, would surely have been congenial to Reade; especially as they too were prone to draw on classical history and the lesson of the fate of empires to give another twist to the analogy between evolutionary biology and that minuscule fraction of it we call history.

Reade was far from being the only exponent of 'hard' evolutionism working through natural selection. 'From *In Memoriam* to *The Woods of Westermain*,' says the historian G. M. Young, 'from *The Woods of Westermain* to the Choruses of *The Dynasts*, we can follow the secular intellect seeking its way to such an apprehension of Being as Process as might hereafter reconcile the spiritual demands of humanity with the rapt and cosmic indifference of Evolution.'[22] Despite Peckham's animadversions, Young is correct in the claim he makes for Tennyson, Meredith and the later Hardy. These do have in common a version of evolutionary theory which is true to the evidence as presented by the Darwinism of the day (that is, their version makes no teleological extension) and yet is simultaneously progressive and stoical. There was no general expression of incredulity when these three prescribed, as Reade did, endurance for the present in the hope that the slow permeation of Spirit (the highest term of Meredith's triad) might after long ages work a transformation; or when they all mentioned Darwinism as a potent inspiration for literary art; or when all gave expression to a guarded optimism deriving from their reading in the biologists. Their audience was not outraged when told that such a series of deductions could with perfect propriety be made from the axioms at the heart of the *Origin*. The Darwinism of a century ago was thought to be able to bear such an interpretation, even if today we know that it cannot.

We have seen in the thought of Wallace that this kind of long-term evolutionism, though of great appeal, was not the only channel of response open to the 'secular intellect' after 1859. Quite a different reaction was the one that confronted directly and rejected as scientifically unsound the evolutionary 'rapt and cosmic indifference'. For Charles Kingsley, naturalist, liberal divine and novelist, the cosmos was far from indifferent. He found no, or little, difficulty in absorbing the new image of a God who creates a single primeval form capable of endless proliferation and variation. He even insisted that the new image was more intellectually satisfying than Milton's God in the seventh book of *Paradise Lost*, who creates by making the grassy clods calve. We notice, however, that Kingsley fits evolutionary theory into his theology, not by importing the supernatural into biology as Wallace did, but by making natural selection itself an agent of divine power. The moral imperative, in the thought of the professional atheist Reade, is very strong; but it is essentially the imperative which orders passivity before the overwhelming force of nature. That is the significance of his structural metaphor of martyrdom. Reade asks us to trust nature, but he is silent as to what is concealed behind nature. Frankly, he knows no more than that natural selection is working in our collective, though not individual, interests. Kingsley, by contrast, finds a very active moral imperative in natural selection. It looms so large in his thought, and in such imaginative products of his thought as *The Water-Babies*, that it is frequently difficult to recall that Darwinism is for Kingsley no more an instrument of the Almighty's will than is, say, gravitation. It took Henry Drummond (1851–97), with his background of Scottish Dissent and agronomics – his father was a ferocious fundamentalist, but also director of a firm of seedsmen – to take the final step. This was the effective deification of natural selection.

Drummond's career was an even odder hybrid than his father's. He mixed active assistance in the Moody and Sankey evangelical revival of 1874–5 with a lectureship in natural science at Free Church College, Glasgow; and he alternated writing inspirational sermons with making a comprehensive and accurate geological survey of the Rocky Mountains in 1879. His evolutionary philosophy is an elaborate and diffuse one, but in so far as it has any central tenets these are to be found in a series of lectures delivered with enormous popular success in Boston in 1894, and published the same year as *The Lowell Lectures on the Ascent of Man*. The pressure for seats at these lectures was so great, records Drummond's biographer, that a flourishing black market in the tickets formed. But apparently the fortunate audience was discriminating. They praised the lectures' lucidity and style, but retained some doubts about the validity of the contents. In their printed form, though, the lectures had a very great sale because they were a relatively sophisticated attempt to mediate between theology and the new biology without denying any of the conclusions of the latter. In process of acting as a mediator Drummond draws heavily if somewhat ingenuously on his considerable fund of information about natural history to piece together his evolutionary history. Like Reade, Drummond travelled in tropical Africa (between June 1883 and April 1884); and, like Wallace decades before, he found the tropics to be an unending source of inspiration about the workings of nature in the violently active raw. Again like Reade, Drummond is anxious, even determined, to insist on that inescapable cruelty of nature before which 'humanity was dumb, morality mystified, natural theology stultified'.[23] Yet he sees no need for a precipitate retreat into the Church of Humanity. God, for Drummond, has not after all absconded. Of even more sanguine temperament than Reade, Drummond did his best to persuade his Boston audience that no excessive price in pain had been paid for progress. His stoicism – he would call it Christian fortitude – is very much in evidence, as it is not relieved by the special comfort which Reade took in man's future ascendancy over the forces of nature, when humanity collectively may become 'what the vulgar worship as a God'.[24] But they do share a Panglossian attitude towards natural selection. For Drummond it 'so discourages imperfection as practically to eliminate it from the world . . . Natural Selection is the means employed in Nature to bring about perfect health, perfect wholeness, perfect adaptation, and in the long run the Ascent of all living things' (pp. 265–6). His refusal to relegate this vision of perfection to some indeterminate but certainly distant future has an odd consequence. Drummond plumps for his own age as exhibiting the fulfilment of guided evolution; and, within his own age, for the domestic cosiness of middle-class English life. At the centre of the family rests the mother, and Drummond conceives nature to be a production line turning out ever more perfected models of mothers: 'the Pisces, then the Amphibia, then the Reptilia, then the Aves, then – What? The Mammalia, THE MOTHERS. There the series stops. Nature has never made anything since' (p. 343).

And never will, is his implication. Again we should notice, not so much the profound dubiety of the biological and anthropological facts Drummond is drawing upon, but how these data, whether true or not, are shaped to fit the conclusion he longs to reach. Beginning with the raw material of the behaviour of lowly organisms, he ends with an amalgam of Marian iconography and sentimental sociology. When he treats of sexuality, the mating habits of tadpoles, aphids and bees all help him to establish his position – so clearly foreshadowed! – that 'maleness is one thing and femaleness another' (p. 328). As he works towards an elucidation of the ethical significance of maternity, seemingly hard biological facts are forced to generate a new natural law, with the very pattern of nature being the Victorian fireside. Of the evolution of the family unit, he concedes that in the state of nature (perhaps he has in mind the tribes of tropical Africa?) the family circle is 'incomplete' – by which he means polygamous, or at least not nucleated. But Drummond's evolutionism encompasses more than the crude accomplishments of nature, even a nature under divine guidance. Social evolution, which for him is merely another name for the linear advancement of civilisation, comes in to seal the leak: 'with the Christian era the machinery was complete; the circle finally closed in, and became a secluded shrine where the culture of everything holy and beautiful was carried on' (p. 404). The religious imagery is not, of course, a mere peculiarity of Drummond's decorated style. Implicit throughout this deification of the family circle at its most suffocatingly claustrophobic ('a secluded shrine') is Drummond's chiliasm. As his lectures end so does history, in those endlessly repeated images of the domestic hearth. The ascent of man and evolution itself close like a three-volume romance with that self-same vision which Orwell humorously evokes in his essay on Dickens: 'a huge, loving family of three or four generations, all crammed together in the same house and constantly multiplying, like a bed of oysters'.[25]

W. H. Hudson and A Crystal Age

We have now surveyed Victorian evolutionism as it was developed by four thinkers – Wallace, Spencer, Reade and Drummond – who were keenly aware of the new dimension given to their philosophies by the unavoidable need to accommodate inside these Darwinian biology. All four managed at the same time to retain their optimism, their essential faith in the rightness of the evolutionary process; though it might well be thought that they are linked by little else. Is there any common ground between the animistic teleology of Wallace and the divinely guided natural selection of Drummond? Between Spencer's highly developed Lamarckism and Reade's total reliance on Darwinism to fashion the future world of his heart's desire? Not at first sight. But the imaginative work may sometimes attain the synthesis denied the purely ratiocinative; and we will continue the discussion of evolutionism with close reference to William Henry Hudson (1841–1922), his novel *A Crystal Age* (1887), and the effect on this novel of his tacit recognition of the advantages and difficulties of all four approaches to evolutionism mentioned above.

'Science and Art to my thinking,' wrote Hudson near the end of his life to a friend, 'are mutually exclusive: and novels may be or not be art or science pure and simple but a sort of oil and water mixture.' Looking back on a long and varied career (his first novel was published in 1885; his last in 1920), he must have felt with some satisfaction that he at least had never written any of those 'semi-pathological treatises, with a story about people's domestic affairs and emotions thrown in to make it a novel'.[26] Perhaps we should be grateful he did not; but in fact few authors were better equipped than he to whip up that elusive emulsion of art and science. On the one side he was a competent and productive light essayist and romantic novelist, though only one of his novels, *Green Mansions* (1904), has indisputably survived the taste of the age. On the other, he earned a reputation after several years of struggle as an accurate and charming natural historian – today we should call him an ethologist – specialising in ornithology. Like Thomas Hardy, who was almost exactly his contemporary, Hudson passed through the whole length of our period as a mature intelligence sensitively alive to the biological debates of the day. Unlike Hardy, though, Hudson, if not quite a professional of the calibre of Wallace or Romanes or Lankester, at least met those doyens of Victorian science on something approaching an equal footing. And, unlike Samuel Butler, Hudson was always felt to be respectable. He refrained from entering public quarrels on technical issues once his writing career had properly begun, preferring to keep his scientific enmities to his private letters. Though strongly inclined to evolutionism he was never (so far as one can judge from his slim literary remains) a Darwinian; and indeed he became in his old age one of the warmest supporters of Butler, that Ishmael of biological speculation. Yet he was singularly discreet about making public his change in allegiance – at least until the time came when the expression of doubts about natural selection could not be construed as an attack on Darwin personally. His long life offered him the advantage of witnessing the total disintegration of Darwin's intimate circle; for, though he was already adult when the *Origin* broke upon the world, he outlived Darwin, Huxley, Romanes, Spencer and Bates. The reward for his discretion was palpable. His field observations were treated with respect and he remained in the confidence of all the greater luminaries of the day. He was never tainted, as was his fellow naturalist-novelist Grant Allen, with accusations of pandering to the non-scientific laity with a slipshod generalisation or a dubious extrapolation. Nevertheless, in spite of all these considerable advantages Hudson's career must remain a disappointment to the student of allied scientific and artistic creativity. He practised very much what he preached about the mutual exclusiveness of science and art. In his novels little enough of his considerable learning manages to penetrate their fictional worlds. Even in the Utopian *A Crystal Age* far too much is left unexplained, and too few of the lengthy digressions on such matters as vivisection and the developmental significance of pain are brought into any organic relation with the tale itself. The 'oil' of the romances and the 'water' of the standard textbook *Argentine Ornithology* (to which Hudson made extensive contributions in 1888–9) remain obstinately differentiated. Yet *A Crystal Age* is still recognisably a future envisioned by a Victorian naturalist steeped in evolutionism: we shall see the truth of that.

Despite Hudson's mania for self-concealment (he systematically falsified many details of his life and destroyed thousands of his personal letters and memorabilia) the various biographical reconstructions present us with a life lived for the most part in obscurity and poverty. His isolated childhood on the Argentinian pampas is poignantly evoked in his *Far Away and Long Ago* (1918) – a highly poeticised autobiography which ends by marking what was clearly the most decisive event of his reading the *Origin* at the age of 19 or 20. (It may have been a little later than that: several details of Hudson's life are still vague.) As Robert Hamilton has shown,[27] there do exist two mildly contradictory accounts of the impact on him of the *Origin*; but the first is the more detailed, rings truer and accords better with the fragments of information available elsewhere. After his first impassioned rejection of Darwinism, he tells us, he tried to look at it again in the light of his own experience as a budding naturalist. For some time Darwinism 'went to bed and got up with me, and was with me the day long' and for several years 'my mind, or subconscious mind . . . went on revolving it', until at last he became convinced of the truth of the broad evolutionary proposition.

The precise extent of Hudson's knowledge of the finer points of the later biological controversies of his day – and we must remember that his day stretched all the way from the first appearance of the *Origin* to the solidification of mathematical, post-Mendelian genetics – is, thanks to his secretiveness, beyond detailed reconstruction. His first biographer laboured under a double difficulty of little real information at his disposal and a lack of sympathy with his subject's angle of vision. He tells us rather condescendingly that his friend had 'no knowledge of physiology',[28] while only a few pages later he is recording Hudson's stimulating and technical conversation with Arthur Keith (1866–1955), the physiologist, cardiologist and populariser of neo-Darwinism. The sole verifiable comment Morley Roberts offers is the truism, apparent to any reader of Hudson's surviving letters, that he was no pure Darwinian. In fact, those letters also give us glimpses of an exceptionally well-read mind which had, in the province of biology, absorbed all of Darwin's major works, Wallace, Vries, Bates, Erasmus Darwin and Samuel Butler. He learnt from all these. His own retrospection is, on the evidence of his later writing, accurate enough: 'insensibly and inevitably I had become an evolutionist, albeit never wholly satisfied with natural selection as the only and sufficient explanation of the change in the forms of life'.[29] Thanks to his temperament Hudson was free to walk in the broad middle of the road, making himself invulnerable to extremist criticism. Reasonable though this self-analysis appears, we must remember that it is Hudson in extreme old age reflecting on what had been highly emotional responses to the discoveries of a science half a century less developed. Who in such circumstances could pledge himself free of distortion? And in fact the sentiments of his first published work, unearthed thirty years ago by David Dewar, do not square too well with Hudson's recollected genial tolerance of Darwinism. Practically nothing is known directly of Hudson's biological (or indeed any other) interests between the end of *Far Away and Long Ago* in the first years of the 1860s and the time when his literary career began in earnest in 1885, long after he had become a permanent expatriate in England. A good slice of his last years in South America was spent collecting ornithological specimens for the Smithsonian, and it was from Buenos Aires that he contributed a series of letters on recondite points of ornithology to the Zoological Society of London. In the third of these, the only one of present interest, on the habits of the pampas woodpecker (*Colaptes campestris*), he takes up the cudgels against Darwin personally:

> the perusal of the passage quoted [Hudson had just cited the fourth edition of the *Origin*] by one acquainted with the bird referred to and its habits might induce him to believe that the author had purposely wrested the truth in order to prove his theory . . . this bird rather affords an argument against the truth of Mr. Darwin's hypothesis . . . in truth, natural selection has done absolutely nothing for our Woodpecker. Its colours are not dimmed, nor its loud notes subdued; but even when it traverses the open country it calls about it the enemies from which it has little chance to escape.[30]

The aggressive tenor of this pompous letter marks out its animus against Darwinism clearly enough; after all, Hudson, even in this toned-down version, is practically accusing Darwin of perverting the evidence to suit his theoretical convenience. Nor can we suppose Hudson placated by Darwin's customarily polite but evasive reply printed later in the same year.[31] This whole episode is a most noteworthy one. It warns us that in the stimulating decade or two after 1860 commitments ran deeper and emotional temperatures higher than the later recollections in tranquillity (which so often must serve as the only primary material) might suggest.

Hudson's metamorphosis from a doubtful Darwinian to a frank and outspoken Lamarckian in his old age was, if the surviving letters tell the full story, in large part a consequence of the reading of Butler. At first it was Butler's satire which attracted his attention. He read *Erewhon* (1871) and plagiarised one of its amusing inversions by making his hero Smith in *A Crystal Age* attract the punishment of thirteen days' solitary confinement for making himself ill by not eating regularly. Much later he also read *Evolution, Old and New* (1880), in which Butler exercises his powers as a scientific historian by proving to his own satisfaction that Darwinism was anticipated by naturalists of the previous century. The book drew Hudson on to investigate Butlerism as a philosophy and to find it at last almost perfectly congenial. His admiration for Butler was real enough, but slow in evolving; for as late as 1918 we find him making perceptive thumbnail sketches of Butler's many deficiencies: 'it was an age of giants; Darwin and so on: their facts were too much for him: drove him into a clever perversity of humour . . . I suppose there is a grain of truth in [his writings].' He came to find rather more than that, for just a year later he is commenting on Butler's 'acute reasoning and splendid independence'.[32]

Hudson's indebtedness to Butler comprises only a small part of his total receptivity to non-Darwinian biology after 1900. In fact there was little real common ground between them. Butler was entranced by theory, the more abstract the better – especially theories of heredity, as a future chapter will show – but he is impatient of the painstaking, descriptive work of the naturalist; whereas Hudson found in the end all abstractions tiresome unless they enlarged on and formalised his strong pantheistic feelings. In this respect his evolutionism is comparable with Wallace's, though he had a far more powerful intuitive grasp of the problem of evil than did Wallace and refused to settle for any easy solution to it. Certainly his objections to the natural selection hypothesis followed the well-worn emotional grooves. Hudson hated his beloved world of nature being transformed into a collection of essentially random events and processes of which the by-products are endless wastage and endless suffering. His hatred of these was so strong that it made him indecisive for many years about the inevitability, or even the likelihood, of biological progress. As an ethologist Hudson's contact with the prolonged and irremediable pain inherent in nature was as observationally direct, as little moulded by the conventional religious pieties, as Darwin's own. Interestingly, in an essay written towards the end of his life, 'Wasps', built around the single incident of watching a swarm of wasps devour a rotting pear, Hudson recalls to mind the habits of the *Ichneumonidae*, an order of wasps which lay their eggs in the bodies of living but paralysed caterpillars. He reminisces about the 'old days when my mind was clouded with doubts, and the ways of certain insects, especially of wasps, were much in my thoughts. For we live through and forget many a tempest that shakes us; but afterwards a very little thing – the scent of a flower, the cry of a wild bird, even the sight of an insect – may serve to bring it vividly back and to revive a feeling that seemed dead and gone.'[33] The description of the wasps' exploiting a singularly unattractive ecological niche is the very one alluded to earlier as causing the normally unemotional Darwin to react with a paroxysm of resentment and horror because it made only too obvious how easily a 'devil's chaplain' could make out a telling case against nature.

Hudson is hardly a devil's chaplain, though the wasps' gorging themselves does thrust forward yet again 'the old vexed question – How reconcile these facts with the idea of a beneficent Being who designed it all?' (pp. 201–2) Instead of launching into abuse against nature Hudson restricts himself here to a gentle debunking of the palliatives invented by Darwin to explain away such disturbing evidences of nature. He finds unpersuasive both the argument that the new concept of the deity who can make new organisms appear consecutively according to natural law is a more noble one than the old, and the further apology that such repulsive instincts are not specially designed but merely a part of a great impersonal law of variation and survival. Such attempted consolations did not appeal to the author of 'Wasps' in 1919 any more than they had appealed to his earlier self of 1900, who at Wallace's request made minutely detailed observations into the power of the cuckoo fledgling to eject robins' eggs from a nest. (The findings were published in *Hampshire Days*, 1903.) In this earlier account he speaks of a 'devilish intelligence' at work and in the later essay he echoes this by finding the most apposite word to describe the adaptations displayed by slave-making ants and the *Ichneumonidae*, as well as by cuckoos, to be, simply, 'diabolical'. This would appear to leave Hudson a long way indeed from any sympathy with the school of evolutionism which linked continued selection and the final attainment of perfection. And yet at the same time there is another side to the detestation so mournfully expressed in 'Wasps', for one can find in Hudson's musings a certain fascination with nature's eternal war for *Lebensraum*. Kindly and pacific in his own nature, Hudson still shares with Reade and Drummond the awareness that there is a certain stimulating quality in pain – for the species as a whole no less than for the individual. Whatever it does or does not mean for the 'goal' (if any) of evolution, pain may be seen at present as a knife carving the rude mass of life into shapes of ever greater significance. For, whatever civilised man makes of it, suffering is there, irreducibly, in wild nature. Much the greater majority of birds starve to death every winter, and every night in the fields multitudes of the higher mammals, no less sensitive than we, die lingering deaths. Hudson agonised over this: few men of his generation, probably, had observed the weaker animal going under at such close quarters. So perhaps it is understandable that he should wish periodically to shift his eyes away from the brutal present of natural history and to let them settle on the distant prospect of fully evolved life when the perfected and vegetarian lion might lie down with a lamb not destined for the butcher's knife. On other more theoretical matters Hudson displays very little consistency, wobbling through the years between the hardest Darwinism and various forms of teleology. His 'mighty scorn of the Neo-Mendelian-Darwinians' with which he is credited by his biographer Morley Roberts was only one of his stances.

These vacillations, though indeed very characteristic of Hudson's thought in his last years, also give his novel *A Crystal Age* its special quality. Though the inventiveness there does exist within a Darwinian framework, and even expressly within the terms of the promised perfection held out at the end of the *Origin*, Hudson's story is ambivalent about that perfection. He allows his hero to develop a deep and personal resentment over the price that the Crystallites have paid for stability; to make in fact his tedium and frustration the indirect cause of his death. When we look closer, however, we see that this is not a deep-rooted rebellion. While it is rather easy for the reader to reject the delicate Crystallites as a degenerate stock in a garden setting devoid of all biological pressure, he cannot read *A Crystal Age* with due attention and find it a dystopia. Hudson is aware that the promise of biological perfection has retreated to the horizon. But it has not yet disappeared over the edge.

A Crystal Age was first published anonymously in 1887. It was only Hudson's second novel, and like its predecessor it proved a complete failure in terms of sales. Indeed, like its successors, too; for not one of Hudson's seventeen books up to 1910 sold over a thousand copies. Seven out of these were works of fiction and five are now utterly forgotten. *A Crystal Age* is still known to the critic of genre precisely because it is so entirely representative of the type of Utopian fiction being written in the 1870s and 1880s. Aldous Huxley once remarked that, to posterity, prophecies are most useful as comments on the age in which they were uttered: Louis Mercier's *L'An 2240*, for instance, shows us 'the ideals of an earnest and rather stupid Frenchman in the year 1770'.[34] As the equally earnest but intelligent Hudson himself noted in the preface to the second and revised edition of his novel in 1906, *A Crystal Age* is riddled with the small concerns, the momentary crazes and the characteristic myopias of its period, and its residual value is exactly that of a period piece. If it was already obsolete in expression and as prophecy by the late 1890s, this is certainly because of the rapidly increasing sophistication in the way its basic assumptions were then being handled. Even for a professed Darwinian, the gesture of using a paragraph from the *Origin* as epigraph to a novel was to invite a very different response in 1860, 1880 and 1900 – especially when that paragraph touched on the biological perfectibility of organic beings and hence by implication of man.[35]

Hudson's own attitude to his creation is hard to fathom. He is self-deprecating to a fault in the 1906 preface and was thoroughly disapproving in his remarks to Edward Garnett thirty years after the first writing, yet at the same time he must have thought it worth revising and reprinting. Presumably he did so with the hope that the ideas of the romance would compensate for its artistic defects. In truth these defects are many, but quickly told. As a description of the far future, *A Crystal Age* has neither the sharp but baffling quality of a dream nor the amassing of enough pedestrian detail to bring the Crystallite society thoroughly down to earth. Too much is left vague, and the first-person narrative makes it difficult for Hudson to knit into the action the long explanations and justifications which any Utopian novel must somehow accommodate. Indeed, he wilfully chooses the clumsiest possible device of having the Crystallites read aloud from books of history whenever they are not haranguing the excusably ignorant Smith in their inexplicably fluent English. The awkwardness of the opening, with the hero preserved by the roots of a tree to awake in the far future as a kind of Rip Van Winkle, and the ending, which has the hapless Smith recording minutely his own sensations of approaching death from poison, are alike absurd. Hudson, like many a science fiction writer since, has no idea what to do with the ingenious world he has spun. The dialogue, particularly that between Smith and the girl of the House, Yoletta, is wooden in the extreme. Smith himself, an impetuous and rather thickheaded youth, is a rather too viscous medium for transmitting the author's philosophical and biological speculations. Fortunately these profound defects need not trouble us. *A Crystal Age* is a feeble novel, but it is admirable primary material for displaying the considerable imaginative potential of evolutionism in the 1880s.

The Cost of Evolutionary Perfection

'As all the living forms of life are the lineal descendants of those which lived long before the Silurian epoch,' says Darwin in his recapitulative chapter immediately before the passage which Hudson chose as epigraph, 'we may look with some confidence to a secure future of equally inappreciable length' (*Origin*, XIV:265). Hudson's reading in the palaeontology of his day supplied him with exactly this same sense of the immense antiquity of life, of the slowness of change; and an awareness, too, that in humanity's extensive future change might be as far-reaching as in its long past. Having awkwardly transported his hero to that future by burying him beneath a landslide and putting him into suspended animation, Hudson spares no pains to create a world as remote from our present as the Palaeolithic. He wisely leaves the exact chronology vague, but we are asked to believe that enough time has elapsed for the most sanguine evolutionist's hopes to have been fulfilled. Natural selection has done its work and a state of equilibrium has been reached which is much the same as the one Wallace liked to describe in his essays and much like Drummond's static dream at the end of *The Ascent of Man*. The world into which Smith wakes is magical, prelapsarian. No species has finally dominated, not even man: all evidence of his dominion has shrunk to a single house, with 'no town, nor village even, and not one solitary spire' (p. 5). This pleasant sunlit landscape which so puzzles Smith is Hudson's carefully deliberated gloss on Darwinian 'perfection'. As an ethologist, conservationist and nature mystic his ideal future is not Reade's technological wonderland but a natural harmonious balance. In fact it is the same harmony which his natural history essays celebrate (not without some inner questioning) raised to a higher power. The Crystallite flora exist only as aesthetic triumphs, as do the wild fauna. The domesticated animals – sheep, horses, dogs – have evolved in intelligence to a point where they work freely unsupervised. (Darwin had spoken in the *Descent* of the mental powers of animals being 'capable of advancement' but he had not predicted their use as unpaid servants and labourers.) The benign influence of natural selection has made the earth over into a garden and a garden devoid of temptations, for the human inhabitants of this paradise have voluntarily turned their backs on pure knowledge, while their morality has contracted into sedulously observed ritual. The evocative power of Hudson's Crystallite landscape, its half-familiarity, is a product of its sources. These are biblical and mythological, obviously; but also literary (Tennyson's Lotos-land 'In which it seemed always afternoon'?); from Hudson's Pampas childhood, later to be offered up with visionary intensity in *Far Away and Long Ago*, which has much the same mood; and certainly from some geological reading. For Hudson's remote future is, thanks to his wide biological grasp, in some respects a mirror image of the equally remote past. After revising his Utopia, Hudson spoke of it as 'a dream and picture of the human race in its forest period',[36] and the nostalgia is indeed strong for a golden age when nursling humanity roamed the woods and savannahs untroubled by the products, abstract and concrete, of thought. In the Crystallite future the nightmare of history is abolished by having a massive portion of all human culture 'politics, religions, systems of philosophy, isms and ologies of all description' consumed, at some time in the very remote past, in a 'mighty Savonarola bonfire' (p. 266). Hudson, who uses this disaster dramatically merely as a way of cursorily whitewashing his canvas before sketching in the new design, fails to gratify our curiosity about it. All we come to know is that man's Faustian side has nearly destroyed him, especially his 'seeking in the living tissues of animals for the hidden springs of life' (p. 70) – a mingled echo of *Frankenstein* and the anti-vivisection legislation of the previous decade.

The Crystallites, then, are the remote descendants of an obscure wandering people who controlled their curiosity about dark biological secrets and diverted their sexuality into the channels of song and dance. Their economic unit is the subsistence farm with a little additional trade; the social unit is the House, which is a mansion, Smith is informed, whose age is believed to be coeval with the earth's. This shelters a single family of twenty to thirty persons in a pattern reproduced everywhere. Hudson allows no larger political grouping, for he despised the socialist Utopias of his time with all the contempt of a naturalist who thinks he can see past the trivialities of temporary human groupings. His sour comment on *News from Nowhere* brings out his distrust of forced change according to a social theory. It is, he told Garnett, a strange delusion to suppose that 'new social laws concerning the price of food and clothes' could possibly bring about the millennium.[37] (Nevertheless he is conventional enough in his own vision to have money abolished, though he adds the sardonic touch that the effect of this is to turn his hero into a serf bound to the House, rather than the happy anarchic spirit the Guest becomes in Morris's later Utopia.) Hudson's strongly patrician tastes, coupled with his genuinely new contribution of having biological forces utterly transforming a species, led him to a much more *drastic* vision than any earlier Utopianist. His Crystallite society has the nature of crystal itself: it is rigidly patterned and totally ordered along every plane of social interaction. Presumably this rigour is the outcome of free choices made in the past, but if so those choices have become so fully endorsed by biology that in the fictional present even the possibility of acting differently does not arise.

In his description of the physiological changes which have overtaken the Crystallites, Hudson prefers to leave muted the question of whether this is due to natural or artificial agencies; whether these distant representatives of humanity have been made over solely by time and natural process, or whether this has been supplemented by eugenic edict. But real physical changes have occurred. The sexes have moved towards hermaphroditism, resembling each other 'in their height, in their smooth faces, and in the length of their hair' (pp. 13–14), with only the uniquely potent Father, the patriarch of the House, retaining his secondary sexual characteristics. Physical beauty and extreme longevity have become the norm. The daily round of the Crystallites is social and inconsequential, as it tends to be in all Victorian Utopias where human diversity has been filed down into bland equality. I. F. Clarke has commented on the 'curiously juvenile streak'[38] in Richard Jefferies' *After London* (1885), as well as in Hudson's and Edward Bellamy's romances of the same half-decade. Clarke's explanation for this is cultural-psychological: that these writers, more sensitive than most to the perplexities of their age, tried to ally themselves vicariously with the simple values of childhood. With Hudson, however, it would seem more convincing to suppose he knew that biological specialisation, including sexual differences, is the result of selectionist pressures. Remove those pressures and the sexes will drift towards resemblance.

But the real originality of *A Crystal Age* rests in its divergence from the pattern of Arcadian myth; and this divergence is most telling in Hudson's handling of the final evolutionary outcome of sexuality. Here personal wish-fulfilment and shrewd naturalist's common sense play an equal part in formulating one set of conditions whereby permanent strife might be set to one side. The exact nature of the reproductive arrangements within the House of the Harvest Melody is shrouded in Victorian reticence but there is no doubt that they accord well enough with Hudson's own beliefs about sexuality: 'the idea that there is no millennium, no rest, no perpetual peace till that fury has burnt itself out'.[39] In this respect his Utopia (which he is identifying with quietude) is a fantasy solution, but it is still securely founded in his study of natural history. He is perceiving unbridled sexual energy almost as a physical force, a driving head of steam impelling along the whole Darwinian mechanism. For natural selection works by population pressure (Darwin imagined this as driving the wedges of species into every vacant corner of the environment) which in turn is fed by the reproductive capacity. Allow that fearsome reservoir to leak away, to dissipate harmlessly into aesthetic expression, and the restless remoulding of life and endless competition between and within the species might slowly come to rest. Such at any rate is the logic of Hudson's dream solution: to debilitate sexuality by restructuring society after that most stable of all life patterns, the insectile. He imagines his House to be a single reproductive unit, with a fertile queen at its centre to regulate the rate of population increase within the human hive and to prevent the Darwinian order from gaining the upper hand. It is in dramatising these curious arrangements that the novel has its only vivid scenes. When the impulsive Smith with his unregenerate sexual habits from the remote past confronts with baffled ardour his love, the passionless drone Yoletta, his mystification is well conveyed. We share it, even if we guess how the Darwinian law has been repealed long before Smith is enlightened by the mortally ill queen, Chastel. She reacts thus to his description of a lowlier stage of evolution when all women might be mothers:

> The human race would multiply until the fruits of the soil would be insufficient for its support; and the earth would be filled with degenerate beings, starved in body and debased in mind. (p. 173)

The Malthusian threat is familiar enough. What is surprising is the angry energy with which Hudson proposes his solution of making reproduction a highly specialised role. He was sufficiently attached to it to refute indignantly Edward Garnett's impression that he had taken it over from the naturalist and sociologist Benjamin Kidd.[40]

Hudson's evolutionism of the late 1880s, then, could be maintained only at the tremendous cost of tampering irreversibly with the fundamentals of human nature. Indeed, Smith wonders whether the Crystallites *are* human still. Even as the realisation of the true price to be paid for working out the beast dawns on him, there comes, too, the suspicion – it never amounts to anything more – that perfection might not be worth the candle. As he confronts the wreck of civilisations, Smith tries to comfort himself with the thought of the great contrast between fragile human striving and the sempiternal natural order. He tries to be a Spencerian, an evolutionary Pangloss. ' "*Passing away*" is not written on the earth,' he assures himself. 'It is vain to ask myself now whether the vanished past, with its manifold troubles and transitory delights, was preferable to this unchanging, peaceful present' (p. 268). Vain, perhaps, for Smith to ask it, trapped in the remote future, but not vain for Hudson – still less for his reader. Certainly Hudson stacks the cards heavily against the nineteenth century, and as the idyll proceeds he grants us little opportunity to object to paying the tax of curiosity and creativity. If the denaturing of man is the price to be paid for entering Elysium to ride a superbly responsive horse through a long golden afternoon, then so be it! Such is the mood of the novel. But what if golden afternoons cannot be indefinitely extended for the race? What if biological theory should come to teach that nature cannot hold on to its precarious perfection but must move onward into a phase of decay? Even the posing of such questions was well beyond Hudson's scope in the 1880s. Much later on he came to recognise that 'the ending of passions and strife is the beginning of decay', as he put it in his 1906 preface; but to make imaginative use of the emerging facts of biological degeneration was a task which passed to other hands. Among these retrogressionists evolutionism ceased to exist in its naïve form after a run of about one hundred years. As we have seen, Darwinism as such was powerless against it, but the post-Darwinian revelations eventually did destroy it for some.

The manner of evolutionism's death and the nature of its antithesis will be the subject of the next chapter.

Notes: Chapter 3

1 Alfred Russel Wallace, *Darwinism: An Exposition of the Theory of Natural Selection, with Some of Its Applications*, pp. 477–8.

2 Wallace's chief biographer, Wilma George, does not attempt in her *Biologist-Philosopher: A Study of the Life and Writings of Alfred Russel Wallace* to connect up Wallace's two roles of her title. She lets Wallace the co-discoverer of natural selection and Wallace the spiritualist remain compartmentalised. Though several historians have tried to trace out the course of Wallace's defection from Darwinism over human evolution, the only reliable accounts are: Malcolm J. Kottler, 'Alfred Russel Wallace, the origin of man, and spiritualism', *Isis*, vol. 65 (1974), pp. 145–192, and Frank Turner, *Between Science and Religion: The Reaction to Scientific Naturalism in Late Victorian England*. The former shows that Wallace's defection cannot be dated to earlier than March 1869 – i.e. after his conversion to spiritualism, which was complete by November 1866. Turner (p. 83) agrees with earlier readings of Wallace's paper 'The origin of the human races . . .' *Journal of the Anthropological Society of London*, vol. 2 (1864), pp. clviii–clxx, which pre-dates his spiritualism, that this does in fact contain 'the latent seeds for all his later departures from scientific naturalism'. Though Kottler seems to me to be correct, there remains the puzzle that Wallace's 1867 paper 'Creation by law' (see n. 3, below), which Kottler does not discuss, dates after his conversion but reads as though Wallace still believed that nothing is exempted from the sway of natural selection. The value of Turner's account is that he describes Wallace's involvement with the ideas of the phrenologist George Combe early in the 1860s, for whom man was 'designed for another and higher destiny' (quoted by Turner, p. 79).

3 In Alfred Russel Wallace, 'Creation by law', *Quarterly Journal of Science*, vol. 4 (October 1867), pp. 471–88. Nominally a critique of the Duke of Argyll's theistic *The Reign of Law* (1867), this review attempts to show how the examples Argyll adduces as evidence of conscious contrivance in nature can be explained by natural selection alone. It is very remarkable that what Argyll takes to be the most telling evidences for his natural theology – non-advantageous beauty, the fitness of nature to man's needs – are vigorously dismissed by Wallace; and yet his later position was more extremely teleological even than Argyll's.

4 This initial statement is contained within a long review by Alfred Russel Wallace published anonymously, 'Sir Charles Lyell on geological climates and the origin of species', *Quarterly Review*, vol. 126 (April 1869), pp. 359–94, which deals with both Lyell's *Principles of Geology* (10th edn) and his *Elements of Geology* (6th edn). The last three pages are devoted to the theory of man's unassisted ascent from animality – a theory which Wallace now calls 'in the highest degree improbable' (p. 391). The following year Wallace expanded these last pages into a complete new paper, 'The limits of natural selection as applied to man', and published it in his *Contributions to the Theory of Natural Selection: A Series of Essays* (1870). His views stabilised at that time, for yet another revision, twenty-one years later, for his new collection *Natural Selection and Tropical Nature: Essays on Descriptive and Theoretical Biology* (1891; new edn 1895), which appeared under the same title as the 1870 version, has very few changes. However, whether the whole evolution of man, or just the evolution of his mental and moral powers, demanded the introduction of an agency additional to natural selection was something which Wallace changed his mind about more than once. In the 'Limits of natural selection' essay of 1870 both body and mind are the result of assisted ascent. By the time of *Darwinism* (1889), Wallace had retreated by handing bodily evolution back to natural selection, retaining spiritual assistance only for the intellectual and moral nature. But seven years later his opinion

was that the human spirit 'is developed *in* and *by means of* the body' – and mental faculties by the brain (quoted by Turner, *Between Science and Religion,* p. 97).

5 Thomas Henry Huxley, 'Mr Darwin's critics', *Contemporary Review*, vol. 18 (November 1871), p. 471.

6 Wallace, 'Lyell on geological climates', p. 392.

7 George Earle Buckle (a journalist and later editor of *The Times*), 'Natural selection insufficient to the development of man', *Popular Science Review*, vol. 10 (1871), p. 22.

8 Alfred Russel Wallace, *The World of Life: A Manifestation of Creative Power, Directive Mind and Ultimate Purpose*. Quotations and examples in the remainder of this paragraph are drawn from pp. 278–84 of the first edition.

9 This interpretation is from Morse Peckham, *Man's Rage for Chaos: Biology, Behavior and the Arts*, p. 13. If we accept Kottler's 'compromise' position that spiritualism caused Wallace's first doubts so that he was led to reconsider the formidable evidence for the excess development of the brain etc., the most important fact is that Wallace never brought paranormal explanations into his public thinking on human evolution. He is concerned exclusively with the technical limitations of Darwinism; and so publicly this was the foundation to his evolutionism, whatever his religious beliefs.

10 Wallace, *Contributions*, p. 398; Wallace, *Natural Selection and Tropical Nature*, p. 205 n. In the last version of his 1870 paper, Wallace denied that even a teleological biology could make any useful statements about the final end of evolution. He rests with these personified forces.

11 This typical judgement is from Joseph Warren Beach, *The Concept of Nature in Nineteenth-Century English Poetry*, p. 6.

12 The Duke of Argyll, for instance, determined to do this. See his correspondence with the more circumspect Spencer between March 1893 and September 1898: *Autobiography and Memoirs*, Vol. 2, pp. 496–500.

13 James Collins, 'Darwin's impact on philosophy', *Thought*, vol. 34 (1959), p. 186.

14 'The filiation of ideas', in David Duncan, *The Life and Letters of Herbert Spencer*, p. 543. Lionel Stevenson, in 'Darwin and the novel', *Nineteenth-Century Fiction*, vol. 15 (June 1960), pp. 29–38, has shown that Spencer's interest in pre-Darwinian evolutionism was reflected in literature, too: less familiar novels by Meredith, Kingsley and Eliot of the mid-1850s make use of the same 'development hypothesis' that Spencer constructed in 1852. Similar, more questionable claims have been made for the pre-1859 poetry of Coleridge, Browning, Emerson and of course Tennyson. Beach makes a critical summary of such claims in his *Concept of Nature*, pp. 415–22.

15 *Westminster Review*, n.s., vol. 11 (April 1857), pp. 445–85. The article is nominally a review of A. von Humboldt's *Cosmos*, the ninth edition of Lyell's *Principles of Geology* and the fourth edition of W. Carpenter's *Principles of Comparative Physiology*. Spencer soon leaves all three behind, however.

16 Morse Peckham, 'Darwinism and Darwinisticism', *Victorian Studies*, vol. 3 (September 1959), pp. 28–9.

17 Charles Coulston Gillispie, *The Edge of Objectivity: An Essay in the History of Scientific Ideas*.

18 Charles Kingsley, *The Life and Works of Charles Kingsley*, Vol. 18, pp. 213–14. This essay, 'North Devon – a prose idyll', first appeared in *Fraser's Magazine* in July 1849.

19 *The Martyrdom of Man*, p. L. All further references in the text are to this reprint for the Rationalist Press.

20 Herbert Spencer phrases it very similarly: cf. *The Principles of Biology*, revised edn, Vol. 2, p. 354: 'Slowly, but surely, evolution brings about an increasing amount of happiness; all evils being but incidental.' There are comparable sentiments in Darwin, too: cf. *Origin*, XIV:266. Out of a long list of scientists whom Reade mentions in his preface as helping him form his views, only three – Lubbock, Tyndall and Tylor – were known for their work in subjects other than biology or related sciences.

21 Charles Darwin, *More Letters of Charles Darwin*, ed. Francis Darwin and A. C. Seward, Vol. 1, p. 94.

22 G. M. Young, *Victorian England: Portrait of an Age*, p. 110.

23 Henry Drummond, *The Lowell Lectures on the Ascent of Man*, p. 256. The quotations which

follow in the text are from this first edition.

24 Reade, *Martyrdom of Man*, p. 423.

25 George Orwell, 'Charles Dickens', in his *Collected Essays*, p. 71.

26 Letter to Morley Roberts, dated 12 March 1920: W. H. Hudson, *Men, Books and Birds*, p. 225.

27 Robert Hamilton, *W. H. Hudson: The Vision of Earth*, pp. 40–1. The two accounts are in *Far Away and Long Ago: A History of My Early Life* (1918) and *The Book of a Naturalist* (1919). The quotation which follows is from the former account, p. 329.

28 Morley Roberts, *W. H. Hudson: A Portrait*, p. 174.

29 Hudson, *Far Away and Long Ago*, pp. 329–30.

30 Quoted from the untitled version, which was the one read by Darwin, in *Proceedings of the Zoological Society of London*, 24 March 1870, pp. 158–60. In W. H. Hudson, *Letters on the Ornithology of Buenos Ayres*, ed. David R. Dewar, Dewar dates the letter to 28 January 1870, adding that it was corrected and its tone modified by the Secretary before publication.

31 Charles Darwin, 'Notes on the habits of the Pampas Woodpecker (*Colaptes campestris*)', *Proceedings of the Zoological Society of London*, 1 November 1870, pp. 705–6.

32 Letters to Morley Roberts dated 22 October 1918 and 21 July [1919; so dated by Roberts]: Hudson, *Men, Books and Birds*, pp. 167–8, 207. Five months later, Hudson speaks of the newly published biography of Butler sending him *back* to Butler's books, though he does not say specifically that he means the 'evolutionary' ones. See the letter to Roberts of 28 December 1919: ibid., p. 214.

33 W. H. Hudson, 'Wasps', in his *The Book of a Naturalist* p. 198.

34 Aldous Huxley, 'Crebillon the Younger', in his *The Olive Tree and Other Essays*, p. 135.

35 Hudson gave his first anonymous edition an epigraph from the fourteenth chapter of the *Origin* 'Recapitulation and conclusion', including in this the familiar sentence 'as natural selection works solely by and for the good of each being, all corporeal and mental endowments will tend to progress towards perfection'. It is indicative of Hudson's very careful reading of the *Origin* that he was able to pick so unerringly on one of the few directly evolutionistic statements in the book. In the 1906 (revised) edition he added his name and a short preface, but removed the Darwin epigraph and substituted four lines (unattributed) from *The Faerie Queene*, bk VII, canto viii, stanza 2, ll. 1–4, which speak of a time when mutability shall cease. The best explanation for this that occurs to me is that by 1906 Hudson's disenchantment with Darwinism was such that he no longer believed in the power of natural selection to generate any kind of perfection.

A Crystal Age poses some complex bibliographical problems, but a cursory inspection of the two main editions shows that Hudson revised quite carefully with a view to curbing the stylistic flourishes of the first edition: 'Universal Mother' (1887) becomes simply 'Earth' (1906 and later), and so on. Quotations throughout are from the first edition.

36 Hudson, *A Crystal Age*, 2nd edn, p. viii. This most readily available second edition went through a number of impressions, all containing the added preface.

37 Letter of 10 June 1917: W. H. Hudson, *Letters from W. H. Hudson to Edward Garnett*, new edn, p. 175.

38 I. F. Clarke, 'The nineteenth-century Utopia', *Quarterly Review*, vol. 296 (1958), p. 86.

39 Letter to Edward Garnett of 10 June 1917: Hudson, *Letters*, pp. 174–5.

40 In his *William Henry Hudson*, John T. Frederick argues that the initial inspiration for modelling the Crystallite society on the habits of the social insects may have come from Thomas Bell's *The Naturalist in Nicaragua* (1874). Certainly this work was well known to Hudson, for he several times quotes approvingly from it; and it does contain a passage relating the social habits of ants to the reformed human society described in More's *Utopia*. However, I have suggested elsewhere that there may have been a more direct source. Hudson's claim to the contrary notwithstanding, a paper by Benjamin Kidd called 'Humble bees' which appeared in *Longman's Magazine*, vol. 7 (December 1885), and deals with the habits of the species *Bombus terrestris* contains such close parallels to the plot that Hudson almost certainly read it before, and not (as he claimed) after, the novel was completed.

4

Laying the Ghost of the Brute: The Fear of Degeneration

when shall we lay
The Ghost of the Brute that is walking and haunting us yet, and be free?
In a hundred, a thousand winters? Ah, what will *our* children be,
The men of a hundred thousand, a million summers away?

ALFRED TENNYSON (*c.* 1892)

On 4 and 11 February 1884, John Ruskin gave two public lectures in London on a strange theme: the alarming deterioration in the English weather which he claimed to have observed over the previous twenty years, and the psychological effect of this depressing change on one sensitive individual – himself. Exercising his rhetorical gifts with the usual skill, Ruskin conjures up the mournful picture of a sun quenched by heavy, vilely coloured clouds and polluted air, and of an earth whose inhabitants are pinched and feeble from a lack of radiant energy. Since the tonnage of coal burnt with no controls whatever had been increasing steadily through the previous decade, there may perhaps have been some kernel of truth in Ruskin's denunciation. He himself certainly insisted that the data he was using had been 'verified with a chemist's analysis and a geometer's precision', and that in the future his own day would be recognised as having seen 'phenomena hitherto unrecorded in the courses of nature'.[1] But describing the manifold shortcomings of the weather does not exhaust Ruskin's indignation. His real fear and outrage is of something deeper: a more catastrophic and permanent snuffing-out of heat and light. By a wild leap of thought he turns his ire on to a book 'by a very foolish and very lugubrious author', Balfour Stewart's *Conservation of Energy* (1872). Stewart, following his mentor William Thomson, offers his reader the prospect of a very distant but certain future when all the energy flows of the universe have equalised; when all the stars have formed from gas-clouds and have radiated away their heat so that all creation is an evenly heated inert mass and all life necessarily extinct. For Ruskin the bad weather and this vision of universal heat-death form some strange mental amalgam, provoking him into a decisive judgement on his personal intellectual history:

> I will tell you this much: that had the weather when I was young been such as it is now, no book such as 'Modern Painters' ever would or

> *could* have been written; for every argument, and every sentiment in that book, was founded on the personal experience of the beauty and blessing of nature, all spring and summer long . . . That harmony is now broken, and broken the world round . . . month by month the darkness gains upon the day. (*The Storm Cloud of the Nineteenth Century*, 137–8)

The five volumes of *Modern Painters* were published between 1843 and 1860. The sublime responsiveness to every mood of nature embodied in the first two volumes (1843, 1846) contrasts violently with the expressive disgust of rural and urban life alike to be found in the final volume of 1860. The *Storm Cloud* lectures more than twenty years later were disastrous to Ruskin's reputation because they, along with others given around the same time during his tenure of the Slade Professorship at Oxford, forced his mental instability into full public awareness. But Ruskin had already suffered three prolonged manic episodes before these lectures, and even in his letters of the early 1870s the same half-crazed illogical linkages are made over and over between the terrible weather, the degeneracy of the natural order and the betrayal of the Wordsworthian vision. In 1871 he complains to Charles Norton of 'this black east wind for three months, this sense of the evil working of nature herself – my disgust at her barbarity – clumsiness – darkness – bitter mockery of herself',[2] and two years later of 'fourteen days of incessant wind and rain . . . Nature herself traitress to me'.[3] These agonised notes, which extend Ruskin's obsessive concern with the weather over at least a full decade, suggest that it is to the literature of psychiatry, not of meteorology, that we should relegate *The Storm Cloud of the Nineteenth Century*. In its confusion of objective and subjective, in its obsessive dwelling on decay and gloom, and in the vague imputation of these to some moral decline it threatens terribly the descent of its author's mind into yet another bout of insanity – realised in the following year. *The Storm Cloud* is one gigantic pathetic fallacy, which, ironically, Ruskin himself coined as a phrase and which he defined as 'a falseness in all our impressions of external things' (*Works*, 5: 205). Yet, as is usual with Ruskin when he is writing out of his own intuition, his hypersensitivity lets him probe and probe away at certain molars which were for others a source of no more discomfort than a dull background ache. The complete inversion of the response to nature between the first volume of *Modern Painters* and *The Storm Cloud* may of course be seen as John Rosenberg sees it: 'one of the crucial experiences of his century: the realisation that the benign Nature of Wordsworth was red in tooth and claw'.[4] But Ruskin's perception was able to distinguish much finer shades than that, even when he was half-deranged. Rather than to his isolated bizarre notions we should give ear to the associative connections he forges; for with these he succeeds in contracting into a very small compass a whole cultural response which in better-aligned minds remained fragmented. The contraction is most evident in *The Storm Cloud* when we examine the movement from 'blanched Sun, – blighted grass'; past the further insistence that harmony is 'broken the world round' and all nature in decay; and finally to the shuddering simile of the whole universe as a candle guttering into extinction. These constantly recurring metaphorical patterns owe little to Darwin directly, though they may perhaps be comprehended as a response to the idea of degeneration within the larger frame of post-Darwinian biology. Instead of nature red in tooth and claw, we have Ruskin's own blighted garden as a concentrated image of the corruption of nature:

> looking over my kitchen garden yesterday, I found it one miserable mass of weeds gone to seed, the roses in the higher garden putrefied into brown sponges, feeling like dead snails; and the half-ripe strawberries all rotten at the stalks. (*The Storm Cloud*, 58)

It is both endearing and pathetic that, for Ruskin, personally experiencing a garden gone to seed was so alarming in its cosmic suggestiveness as to require an immediate public statement. But even for saner writers the same image, when transformed into an impersonal metaphor, was a favourite device for making concrete a group of connected biological abstractions. It was used thus quite consciously by writers as diverse as Mill, Hardy, Gissing and, above all, by T. H. Huxley.[5] As Oma Stanley has shown in detail, the young Huxley tended romantically to personify nature and to come perilously close to a teleological reading of it.[6] In his essay 'On the formation of coal' (1870), Huxley does not go quite so far as Wallace was later to do and suggest that the coal measures were laid down *specifically* for man's future use. However, when he describes how the elements of burnt coal, if collected, would be identical with the nutritive elements taken in by the organisms which formed it Huxley does not scruple to apostrophise nature as no prodigal, but the most efficient of housewives. Clearly, he is trying to rectify the impression left by Darwinism in most minds of nature as a wanton. Huxley afterwards changed his ambitions on that score; possibly to correct certain late eulogies of nature by Swinburne, Meredith and Jefferies which he regarded as unhealthy; or possibly (as Oma Stanley argues) because he had read J. S. Mill's influential essay 'Nature' when it appeared in 1874. Be that as it may, ten years after Ruskin's *Storm Cloud* lectures Huxley may be discovered in his 'Prolegomena' (1894) to his Romanes Lecture of 1893 seeking a metaphorical structure which describes possible responses to the inevitable decay of the universe towards maximum disorder. (The concept of entropy was first presented by Clausius in the mid-1860s.) A year earlier Huxley had moved in his lecture restlessly through three complex metaphors without pausing long at any of them. In 'Evolution and ethics' we skip from Jack and the Beanstalk to the ape and tiger to Tennyson's Ulysses in a confusing manner. In the 'Prolegomena', however, the parable of the weeded garden – that is to say, a carefully maintained artificiality carved from a wilderness to which it will rapidly revert unless attention is unremitting – is a vivid expression not so much of Huxley's belief that man should combat the natural order but, rather, of his fear of degeneration transformed into a homely image. For Huxley's weeded garden soon gets overrun by Ruskin's unweeded one. As his train of thought runs on it emerges that what really interests and appals the retired Huxley as he looks out upon his newly laid-out flowerbeds at Eastbourne is not how the garden may best be weeded, nor what it looked like in a state of nature before Caesar first came to Britain, but what would happen almost instantaneously if the hands of Caesar's modern descendants were withdrawn.

> The walls and gates would decay; quadrupedal and bipedal intruders would devour and tread down the useful and beautiful plants; birds, insects, blight, and mildew would work their will; the seeds of the native plants, carried by winds or other agencies, would immigrate, and in virtue of their long-earned special adaptation to the local conditions, these despised native weeds would soon choke their choice exotic rivals. A century or two hence, little beyond the foundations of the wall and of the houses and frames would be left, in evidence of the victory of the cosmic powers at work in the state of nature, over the temporary obstacles to their supremacy, set up by the art of the horticulturist.[7]

This fine musing passage is the equal of the Ruskin piece in its dramatic power, and its half-clarified intuition that all life, whether the product of nature, horticulture or human culture, is only an example of entropy running backwards for a while and in the long run is certainly no bastion against the universal ruin. Similar in theme they may be, but in tone they are quite distinct. Ruskin's is that of a sunken and immobilised despair of the artist before a nature turned treacherous. Huxley's is the noble and actively stoical pessimism of a biologist with no romantic illusions about nature left to be shattered. Both do agree that in the very long term man's future is finite. Ruskin allowed this knowledge to poison his present. Huxley was content to clarify the issues and to offer the temporary consolation (borrowing some lines from Tennyson's 'Ulysses') that before the end 'some work of noble note' could well be done. We shall find that the imaginative response to these unsettling truths about a nature on the wane followed the pattern laid down by these two preliminary examples.

Meanwhile, our central concern is to bring into exact focus the vague connotations hovering around the late-Victorian understanding of *biological* degeneration (otherwise retrogression, or devolution). For a simple concern with moral or social degeneration is a perennial cultural need, feeding off different evidence at different times. For Western pre-scientific cultures a belief in the progressive decline of man and nature found its explanation in the Genesis myths. Even earlier, the Greek legendary history retailed by Hesiod and others already told of an ideal Golden Age in even remoter antiquity which had long since relapsed into the contemporary Age of Iron. And much later, in the seventeenth and eighteenth centuries, the literary battle over the respective virtues of ancient and modern writers raised some quasi-biological issues. Are nature's powers limited or infinite? Has she exhausted her powers with the passing of the classical age? Defeated spirits agreed with Sir Thomas Browne that ' 'tis too late to be ambitious, the great mutations of the world are over', and resigned themselves to rapid universal decay. Others of more elastic temperament preferred to believe that nature is elastic too, and capable of regeneration and renewed creativity. Whatever course such speculations followed they were inevitably amorphous and a matter of blind prejudice or temperament, as when Dr Johnson mocked Milton's proud boast of being a giant in a degenerate age when not even the trees grew as high as they were wont to do. They took on a sharper outline only when it seemed to become clear, as the first enthusiasm for the new biology subsided, that certain established assumptions about man *as a species* (and not simply about certain races, nations or cultures) might need to be abandoned. The warning came most distinctly from the direction of astrophysics, but that in turn was a direct reaction to a specially bold claim by the biologists. In short, certain writers had to face up to the frightening prospect of degeneration in the future because we are already living in the bright noon, and not the dawn, of man's day. These fears co-existed uneasily with the progressivism inspired by another selection of evolutionary facts.

The Victorians' Parochial Future

When he came to write on 'The scientific movement and literature' in 1877, the critic Edward Dowden felt a need to distinguish carefully between the literary possibilities of the ideas of Progress among the Romantic poets and in Tennyson. The difference – Dowden finds it obvious – is the new post-Lyell, post-Darwinian attitude to time. His metaphor is not perhaps the freshest he could have chosen, but it serves:

> We ascertain that the waters have come from some mysterious source among strange mountains a thousand leagues away, and we are well assured that they will descend a thousand miles before they hear the voice of that mysterious sea in which they must be lost.[8]

Perhaps it is taking Dowden rather too literally to press heavily on his ratio of three to one between the availabilities of past and future time for man. Nevertheless, his general proportions are clear. Writing in the late 1870s, Dowden could readily have known that, according to the best estimates of his contemporary palaeontologists, *Homo sapiens* and his related but now extinct species had been active for at least a million years; that is, if one were to accept the persuasive calculations of William Thomson discussed in an earlier chapter, as much as one-tenth of the whole span of the past. A by-product of Thomson's attempt to fix quantitatively the age of the sun was that it also placed a very restricted limit on its future output of energy. That limit came farther down to the present and encroached more and more upon Darwin's early confident expectation of 'a secure future of . . . inappreciable length' (*Origin*, XIV:265 – and in the sixth and last edition Darwin changed the last phrase, in deference to Thomson, to 'great length'). Thomson's supposition that the sun's heat is produced by chemical reactions alone led inevitably to his warning that Time must have a stop; that even if the sun were made of a solid lump of high-grade coal it could not be expected to burn for a million or so years longer. Taking all other possible factors into account and interpreting the evidence as generously as possible, Thomson still placed his absolute upper limit at 'five or six million years of sunlight'.[9]

These now thoroughly outmoded opinions help us towards two important generalisations about the Victorian image of the future. First, that future was a relatively parochial affair. The geology of Hutton and Lyell had already broken through Archbishop Ussher's temporal boundary (in 1658 biblical chronology had enabled him to date the universe's age to 5662 years, plus or minus one year); but even these Uniformitarians, teaching that no radical change in the nature, only in the degree, of geological processes had ever happened, were, thanks to Thomson, Balfour and others, able to form no real conception of the enormity of time. With a very few exceptions, such as Darwin's own quickly retracted figure of 300 million, they measured the age of the earth itself and therefore of the life upon it only in scant millions of years. To conceive of a past thousands and a probable future tens of thousands of times longer than this was, between about 1865 and 1890, outside the imaginative range of even the most astute intelligence. Second, this prospect of a limited future had great human immediacy, for when the sun flickered out a few million years hence man, given the slowness of biological change, might still be recognisably man. So Darwinian evolution offered the almost unendurable prediction, after the passage of some millions of years, 'with every continent swarming with good and enlightened men', of everything having to be forfeited at the very acme of achievement. Darwin himself felt that the humiliating slowness of natural selection in bringing about improvement added to the horror of this vision of heat-death.[10]

The Thomsonian future, therefore, if his calculations were to be accepted (and there were few to question them: given the premiss of molecular energy transformations alone, they are mathematically impeccable), was one of continuous degradation, failing light, of gloom descending over the whole of creation like a storm-cloud over central London. Now, it has been something of a cliché to describe the impact of Darwinism as being equivalent in intensity to that of Copernican cosmology in the sixteenth century. In one respect, however, the comparison is a misleading one in that at least part of the human significance of heliocentrism entered the public domain only with Darwinism itself. Before that it was possible even in a heliocentric solar system to imagine man still within the warm and snug confines of a cosmic household. But as Thomson's warnings about the sun – the sole source of biological energy – dissipating its force took effect F. W. H. Myers's poetic description of that system held a chastening power: 'an unroofed and fenceless plot, from whence every moment the irrecoverable sun-rays tremble out into the blackness and are squandered in the gulf of heaven'.[11] Astrophysics and biology united in recognising waste as inherent in the natural order. Degeneration out in the organic world is equatable with degeneration in the organic world: that is the implication of what Ruskin saw in his rotting garden.

The Darwinists and Degeneration

Are all species (including man) perpetually either evolving or devolving? Few biologists of the day felt impelled to seek a *tertium quid* on this emotive issue. One or two argued that, under certain very rare conditions, perfect adaptation to a static environment can result in permanent equilibrium: what ecologists now call the 'climax' condition. They believed that there is no inner force making species evolve; that they do so only when new niches in the environment are opened up, into which the selection mechanism can force the 'thousand wedges' (Darwin's own metaphor) of genetic variations. Several such 'living fossil' species were known to the Victorians which have continued since the very earliest times without manifest change in either direction. But what exactly these examples might mean for the human future was strongly disputed.[12] Herbert Spencer took them at face value. Arguing from economics back to biology as was his wont, he taught that when *laissez-faire* conditions prevail more generally the human population will eventually stabilise, since the more highly organised the intelligence (a trend which he took to be inevitable with free-running natural selection in society), the lower the fecundity. He anticipated a general uneventful harmony where quiet philosophical men like himself might flourish; when 'pressure of population and its accompanying evils will disappear; and will leave a state of things requiring from each individual no more than a normal and pleasurable activity'.[13]

This dream of an Eden where the only remaining serpent might be yawning boredom certainly had its literary adherents, as we shall see; but it had far more, and more persuasive, scientific critics. Evolutionary stasis as a permanent feature in nature fell from notice when the neo-Darwinians tried with some success to show not only that natural selection alone can make for advance, but also that any tampering with its policing power would mean instant decline. To survive unaltered, to avoid the effects of Weismann's 'panmixia' (his term to describe degenerative change consequent on disuse had a brief general currency after 1890), a species must be kept well chiselled to shape. But even those who were stoic enough to be transformed because only struggle and competition bring progress were nonplussed to find that degenerative adaptation can produce the most dramatic and ugly results. The blinded fish living in the dark Kentucky caves, for example – what had selection done for them? Certainly they were adapted well enough – to cave life. The moral could hardly be more pointed. Or what of the habits of the slave-making ants, *Formica rufescens* and *Formica sanguinea*, so abjectly dependent on their household captives, *Formica fusca*? What of the rearing habits of the cuckoo? Here, for the liberal Darwin, belief in the goodness of his principle collided with his data: in the first case he finds the instinct to be at once 'wonderful' and 'extraordinary and odious' (*Origin*, VII:115, 141). Natural selection has produced something morally detestable. Are not such adaptations genuinely degenerate? Darwin's anthropomorphism, as he wrestles with these dissonances (he concludes that the British species of ants are less slavish than the Swiss) is quite muted: it is only slightly doubtful if his mind is fully on the ants. At no point, at least, does the discussion expand overtly out of natural history. It is the same with his struggle to devise an undistorting symbol of the evolutionary process (as he moves in the *Origin* from branching tree to tangled bank) which does not beg too many questions about 'higher' and 'lower' species. He admitted in 1854 that even the useful guide of morphological differentiation fails to work in every tricky case of degeneracy; that 'my ideas are no clearer than those of my brethren'. Even four years later with the *Origin* nearly completed, all he *is* clear about is that 'highness' in the special biological sense is 'different from highness in the common acceptation of the word. It might be connected with degradation of organisation: thus the blind degraded worm-like snake (*Typhlops*) might supplant the true earth-worm.'[14]

But despite his hesitancies and semantic caution, in the less meticulous hands of Darwin's 'brethren' his work eventually gave a very fine edge to the broader concept of degeneration. Inevitably so; for any layman of imagination could perceive that devolution was but the dark reverse of a coin whose bright obverse was biologically endorsed progress. A man who does not believe that the working of some natural law may underwrite human advancement can hardly be troubled by fears that nature may be planning man's obsolescence, too. Only a culture which has implicitly accepted evolutionary theory can find troubling the warning that 'in no case does the record of the fossils show a really dominant species succeeded by its own descendants'.[15] It was those very writers most anxious to promote the tenets of evolutionism who were most troubled by warnings of decay and who were the most eager to respond with moral exhortation.

Only some of the manifestations of Victorian pessimism about the future may be found to have a biological base, but those that did are certainly incomprehensible in any other terms. One of the earliest explicit formulations, which later served many more writers as a convenient mine of examples, was Ray Lankester's *Degeneration: A Chapter in Darwinism* (1880). For most of this short book Lankester dutifully recites the evidence for believing that evolution does not always run in the direction of greater heterogeneity. Only in the last chapters does he turn to the wider issues: the relevance of these examples to the individual's and nation's moral life. His conclusions are restrained but decisive. Collapse into 'a contented life of material enjoyment accompanied by ignorance and superstition' (p. 61) must be the fate of anyone of any race who refuses to heed the dreadful lesson of the tapeworm. Nature is not mocked. If in our idleness and stupidity we do not obey her dictates of enterprise and aggressive puritanism, then we shall soon be at one with the dinosaurs.

Such extensions from certain indubitable facts – extensions which in Lankester's monograph form no more than an afterthought – became, as the 1880s passed by, the issues at the centre rather than on the periphery of the debate on retrogressive evolution. It followed inevitably that the conclusions of scientists with various moral axes to grind became more and more detached from the hard core of evidence. Andrew Wilson, lecturer in zoology at the Edinburgh Medical School, pressed forward in an essay written in the wake of Lankester's book with the concept of degeneration as a veritable skeleton key to history, able to penetrate

> fields of thought often widely removed from the original topic which interests the reader. The present subject of degenerative changes, regarded as part and parcel of the living constitution, can readily be shown to possess applications far removed from zoology and botany, and extending into the most intimate spheres and phases of human history itself.[16]

While few readers of the *Gentleman's Magazine* would probably be very intrigued by details of the tapeworm's life-cycle, Wilson's opening up of a peephole on degeneration in history held a spicier promise. Gibbon's pages on Caligula make livelier reading than those on Augustus, even though the historian is more bound by the proprieties than a biologist. For it is part of the latter's job to gaze on vice with a cold eye and report what he finds without shrinking, and Wilson took his professional obligations seriously. Without breaking stride even momentarily to consider how far analogy can fairly be made to stretch in this case or what the precise force is of certain pejoratives he uses so glibly, Wilson marches from the habits of the male rotifer and the linguatulina which in adulthood lose their limbs to become little more than digestive tracts, to sonorous philosophising on the decay of civilisations. Certain lowly crustacea, in losing 'eyes, legs, digestive system and all the ordinary belongings of animal life "go to the bad," as a natural result of participating in what has been well named "the vicious cycle of parasitism," ' he tells us (p. 480), having the grace at least to mark his recognition that it is a fair step from shellfish to Newgate. But the awareness that he is employing a dubious figure of speech does not restrain Wilson from his foregone conclusion that retrogression is immutable law, to be numbered among the sternest realities of organic existence. Even here, though, the tendencies we are trying to exhibit still remain partially latent. Analogies are indeed grossly abused, but the author does retain some rudimentary insight into the difference between a biological fact and an ethical prescript. In two long chapters of his immensely popular *Natural Law in the Spiritual World* (1883), Henry Drummond smudged the two beyond redemption and earned plaudits from his audience for doing so.[17]

As we saw in the previous chapter, Drummond's textbooks of imitation biology are just manuals serving up the morality of evangelism in the guise of so-called 'laws' extracted from the world of life. The method is hardly more sophisticated than in a medieval bestiary. It is only in observing Drummond's method here, in his role of a post-Darwinian Paley looking out for fresh Evidences and determined to find them, that we can gain at last a sufficient idea of the sheer richness of the metaphor with which biology had supplied the none-too-scrupulous rhetorician. Here is Drummond on the habits of *Sacculina*, an organism which is parasitical on the hermit-crab. In its youth this creature is a lively individualist: 'it paddles briskly through the water. For a time it leads an active and independent life' (p. 343) – and is practically a disciple of the *Self-Help* creed. But *Sacculina* is doomed to star in a Rake's Progress of submarine life:

> The hereditary taint of parasitism is in its blood, and it proceeds to adapt itself to the pauper habits of its race . . . The Sacculina sets out in search of a suitable host, and in an evil hour, by that fate which is always ready to accommodate the transgressor, is thrown into the company of the Hermit-crab. Whereupon the whole of the swimming feet drop off, – they will not be needed again, and the animal settles down for the rest of its life as a parasite. (p. 343)

In its youthful form *Sacculina* resembles a shrimp or lobster, but in adulthood it declines to seize its opportunities. Its 'sin' therefore for Drummond is a double one:

> It was first a disregard of evolution, and second, which is practically the same thing, an evasion of the great law of work. And the revenge of Nature was therefore necessary. (p. 344)

These passages on the 'physiology of backsliding' are not introduced to labour their absurdity – that is hardly required – but because they are extreme representatives of a common logic. Such evidences from nature were still thought to be highly convincing by a substantial cross-section of the literate public; and, whatever we might think of them, they do demonstrate the absurdity of assuming that Darwinism instantly reduced nature to a mere backcloth to human concerns. When Drummond pointed an accusing finger at the decadent social morality of animal parasitism, and drew the human lesson that 'beggars of the market-place . . . are living and unlying witnesses to the unalterable retributions of the law of parasitism' (p. 350), he was touching a very responsive chord. His use of 'unalterable' suggests, perhaps, something of the deeper logic at work. If the laws of biology could be granted the universal applicability and inevitability of physical law, then so many social sores could be sterilised without resorting to the surgery of economic reform. The easiest way to deal with an uncomfortable problem is to conceptualise it afresh in a more appealing form. If beggars, instead of being the casualties of an inadequate man-made system, could be seen instead as domestic pigs reverted to a worthless feral form by their own built-in retrogressive tendencies, then this identification removed much perplexity. Though they might disturb, atavisms in the social structure like paupers and criminals were not ultimately things one had to worry about, for they could safely be left to nature with her complex but infallible system of checks and balances. If Drummond's complacency repels, it is as well to reflect that the biology of his time was able to be forced to yield grounds for complacency. That alone enabled Drummond to see frailties as throwbacks rather than as rising from the strained human position of being 'plac'd on this isthmus of a middle state'. In this new, popular conception of original sin, guilt, rather than reaching onwards even to the seventh generation, is just something to be bred out like an undesirable characteristic in a pig.

Is Degeneration Really Perfection?

It takes very little reflection to appreciate just how superficial were Victorian biology's deductions from parasitology. Those degenerative changes which through natural selection have reduced the physiological machinery of a liver fluke to the simplest level compatible with survival in a highly protective environment cannot seriously be called deleterious, unless one is prepared to praise mere superfluity as a virtue. Devolution in form should properly be seen, not as backsliding, but as a new and successful adaptation to a fresh ecological challenge. There is a spectrum of achievement even among parasites; and, though there may seem to us to be an undeniable element of progress in the evolutionary movement from amoeba to philosopher, would the amoeba, if consulted, agree? This is to put facetiously a profound difficulty which remains after more than a century to vex the theoretician. As one of the most thoughtful of these has put it:

> one hardly knows where to begin, in ranking such widely divergent organisms as butterflies and orchids. The perfection, even in the sense of engineering criteria, differs from one structure to another . . . An attempt to create a calculus or scale of perfection must be founded upon the assumption that perfection in one structure is equivalent to perfection in another. But this is absurd.[18]

Absurd, but difficult to resist, as Darwin himself found, despite having made much the same point about the perfections of cuttlefish and bees. Where he erred was not in failing to limit his use of such labels as 'perfect' and 'degenerate' to precise contexts, but in being insufficiently sensitive to the nuances of language and the auras of meaning which surround these terms. It is not that Darwin was unwilling to examine his assumptions about progress. He was well aware, as his letters if not the *Origin* tell us, that 'perfection' might mean one thing in a biological context and quite another in a social. Rather, he too ingenuously believed that people would let him speak out as a biologist only. But of course they did not. Wilson, Drummond and (later) Benjamin Kidd compounded his error out of a pressing need to find some moral significance in the devolved organism. For the point that Ghiselin makes does not derive at all from any fresh information available only to us. To understand it takes only a slightly adjusted angle of vision, a little less anthropocentric a stance. Why no popular Victorian interpreter of biology was willing to make that effort must be buried very deep in cultural history.

The retrogressionists, then, took human societies to be inherently and naturally unstable, requiring only the jolt of some disaster to make them collapse back into simpler autonomous units. Richard Jefferies presents such a retrogressed future in *After London* (1885), where a feudal system is shown as arising with internecine strife and slavery from the ruins of the devastated metropolis. Or, alternatively, the future society could be imagined as being in the iron grip of its institutions, hopelessly inflexible in a changed environment, dwindling down into senescence and final dissolution. Either way there was the suspicion that certain insensate forces in life have the future well programmed ahead of time; or at least that our destiny may be roughly predetermined. *Amor fati* at all times has a strong appeal to the literary imagination; nowhere more so than in the 'tale of the future' – the medium which carried most plausibly the Victorian devolutionary eschatology.

Degeneration and the Utopia

To arrive at a satisfactory estimate of the quality of the interaction between retrogressionist biology and speculative fiction, it is convenient first to establish some definitions and distinctions. The 'tale of the future' (I. F. Clarke's handy phrase) may best be classified as a sub-genre of Utopian fiction generally. Like other literary Utopias, the tale of the future offers us an imaginary world created for the purpose of exhibiting a writer's schemes for an improved society or else for satirising existing trends in society. But this definition is altogether too broad a one, although it does permit some useful generalisations. Clarke's demonstration that there was after 1872 an unprecedented flood of Utopian fiction, including futuristic Utopias, leads on to his important formulation that in any period the number of such fictions published is inversely proportional to the acceptance of the prevailing order. He notes further that 'the seventies mark the beginning of a crisis in the nation's way of life':[19] presumably he would have this judgement understood as a wide comment on the social and intellectual ferment of that decade and its immediate successors. Nevertheless, Clarke's reference to the Franco-Prussian War (1870–1) does tend to concentrate attention on the political and economic elements of the Utopia at the expense of its scientific and technological content. One needs to read only a few of the Utopias written between 1870 and 1900 to appreciate that the major recurring themes – female emancipation (pro and contra), religious toleration, state-manipulated breeding and child-rearing, a broad-based judicial system and 'rational' town-planning – are merely variants on their Greek and Roman prototypes. Here there is nothing really *new*.

Let us introduce some further distinctions.

We will first distinguish between 'geographical' and 'temporal' Utopias: between those imaginary societies detached from the locations of the writer in space, and those detached in time. We may further notice at once that both spatial and temporal Utopias may be written either in the dystopic mode (concerned, that is, with offering a critical or condemnatory reflection of the real-life order of things) or in the eutopic mode (the imaginary paradise of social and personal contentment). This gives a suitable two-dimensional scale which will accommodate most late-Victorian examples, and is adequate for the purposes of the present discussion. On such a scale, extreme examples might be: of the geographical eutopia, Ellis Davis's *Pyrna: A Commune* (1875); of the geographical dystopia, Butler's *Erewhon* (1871); of the temporal eutopia, *A Thousand Years Hence* (1882) by 'Nunsowe Green'; and of the temporal dystopia, Wells's *The Time Machine* (1895). Once this rather mechanical scale has been drawn up, there is one very clear development to be accounted for: the progressive displacement, as the century went on, of the Utopia in space by the Utopia in time. Thus Bulwer-Lytton's *The Coming Race* (1871) is set in gloomy underground caverns; *Erewhon* is not precisely located in space; the anonymous *Etymonia* (1875) is set on an island in the North Sea; *Pyrna*, somewhere under an Alpine glacier; *A Thousand Years Hence* is a series of vignettes reaching that far into the future; *A Crystal Age* (1887) is set at a very distant but vague period; Morris's *News from Nowhere* (1890) in the opening years of the twenty-first century; and *The Time Machine* in the exact year *AD* 802, 701. At the same time the ratio between eutopias and dystopias remained fairly constant, though it became progressively unbalanced in favour of the latter after 1900. If we again turn to Clarke for an explanation (this time to his exhaustive bibliography of temporal Utopias), he suggests that this sub-genre arose from the older tradition of placing fictional speculation in a remote but identifiable part of the globe.[20] By the early 1870s, therefore, Butler's use of the hinterland of New Zealand represents a geographical scraping of the barrel, and *The Coming Race* an intermediate phase in the progress of Utopian fiction. All the concerns of the futuristic Utopia are present in the latter fantasy. The Vril-ya are evolutionary supermen, destined to conquer the world as the title suggests; yet by siting their world in deep caverns Bulwer-Lytton does not have to cope with the artistically tricky leap into the future. But there is another way of looking at this; for, as F. L. Polak puts it, placing a Utopia spatially – for example, in the middle of the Indian Ocean (Etienne Cabet's *Voyage en Icarie*, 1840) – is to admit at least the possibility, in an age clearing up the odd unexplored corner, of accessibility and hence of discovery.[21] Psychologically, the shift into the future would appear to satisfy either the need to distance the fear of current trends or else a hopeless yearning for perfection which the present cannot satisfy. This sounds persuasive in that the indeterminacy of time as well as the conscious revocation of the past in *A Crystal Age* did indeed help Hudson to realise his obvious wish fulfilment. Yet when Edward Bellamy placed the Boston of the twenty-first century in painful juxtaposition with that of the nineteenth in his eutopian *Looking Backward* he did so in a deliberate attempt to make the future seem as accessible as possible and so worth striving for. Yet both these Utopias were published in the same year!

An alternative partial explanation is that the temporal Utopia, whether eutopic or dystopic, may readily have incorporated in it biological processes which can be shown as coming to fruition only in the long term. After all, it is one of the charms of the Utopian literary genre as a whole that it is able imaginatively to digest and synthesise a whole sweep of information. It is almost unnecessary to remark that post-Darwinian biology entered in some form into the fabric of practically every one of the temporal Utopias written after 1870. In fact, the specific tags of 'evolutionary' or 'physiological' have been attached to some Utopias of this class.[22] But one finds almost everywhere the dominant motif of biological change – change much more radical than the minor shifts in technology or in political institutions which bounded the imaginary horizons of the earlier Utopias. More in *Utopia* (1516) and Bacon in *The New Atlantis* (1626) introduced the benefits of selective breeding; but only of breeding directed towards the clearly limited end of improving the proportion of 'worthy' men already making up the ruling clique. To select artificially for a being who existed only as an ideal would have been viewed, to say the least, as impious.

Biological plasticity, infinite in extent and incessantly working, spewing out bizarre and unprecedented forms; change through the whole organic world, completely overthrowing the established order: this, for the Utopian, was the vital message of the new biology. Certainly the latter did not just suddenly intrude 'bloody images' which made it that much more difficult to write a convincing temporal eutopia – at least, not universally.[23] The paradisal futures of William Morris and of Edward Bellamy are quite innocent of any biologically inspired dark vision. Though it would hardly be possible to exaggerate the differences in orientation of *News from Nowhere* and *Looking Backward*, both of these eutopias share a comfortable attunement to man's new status. The citizens of London and Boston around the year 2000 seem as little troubled by their simian ancestry as by the prospect of undergoing physical and mental disruption, and perhaps decay, in the yet remoter future. Even a futurist like Edward Maitland (1824–97), who looked to human engineering to solve social problems, went out of his way to dissociate himself from any dystopian 'literature of despair',[24] even though he relied heavily in his following pages on a eugenic programme and on relocating nuclear families in state phalanxes.

It might be argued that if *News from Nowhere* is free from all qualms about biological degeneration it is equally free from the taint of evolutionism. Yet part of its sub-title is 'an epoch of rest' and that does closely define its mood, which is one of a yearning for uneventfulness. The events are locked into a tight frame of a succession of sunny and placid days. Morris's didacticism is essentially eliminative, in that his lesson is that men can be happy by doing without, and that they can gain in freedom by refusing to be hag-ridden by progress. This very concern with fixity and stability may be regarded, then, as a perverse response to the new Darwinian fluidity. It is as though certain sensibilities had frozen into one posture. The same response takes on another aspect in the nostalgic juvenility, the 'prostration before the child',[25] which is so dominant a note in all these late-Victorian fantasies. In *A Crystal Age* and *The Time Machine* the heroines Yoletta and Weena are both mature women but appear as sexless children; in the former Utopia, as we have seen, there is a more definite shuddering away from sexuality, with the hero, Smith, unmanned before the eunuchoid Crystallites. Similarly, in *The Coming Race*, where there is a more explicit debunking of feminist ideals, the hero's masculinity is defeated by the predatory Amazonian Vril-ya women. What we have here is the objectification of a wish for a society and a species insulated from all the Darwinian pressures and frozen for ever in one crystalline form. The dream of icy rigour is present with startling literality in Ellis Davis's forgotten *Pyrna: a Commune; or, Under the Ice* (1875), a society which the narrator stumbles across beneath the Rhône glacier. Most of the detail of this Utopia is pedestrian enough: the great city with its massive architecture, stocked with statuary; the flawless citizens of identical physical type in flowing robes; the twin museums of Science and Art; the abolition of all sexual distinction 'except that required by modesty'[26] – all these are the traditional props of the Utopia of this time. Davis allows no hint of ironic condemnation to creep into his fantasy of social perfection at the expense of all passion, vitality and spontaneity. Only in his long description of the mortuary habits of his people (their dead are lodged incorruptibly in personal icebergs), and in the fascinated repugnance his hero at one point expresses for his deep-frozen hosts, do we sense some ambivalence in his attitude to a race which defies the sternest injunctions of biological law:

> I began to feel a fear and a dread creeping over me. I was fatigued: I was in a strange, cold world, among beings with whom I had no sympathy, and who appeared scarcely mortal. Beside me was one of them; I was at his mercy. His cold, clear features looked colder and clearer in repose, and his glittering eyes lost none of their lustre. He was as white as a corpse. The idea and dread of death stole over me. I felt as if I were in the power of some malignant spirit who was chilling and freezing me into the grave with an icy breath and a glassy stare. (p. 62)

Those who, unlike Davis, chose to confront the new reality of perpetual change found themselves forced to wrestle at some level with several related questions. Are the consequences of evolution likely to be 'progressive' in the human sense, or even pro-human at all? Does Darwinism indefinitely extended, with its disquieting licence of slavery and cannibalism, supply an emotionally pleasing version of the future? If not, what more tolerable alternatives are available, and how can they be made persuasive? How these questions are handled in the temporal Utopia is largely a question of tone. Even *A Crystal Age* has odd ironic moments, but these are always subdued to Hudson's desire to create an Arcadian eutopia. Evolutionary biology is therefore not allowed to play any really nasty tricks on the Crystallites. But an equally committed artistry could produce very different results from much the same kind of material. H. G. Wells, backed by an elaborate formal training in biology and filled with concern about the effects on humanity of degenerative change, put into *The Time Machine* (1895) his version of *Homo inferior* amid an Arcadian landscape, the better to contemplate his deterioration; the artistic product which results is a satirical, even a tragic, evolutionary dystopia. Later on in the decade, this *Homo inferior* appears again and again in a variety of disguises before being decisively banished with Wells's advocacy of a wide-ranging programme of eugenics. In the rest of this chapter and the next we shall survey this development and try to estimate at each stage what was the imaginative effect of this biological solution to a biological problem.

The Biological Vision of H. G. Wells

Unlike the other authors discussed in these pages, H. G. Wells was a member of the first post-Darwinian generation. As he was not born until 1866, *The Origin of Species* was simply for Wells an immutable fact of existence and the evolutionary hypothesis just another unquestioned item in his intellectual baggage. But although he was detached in time from the momentous events which so troubled his parents' contemporaries, Wells did not reach maturity so late as to miss altogether the key personalities of the drama. Darwin himself died while Wells was still a wretched apprentice at the Southsea Drapery Emporium; but Darwin was an indistinct, barely human figure who for the public at large had long since dissolved into his reputation. But T. H. Huxley, infinitely the most romantic member of the Darwinian circle, lived to see Wells firmly established as a successful novelist; and, at the age of 59, was at the pinnacle of his fame when Wells came under his tuition at the Normal School of Science in the academic year 1884–5. He, the man who coined the word 'agnostic' as a label for himself; he who had harnessed together uncompromising free thought and moral rectitude so cleverly that even his mortal enemies could not prise them apart; he who had engaged in almost lengendary combat with Owen and Wilberforce before Wells was even born: Huxley, it almost goes without saying, was the only fitting deity for his young and often poverty-stricken students.

In their excellent biography *The Time Traveller: The Life of H. G. Wells* Norman and Jeanne Mackenzie have pointed to the close similarities of background and temperament of the great biologist and the future novelist and prophet – similarities which, they suggest, made Huxley even more the object of adulation for Wells. They point particularly to the ambivalent attitudes of master and pupil alike to the demands of pure biology as a full-time research occupation, and the allure for both of the role of sage social commentator.[27] The strong egoism of both men accorded badly with the modest reputation which might be earned in the relative anonymity of the laboratory. One wonders, however, just how far these similarities were apparent to the young Wells. He and his teacher achieved no personal relationship whatever, and Wells probably knew little of Huxley's earlier private life, comparable though it was in its first years with his own. It is perhaps more fruitful to see Wells's idolisation of Huxley as being the reverence due to a living and successful symbol of a professional scientist whose views were respected in the worlds of letters and politics because he *was* a scientist. That was a cultural novelty. And a scientific education generally, and a biological training in particular, had powerful emotional overtones for Wells. Such a training was a synonym, to his mind, for freedom – freedom not only intellectual but social; freedom from the sordid restraints of the genteel poverty of his youth and the middle-class morality that went with it. In these terms, Huxley may have appealed to Wells as the first model for his Samurai, the detached overlords of *A Modern Utopia* (1905) who practise plain living and high thinking as they exercise their benevolent dictatorship.

Despite the expansion of his mental horizons which was induced by the zealous study of zoology, Wells proved in the following year that he did not have sufficient capacity for sustained work in the physical sciences, and he eventually withdrew from his course. Perhaps it is not too fanciful to suggest that his permanent association of the life sciences with social reform was to some degree adventitious. It is somewhat misleading that, looking back on his *annus mirabilis* half a century later, Wells should have referred to his studies at this time quite uncompromisingly as organising a 'vision of life that was growing in my mind'.[28] For that carries the implication that biology supplied him with a dogma or self-contained structure of belief to substitute for the ingrained evangelicalism of his upbringing. In truth, Huxley's presentation of zoology must have had the effect of smashing certainty, not of creating it; of forcing Wells to recognise the multiplicity and uncertainty of the prospect opened up by the Darwinian vision. Geoffrey Wells, whom Wells supplied with much unpublished material for his study, gives us an admirable corrective which was almost certainly written closer to the events than the *Autobiography*'s account. Here Wells muses on his education in these rather different terms:

> I had man definitely placed in the great scheme of space and time. I knew him incurably for what he was, finite and not final, a being of compromises and adaptations. I had traced his lungs, for example, from a swimming bladder, step by step, with scalpel and probe, through a dozen types or more, I had seen the ancestral caecum shrink to that disease nest, the appendix of to-day, I had watched the gill slit patched slowly to the purposes of the ear and the reptile jaw suspension utilized to eke out the needs of a sense organ taken from its native and natural water.[29]

'I knew him incurably for what he was, finite and not final . . .' More than any other this sentence captures the essence of Wells's reading of man's fundamental condition in his scientific romances, and it surely is the essence, too, of that hard biological vision which Wells so thoroughly absorbed from Huxley's lectures, which were so luminously replete (if the published remnants are representative) with asides on matters far removed from the immediate point under demonstration.[30] In those lectures Wells found the germs of a whole string of articles and stories, and he especially took to heart Huxley's careful and much-elaborated disarticulation of progress from evolution.

At the same time Wells's passionate determination to make the evolutionary process yield its last dregs of human significance did not extend to speculation on the evolutionary mechanism itself. There he was quite unquestioning, perfectly content to follow his teacher's overt acceptance of classical Darwinism. It is true that Wells, unlike that other literary pure Darwinian Thomas Hardy, was later drawn to eugenics; but eugenics, at least on the Galton–Pearson model, was nothing more than applied Social Darwinism. In fact its main appeal was to those, like Wells, who sought an escape from the impasse of Darwinism but who never felt the pull of romantic Lamarckism or any other teleological brand of evolutionism. The reason for this highly unusual, though not unique, uncritical acceptance may be that theories in general bored Wells unless they could be given the human interest of an article or the plot and dialogue of a story. Or there may be another explanation. In 1892, when Wells had written nothing but a few columns for periodicals, he got together a *Text-Book of Biology*[31] while scratching a living teaching that subject for the Universal Tutorial College. Much of the *Text-Book* is a crammer pure and simple, though always lightened by Wells's inimitable vividness of exposition. It is noteworthy that the brief account of evolution which it contains is utterly Darwinian, failing even to mention any of the alternatives. In this respect the *Text-Book* is very different from the much more ambitious, encyclopaedic survey of biology which Wells wrote in collaboration (if that is the right word for a relationship where the collaborators were ruthlessly forced by Wells into the role of mere data-processors) with his son and Julian Huxley, *The Science of Life* (1929–30). Wells included there a most carefully and fairly written section, 'The essence of the controversies about evolution', which reviews not only neo-Darwinism as it then stood but also in turn neo-Lamarckism, orthogenesis and Bergson's *élan vitalisme*, as well as making some illuminating remarks on the types of personality among biologists to whom these various alternatives might appeal. Thus, neo-Darwinism is said to appeal to a cold, hard and clear type of mind; neo-Lamarckism to 'those combatant spirits who would figure man in a Promethean and finally hopeful conflict with the universe'. There is the particular recognition that 'several distinguished living biologists are neo-Lamarckians' and that 'various shades and blending of opinion are hotly debated'.[32] Bearing this in mind, the omission in the earlier *Text-Book* may be nothing more than a concession to the examination pressure on his students; but it is perhaps more likely that the bias is due to the rise, at the exact time when Wells was learning his biology, of the neo-Darwinian school to a dominating if temporary ascendancy. Certainly Wells was ready in 1896 to accept August Weismann's absolute rejection of use-inheritance, at least as a factor in human evolution since the Palaeolithic.[33] In *The Time Machine* published the year before, at any rate, Wells had no reason to introduce any evolutionary hypothesis other than Darwin's. By setting his story at a point in future history as remote from us as we are from the origin of our species, he gave natural selection all the elbow-room it needs to work the dramatic transformation Wells presents.

The Time Machine: Social or Biological Allegory?

It should be unnecessary to insist, after the close critical attentions paid to it by Bernard Bergonzi, Mark Hillegas and others, on the very substantial literary merits of *The Time Machine* (1895).[34] Far from objecting any longer to a certain 'creaking in the explanatory passages, a tentativeness, a lack of assurance',[35] recent criticism has instead pointed out the novel's impressive mastery of detail and the carefully constructed illusion of theoretical and technological convincingness which make all preceding tales of the future look the merest flim-flam by comparison. Others have gone yet farther, finding archetypal patterns in the novel's light–dark and other colour symbolisms; or even discovering mythic allegories of a cosmic Apocalypse. It is a simple but elusive novel, interpretable in many ways; but with respect to Wells's biological studies these two points are the ones which seem to be crucial. First, that *The Time Machine* is a key document in the *fin de siècle* history of scientific ideas. Not only does it condense and exploit many of the gloomy attitudes towards a degenerated future charted earlier; it also, when set alongside Wells's other writings of its decade, shows a way forward without violating its key premiss of Huxley's and Thomson's ultimate cosmic despair. Second, internal references prove *The Time Machine* to have been written with the playful yet quite deliberate aim of overarching and assimilating other fiction in the same genre, including *A Crystal Age*, by showing up the inadequacies of such temporal eutopias.

Wells's success in achieving these considerable goals rests at bottom on the realism of his art and its teasing association with mundane Victorian scientific prose – his self-acclaimed power 'to *domesticate* the impossible hypothesis'.[36] Though he shared the common lack of confidence about plunging fictionally into the future without a preliminary journey, he triumphantly overcomes all the awkwardness inherent in describing such travel. The absurdities of the minutely detailed dream or the long sleep in suspended animation are not for him. The clumsiness of the device precipitating Smith into the future in *A Crystal Age* and Hudson's barely concealed boredom with such strategies are alike quite alien to Wells's zestful art. Over and over, little hard nuggets of fact are scattered through pages of nearly preposterous scientific extrapolation. In *The War of the Worlds* (1898), an airy reference to 'that distinguished anatomist Professor Howes' – a real acquaintance of Wells's from his tutoring days – brings a close description of the physiology of the Martians as close as the next issue of *Nature*. Similarly in *The Time Machine* the anonymous Traveller serves up a bogus philosophy of time under cover of a calm citation of the work of the contemporary American mathematician Simon Newcomb. Next, the Traveller delivers a lecture, complete with working model, with all the aplomb and theatricality of Huxley before a Mechanics' Institute audience. We are seduced by the cool legerdemain of the laboratory demonstration.

Only then, having numbed the reader's critical faculties, does Wells introduce the time-travelling super-bicycle. Unlike the description of the theory behind it which concentrates on the broad effect but skips over the paradoxes time travel involves, the machine is real only in the magnificence of its hardware: 'parts were of nickel, parts of ivory, parts had certainly been filed or sawn out of rock crystal' (p. 15). But the effect works: when the Traveller says gradiloquently 'upon that machine . . . I intend to explore time. Is that plain?' (p. 15) we are no more ready to dissent than the members of the stolid, gaping circle around him.

Despite the felicity of this scene and many others, Wells was no congenital novelist. His own judgement on himself as an advanced journalist is, in the whole context of his career, substantially correct. Wells the romancer soon gave way to Wells the reformer of society – in his early days a less-than-adequate role because of his irresolute political sympathies. Nominally a socialist in the early 1890s (although his stormy interlude with the Fabians did not begin until 1902), Wells's romances are paradoxically riddled with what A. L. Morton with his Marxist perspective has called 'an active, if half-suppressed, fear and hatred which assumes curious forms'.[37] The suppression is indeed better than half-effective in the social comedies like *Kipps* (1905), but in *The Time Machine* and elsewhere the result is more than curious if we try to read these romances as mordant critiques of trends in late-Victorian unbridled capitalism; as being about an economic system in degeneration rather than a species. If we glance briefly at a short story Wells wrote in 1897, 'A Story of the Days to Come', we find that there Wells has attempted to fill in imaginatively one of the possible ways in which the transition from England of the nineteenth century to the world of AD 802,701 might come about. 'A Story' is set in a garishly mechanised London a few centuries hence, when an uninhibitedly *laissez-faire* economy is already producing symptoms of the Eloi–Morlocks syndrome. 'The new society', we are told, 'was divided into three main classes. At the summit slumbered the property owner, enormously rich by accident rather than design . . . Below was the enormous multitude of workers employed by the gigantic companies that monopolised control; and between these two the dwindling middle class.'[38] The aristocratic and proletarian languages are already beginning to diverge. (Dramatising this allows an unusual blending of Wells's powers as a comic novelist of manners and scientific romancer.) Soon their tongues will have become mutually incomprehensible. The action of the story describes Denton and Elizabeth, a well-born couple, toppling from the ladder of status into the toils of the Labour Company, a monolithic private enterprise which already has a full third of the population as its serfs and debtors. The couple are forced into the workers' army and set to labour in subterranean factories, tending giant machinery and other mindless tasks of mass production. The contempt and disgust felt by hero and heroine for their stunted and semi-moronic workmates are hardly less than the Traveller feels for the Morlocks in the earlier novel. Like him, they recoil with horror from the 'offensive inferior animals' (p. 773). Though they come at last to an uneasy truce with the proletarians, the story ends, significantly, with their miraculous return to the class of their birth, shaken by their descent into the underworld. We are reminded of that briefer scene in *The Time Machine* where the Traveller descends the Morlocks' ventilation shafts, with its active imagery of detestation, blood-lust, noisome darkness and carnivorousness. It is most tempting to read this, as Morton insists that it should be read, as an expression of Wells's own class hatred (he was, after all, born the son of a struggling petty bourgeois); of his own, and therefore his characters', fear of dropping through the bottom of their class or (worse) having to beat off a proletarian invasion. But this is not the pattern of *The Time Machine*, where we are not permitted to side with the Eloi, the effete aristocracy, in any way at all. For, if the Morlocks are the proletariat grown bestial, the Eloi are the *haute bourgeoisie* gone to seed. With its final revelation of the inverted master–slave relationship, the novel is an ironic and masterly treatment of a possible outcome of the class struggle; except that in Wells's version every class is vanquished by biology's blind workings, which have so totally endorsed the exigencies of economic structures that behaviour has set into an unbreakable mould. When compared to the power with which Wells realises *that* vision, the novel as a sociological allegory or an existential tract is as nothing.

The point of this digression is to insist that in his appreciation of degenerative change Wells has his imaginative focus set on a biological phenomenon which only at a subordinate level has cultural overtones. Perhaps there is something to be gained from reading *The Time Machine* as a product of the 'yellow nineties' but it is not, on inspection, very much. There is a world of difference between the fashionable lethargy of the literary decadence and Wells's essentially constructive pessimism. There is none of that sensuous depravity for its own sake which marked the union of the cruelty and beauty of nature in works by Pater, and was expressed in the previous decade (with the Darwinian analogies well to the fore) by several Iago-like figures in Tennyson's 'By an Evolutionist', 'Despair' and 'The Promise of May'. Even in its incidental trappings *The Time Machine* does not owe a great deal to the *Savoy* magazine. The muted sexuality and the hermaphroditic beauty of the Eloi are probably less the 'clear reflections of *fin de siècle* visual taste'[39] than the product of Wells's Darwinistic awareness – like, eight years earlier, Hudson's – that the reduction or elimination of selectionist pressure coupled with some deliberate breeding policy would be likely to lead to sexual homomorphism. Nor do such gloomy previews of the decline of the West as Max Nordau's *Degeneration* (published in English in 1895) stand behind the novel as much as Wells's own excursions into popularised biology like the very early article 'Zoological retrogression' of 1891. Capitalising on the still acute interest in this subject, Wells there uses, like Lankester and Andrew Wilson before him, the familiar examples of parasites and organisms which have 'retreated' into a less specialised condition of life to expose the inadequacies of what he calls 'Excelsior biology'. How can we, he asks with all the indignation of a proudly rational naturalist confronting the delusions of his literary brethren, spout such clichés as 'the march of Mind' when we look carefully into the *modus vivendi* of the ascidian or mudfish? All we can be sure of is the fact that modifications go on constantly – only our quite unjustified vanity lets us call that progress. 'The Coming Beast', Wells tells his readers, reminding them of Bulwer-Lytton's evolutionistic fantasy, 'must certainly be reckoned in any anticipatory calculations regarding the Coming Man.'[40] It is not the tone of the hedonic aesthete which dominates either here or in *The Time Machine*, nor that of the prophet of incipient cultural collapse, but rather the anxiety-making tone of the speculative biologist and moralist. Wells and Henry Drummond, miles apart in their creeds, nevertheless were exploiting the same public sensitivities. 'Zoological retrogression', like many of his pot-boiling articles and stories of the very early 1890s, displays Wells's scientific imagination working at low ebb and exercising itself solely in producing fear, awe, or thrilling amusement. In the short story 'The Empire of the Ants' of this same period, a community of intelligent ants have in some way, which Wells wisely leaves vague, evolved a culture aggressive to humans. Nothing of substance happens in this tale at all: it simply creates a vague atmosphere of menace to convey the message that natural history can indeed be very frightening.

The Time Machine also conveys fear and awe, but Wells's handling of the theme of retrogression there is stopped from being crudely hectoring by the presence of his spokesman, the Traveller. The longer the hero stays in what he assumes is a new Eden, the more the air of decay forces itself on him. His first impression of the Eloi themselves is of their rather unhealthy beauty, hectic like 'the more beautiful kind of consumptive' (p. 33). Not only is the species devolving; so is the human culture, that other channel of inheritance down the generations. The winged sphinx, symbol of arcane knowledge, is the first object the Traveller glimpses through the hail on first arriving: it looks as though it is afflicted with leprosy. A little later the smashed windows of the living-halls and the fractured corner of an opulent marble table catch his eye. Everything is in a condition of 'ruinous splendour' (p. 44). The Palace of Green Porcelain, repository of the genius of the ages, is ravaged and deserted. When we reach the penultimate chapter, 'The further vision', in which the even more distant future is telescoped into a series of snapshots millions of years apart, the continuous regression of life is the only uniting theme. It has been shown that Wells has here inverted the common line of evolutionary ascent with some exactitude. In this further vision the traditional branching tree with its great trunk of lowly organisms gradually rising into the ever more finely divided branches of complex forms (the orthodox seriation with which Wells must have been very familiar as diagrams in the textbooks at the Normal

School) is fully inverted. Relative complexity diminishes over time; the fine branches merge into the great boughs of the genera and thence into the undifferentiated trunk. The monkeyish Morlocks give place to the arthropods, to crustacea, and finally to some invertebrate animal akin to the medusa. Life is running downhill all the way into extinction. This 'folk biology', as Darko Suvin has termed it,[41] which simply describes the disintegration backwards and downwards into unicellular organisms, is neatly contrived; but it does not exhaust all that Wells has to say about degeneration, particularly if one takes into account the whole series of his scientific romances of the 1890s. We shall see then that these 'romances' are singularly ill-named. At their best they are dystopias, not melodramas, offering not mere sensationalism but constructive admonishment. The ideas they contain are remarkably similar to those of Wells's articles of the same time.

It may be demonstrated in fact that the Wells of this period had but *one* conception of the degenerative consequences of selection working uninterruptedly on our species over aeons of time: he expected it to result in the whole body becoming an organ as vestigial to the brain as the present-day appendix is to the intestine. Though his handling of this hypothesis is very thorough and deeply felt, Wells did not invent it. Rather, it was implicit in the Darwinian view of biological perfection. For the perfection of natural selection is not what had traditionally been understood to be the perfection of form and function: it is the *minimum* refinement compatible with survival. All the ills that flesh is heir to are not divine punishment inflicted on a fundamentally balanced frame, but the consequences of imperfect adaptation. There were two ways of escape from this uncomfortable recognition. For the optimistic Lamarckian, there was no real difficulty. Richard Jefferies, the naturalist and mystic, having pointed to the body's 'unsuspected flaws, handed down it may be for thousands of years', could still hope that physical immortality might one day be realised, by means of the human will itself repairing the defects of its cage. The physical excellence of the Spartans proved what might be done; and surely 'each individual labouring in his day through geological time in front must produce an effect'.[42] (We may notice in passing that Jefferies' argument is very muddled biology. The Spartans in developing their eugenic society certainly did not rely on 'soft' factors of inheritance – their methods, such as child exposure, were harshly Darwinian.) But those who denied the infinite power of the will to perfect the body internally had more of a problem than Jefferies. They had rather to fall back on the practically infinite possibilities of biological engineering. Without too closely specifying the means, Winwood Reade excitedly told his readers that 'these bodies which now we wear belong to the lower animals; our minds have already outgrown them; already we look upon them with contempt. A time will come when Science will transform them':[43] Reade, good Darwinian though he was, could not expect his readers to wait on the Darwinian machinery. Both Jefferies' and Reade's versions of this contempt for the body suggest the pass to which Paley's natural theology had been brought: no longer a sophisticated product of engineering design by a Divine Craftsman, the body had to be viewed instead as a ramshackle structure where make do and mend is the only guiding inspiration; a body tacked together, like Frankenstein's monster, from a variety of ill-fitting animal parts.

Wells, too, 'had watched the gill slit patched slowly to the purposes of the ear'. He shares the odd blend of repulsion and attraction for this process, and it informs his handling of the same theme, 'The man of the year million. A scientific forecast' (1893), where he attempts to chart out in detail Reade's vague prediction by pushing forward in strictly logical sequence from the plasticising influences even now at work on us. Like Reade, he insists that our animal inheritance is steadily being eroded. Our musculature is degenerating through lack of use. Our digestive system is failing to the point where it can no longer cope with raw meat and vegetables: soon its incompetence will include raw fruit. The primitive function of the mouth – a seizing and tearing organ – has already been usurped by cutlery: 'a mandible not of irreparable tissue, but of bone and steel'.[44] This trend, this annexing of organic tasks by artificial aids, must continue indefinitely, for that is the evolutionary logic. (It was certainly a logic of personal appeal: Wells's vision of the 'perfected' body, like those of Reade and Jefferies, is that of a sick man. Reade died in 1875; Jefferies in 1887; Wells in the early 1890s was several times critically ill with diabetes.) The labours of salivary glands and teeth will be arrogated by vats of digestive chemicals, and natural selection will speedily smooth away the redundant organs. What will be the end of this assisted evolution? It is hard to know whether Wells finds his conclusion tragic or comic as he proffers a large, smooth hairless globe supplied with a single pair of giant hands for manipulation and transport; a creature without intestines which will feed by bathing itself in a tub of nutritive fluid. Though Wells is determined to astonish and to evoke the cosmic shudder, he retains the measured periods that he had learnt from Huxley to be appropriate to biological discourse. Further, the Olympian stance is striking: the careful balancing of judgement and the refusal to see this outrageous transformation either as inevitable progress or as an obscene degeneration of humanity. Wells is capable of a relativism on this issue akin to that of Darwin himself, and quite foreign to the biological moralists of a decade earlier. His purpose is the final deflation of Paley's evidences for divine mechanism, and indeed the final abolition of the Divine Watchmaker himself. 'Not two guests are choked in a century,' was Paley's triumphant observation after a very close description of the supremely ingenious functioning of the human epiglottis; and here he is merely adding to the labours of William Derham who, in his *Physico-Theology* of 1713, had pointed as examples of conscious design to the muscles of the eyeball and the mathematical precision of the cells of the beehive, built with the minimum of wax to contain the maximum of honey. The late-Victorian imaginative interest in bionic men was a reaction to the destruction of this smooth-meshed harmony by physiologists like Helmholtz (who dealt abrasively with Paley's claim that the eye must have been designed for vision just as much as the telescope has been designed to aid it, by proving that even a shoddy telescope has less optical defects than the eye) and by Darwin himself (who showed, in the *Origin*, VII:154–229, that the accuracy of beehive cells is not especially remarkable).

Having established his image of the degenerated man, Wells returned to it over and over again. In *The War of the Worlds* (1898) the planet Mars suggested itself to him as an ancient and exhausted earth.[45] The distance of Mars from the sun; the far-advanced cooling of its core; its mostly evaporated oceans; its frigid climate even at the equator: these facts and suppositions supplied Wells with the opportunity of imagining a native race so toughened by the savage environment of a dying planet that 'the immediate pressure of necessity had brightened their intellects, enlarged their powers, and hardened their hearts'.[46] Yet the Martians, no matter how loathsome and alien their appearance, are presented quite explicitly as having already run the same course of evolution as that destined for man. Under the fierce pressure of selection the Martians have evolved (or should it be devolved? Once more Wells takes advantage of the prevailing uncertainty) into soft globes with simple tentacular appendages, everything else having been sloughed off in the adaptive battle. The Martians, with their 'large round bodies – or, rather, heads – about four feet in diameter' (p. 99), clearly stand at the end of the same line of development as the terrestrial man of the year million. So again does the Grand Lunar, ruler of the Selenites in *The First Men in the Moon* (1901), with his minute eyes and his shrivelled white body overhung by a featureless brain-case yards across. Here, as in *The War of the Worlds* and elsewhere, the tone is one of studied detachment. The Martians' natural food, for instance, is humanoid: dried remnants from the rocket capsules, the narrator tells us, have shown that they feed by injecting the blood of 'bipeds, with flinty silicious skeletons . . . and feeble musculature, standing about six feet high' (p. 100): the Martians, too, have divided into two races with the relationship that of Morlocks to Eloi. Inversely, in *The Time Machine* the Morlocks are not to be taken as just the pygmy remnants of the proletariat but the same model of evolved or devolved humanity some time before the end has been reached. They are even akin to the Martians with their 'filthily cold' touch (p. 79); their 'bleached look' (p. 74); their 'nauseatingly inhuman' look, with 'pale, chinless faces and great, lidless, pinkish-grey eyes' (p. 86), and so on. Just at the point where the Traveller crosses these creatures' long sloping line into extinction, the Morlocks are beginning to reassert themselves: cruelty has grown into a common passion, just as the Traveller suspects it may come to do before he voyages forth. But for the Morlocks it is too late. The ease of farming the Eloi, the warm climate and the extinction of all other predators will never serve to spark their degraded mentalities back into true intelligence. As the Traveller moves yet farther forward in time until the very sun is guttering out, the freezing thin air and darkening landscape again suggest another and inhospitable planet. All that is left of vertebrate evolution and the March of Mind is some nameless 'round thing, the size of a football perhaps, or, it may be, bigger, and tentacles trailed down from it; it seemed black against the weltering blood-red water, and it was hopping fitfully about' (p. 133). Such is the loathsome heir to all the ages: the man of 30 million AD.

The Time Machine and the Garden Metaphor

Darko Suvin, in the essay already mentioned, advances the following as the most general statement possible of how Wells uses his raw biological material in *The Time Machine*:

> A Wellsian structure is oriented toward cognitive horizons that it shares with any good handbook of sociology, biology, or philosophy of science. But the orientation is achieved in its own way, following the autonomy of a narrative, fictional aesthetic mode. ('*The Time Machine* versus *Utopia*', 341–2)

This is perhaps an unnecessarily broad formulation. The 'mutation of scientific into aesthetic cognition' to which Suvin refers immediately before the passage above is hardly one restricted to Wells's art: Hardy, Butler, Allen and other contemporary imaginative writers all offered their own mutations with greater or lesser success. Wells, it is true, did know more biology than all these, and he did allow it to penetrate his sensibility more in his earlier career; but he could not do anything qualitatively different with his knowledge than they: he could only literalise the abstract more persuasively, and expand metaphors into myths more energetically. What Suvin does not include, in his list of the 'cognitive horizons' to which Wells was oriented, is the ever-present one in *The Time Machine* of the preceding tradition of Utopian fiction. Constant indirect allusion is made to this. Though we are hindered somewhat here by having no exact documented evidence of Wells's reading in the 1890s in the genre which he came so thoroughly to dominate,[47] enough can be discovered to make evident the informed way in which Wells's version of a devolved future successfully assimilates and then transcends other Victorian temporal Utopias.

The Traveller, in finally arriving at a true conclusion about the relations between the Eloi and the Morlocks, takes up in turn no less than four partial explanations. Moreover, he does not reject the first three, but absorbs them each in turn into a more inclusive supposed explanation. At each stage we can see (even if Wells himself in every case did not) that a parallel may be drawn with earlier speculation. As he watches the Eloi about their childish business soon after arriving, there arises in the Traveller's mind 'with irresistible merriment' his fearful anticipations before setting out of 'a profoundly grave and intellectual posterity' (p. 37). The Traveller, and behind him Wells, is obviously making mock here of the cold and calculating Vril-ya of *The Coming Race*, or the cerebral citizens of any of a dozen other temporal and geographical eutopias offering up identical models of futurity. The Eloi, instead of being supermen whose intellects have expanded beyond the warm constraints of the emotions, expose the futility, in Wells's estimation, of imagining a society where biological competition has been eradicated or suppressed. All about, in the Thames valley of nearly a million years hence (ironically, the same location as the cheerful anarchic paradise of *News from Nowhere*), lies the rusting machinery and crumbled Cyclopean architecture as mute evidence that biology cannot be subverted or frozen still. Here *The Time Machine* deals harshly with the parallel dream of attaining equilibrium with the natural world by ecological engineering. Superficially the Eloi's world might be interchanged with that of Hudson's Crystallites. Huxley's 'weeded garden' has become a universal condition – 'The air was free from gnats, the earth from weeds or fungi; everywhere were fruits and sweet and delightful flowers; brilliant butterflies flew hither and thither' (p. 47) – but it is a sinister Elysium to the Traveller, who puts together this equation as his first hypothesis to explain what he sees: that monotony of environment plus a stabilised population equals cessation of competition, panmixia, and stagnation of body and mind. Apparently as well read in Drummond and Reade as in Bulwer-Lytton and Hudson, the Traveller draws the same conclusions as they: that 'we are kept keen on the grindstone of pain and necessity, and, it seemed to me that here was that hateful grindstone broken at last!' (p. 49). Above ground the Darwinian law has ceased to run, and all the steam has gone out of the race.

But, by introducing the troglodytic Morlocks, Wells pushes the dialectic of the biological Utopia forward to a second stage: that humanity has become separated, not merely into the Two Nations of the rich and the poor, but into two distinct species. Now he takes the common fear of the time, given forcible expression in the writings of W. R. Greg and others, that human institutions may in certain phases of their development mirror morally objectionable relationships in nature – parasitism, slavery, cannibalism – and further proposes that, if such a relationship is not to be found now, then human greed and laziness may well permit it to emerge in the future. In this second phase of its argument, *The Time Machine* rests on the analogy of two species in symbiosis, in the same way that Hudson makes *A Crystal Age* rest on the analogy of the apiarian society. But a continued and bitter acquaintance with the stupidity of the Eloi leads on to a third, modified hypothesis: that, while the Eloi have subsided into a beautiful futility, the roles have become so instinctive that the Morlocks continue to serve them out of irresistible habit, as 'a standing horse paws with his foot' (p. 89). The biological model changes from the mutually beneficial association of symbiosis to that of frank parasitism – a common one in other Utopias.[48] In time that model fails, too, and the Traveller comes to his final monstrous conclusion: that it is the Morlocks who are the real masters, and their way of living a selective cropping of their herds of Eloi. Here that ubiquitous image of the cultivated garden, the product of a cautious and enlightened husbandry, is given a conclusively ironical twist. Inside the new Eden survives one degraded fragment of humanity preying mindlessly off another. All of these early, highly contrived parables force home to us just how strong Wells's rejection of evolutionism was. He is, in this phase, as pessimistic as Hardy. Unlike Hardy, however, he found too much strain in his vision long to tolerate it; and it is at exactly this point that he parted company with his mentor. If he went back to Huxley, searching for a more apposite image than the weeded garden, he must have been disappointed. About the time that his student was working on the earliest draft of the novel which effectively writes finis to a fifty-year sequence of evolutionary Utopias, Huxley was comparing human evolutionary 'progression' to the trajectory of a mortar shell which must turn eventually downwards; and was driving home this analogy by telling his contemporaries that 'the sinking half of that course is as much a part of the general process of evolution as the rising'.[49] Against this, even the pessimism of *The Time Machine* sounds positively constructive. It is at least the pessimism of someone who has soberly weighed the odds against long-term survival and found them great, but who nevertheless sees a slim but real chance. That chance is for man to seize conscious and far-sighted control over his own evolutionary potential. The lesson of *The Time Machine* is, in the context of Wells's later career, a demonstration of what will inevitably be the consequence of a failure of nerve in this high task.

Wells's Controlled 'Inductive Future'

In artistic terms it may well be that Wells travelled farther and farther away from that wellspring of his inspiration, his biological training, as he became more interested in radical social reforms. Given his ebullient personality, it was inevitable that he could not be weighed down for long with the ballast of his Huxleyan 'cosmic pessimism'.[50] It was equally inevitable that his optimistic solution for preventing the world of *The Time Machine* ever being realised should be based on applied biology, just as his gloom was based on 'natural' biology unimpeded by human controls. By the end of 1900, Wells had published both *The First Men in the Moon*, the last of his elaborate scientific romances, and *Anticipations*, which marks his entry into his long career as a polemic journalist. The first of these works hints at a solution to the devolutionary nightmare which is made explicit in the second: both are briefly taken up in the following chapter.

The new trend in Wells's thought which set in right at the end of our period is best considered from the vantage-point of an important lecture which he delivered before the Royal Institution in February 1902, 'The discovery of the future'. By the time of this lecture, in which Wells reflects upon some of the techniques available to the would-be student of the future in the last decade of the century, his reputation as a prophet was already secure. He takes, therefore, this highly public occasion – it marked Wells's entrance into the Establishment[51] – to review the attitudes behind all his creative work in the 1890s. He reflects on the two possible ways of visualising the future. One way is to see it as a kind of 'black non-existence' on which the present writes as it advances. Or – and this is of course the tenor of his own writing – one can see that 'the present is no more than material for the future, for the thing that is yet destined to be' (p. 326). Is it possible, he goes on to inquire, to induce the future from the present, just as we infer the past from geology? He is in no doubt that it is possible: and one of his inductions, as we might guess, is that the future cannot possibly come into one condition and rest there. He now explicitly mocks the stable Comtian state and the static eutopias of his youth, plagued by Darwinian fears. We must abandon, he says, 'any thought of finality, any millennial settlement of cultured persons' (p. 330) – and therefore, by implication, a whole Utopian tradition. By this time even the cosmic heat-death itself, which he says 'of all such nightmares is the most insistently convincing', has no paralysing effect on Wells's imagination: he suddenly reverses with a brisk 'and yet one doesn't believe it' (p. 331). He refuses to speculate beyond the human future; he will work only for the birth of those beings who 'shall laugh and reach out their hands amidst the stars' (p. 331).

These proposals – Wells's whole stance, even – are revolutionary in the context of the preceding discussion. It is as though Wells had turned again to his mentor for encouragement and taken to heart the words from Tennyson's 'Ulysses' which close with some dignity the sad arguments of 'Evolution and ethics':

> Death closes all: but something ere the end,
> Some work of noble note, may yet be done,
> Not unbecoming men that strove with Gods.[52]

They suggest that we can rescue ourselves from the earlier need to rationalise away the disagreeableness of the biological future if we become aware that the image we form of that future creates the present. None of the earlier speculative writers living that much closer to the origin of their difficulties could release their future from the shackles of the evolutionary past. It is no wonder, then, that they tended to be deterministic even in their optimism; and no wonder that the best they could offer was different modes of accepting an unpalatable truth, ranging from the stoical to the defiant.

But in Wells's talk of an 'inductive future' it becomes evident that to think in this new way is itself evolutionary. It expresses a determination to confront a self-chosen future rather than the one nature will otherwise foist on us. In concrete terms (though this does not emerge in the lecture) Wells had eugenics in mind as offering the best chance of enlightened control. It was only to certain jaundiced eyes much later that his remedy for degeneration started to look just like Control.

Notes: Chapter 4

1 John Ruskin, *The Storm Cloud of the Nineteenth Century: Two Lectures Delivered at the London Institution . . .*, pp. v, 43. But notice that Ruskin is careful to distinguish between his plague-wind and a common smog where 'the air itself is pure, though you choose to mix dirt with it, and choke yourself with your own nastiness' (p. 60). It is just possible that the stupendous volcanic discharge of Krakatoa in August 1883 may have precipitated the writing of these lectures. The explosions hurled five cubic miles of dust into the atmosphere and intercepted some sunlight worldwide for several years. But, as Ruskin's letters of the 1870s show, this certainly does not preclude a psychogenic explanation.
2 Letter to Charles Eliot Norton, 3 April 1871: *The Works of John Ruskin*, ed. E. T. Cook and Alexander Wedderburn, Vol. 37, p. 30. Hereafter cited as *Works*.
3 Letter to Charles Eliot Norton, 15 January 1873: *Works*, Vol. 37, p. 57.
4 J. D. Rosenberg, *The Darkening Glass: A Portrait of Ruskin's Genius*, p. 216.
5 On the use of the garden metaphor by Mill and Hardy, see the chapter 'Hardy's evolutionary meliorism: nature, the garden, and "God's gardener" ' in Roy Morrell's *Thomas Hardy: The Will and the Way*. Morrell analyses there the scientific accretions which formed around the basic image from *Hamlet* and *Richard II*.
6 Oma Stanley, 'T. H. Huxley's treatment of "Nature" ', *Journal of the History of Ideas*, vol. 18 (January 1957), pp. 120–7.
7 T. H. Huxley, 'Evolution and ethics: prolegomena', in his *Collected Essays*, Vol. 9, p. 10.
8 Edward Dowden, *Studies in Literature, 1789–1877*, p. 104.
9 The lower figure is from his 'On the age of the sun's heat', *Macmillan's Magazine*, vol. 5 (March 1862), p. 390; the higher from 'On the sun's heat' (dated 1887), in his *Popular Lectures and Addresses*, 2nd edn, Vol. 1, p. 397.
10 Quotation and paraphrase from a letter to J. D. Hooker, 9 February: *More Letters of Charles Darwin*, ed. Francis Darwin and A. C. Seward, Vol. 1, pp. 260–1. The editors date the letter 1865.
11 F. W. H. Myers, 'Tennyson as prophet', in his *Science and a Future Life; with Other Essays*, p. 153.
12 Henry Adams, told of the stable *Terebratula* (lampshells) by Lyell in 1867–8, suspected that these proved Darwinian evolution had never occurred at all – but he was an exception: *The Education of Henry Adams*, pp. 226, 228. The best known of the living fossils, *Coelacanth*, was not discovered until 1938.
13 Herbert Spencer, *The Principles of Biology*, revised and enlarged edn, Vol. 2, p. 506. This passage is unaltered in the first edition of 1864–7.
14 Taken from letters to J. D. Hooker, the first undated, the second of 30 December: Darwin, *More Letters*, Vol. 1, pp. 76, 114. The editors date these letters to, respectively, 1854 and 1858.
15 H. G. Wells, 'The extinction of man: some speculative suggestions', *Pall Mall Gazette*, 25 September 1894, p. 3.
16 Andrew Wilson, 'Degeneration', *Gentleman's Magazine*, vol. 250 (April 1881), p. 491.
17 Henry Drummond, *Natural Law in the Spiritual World* (1883), though now forgotten, was

in its day a bestseller. It sold at least 70,000 copies within five years of publication, and went through thirty editions up to 1890. By the year of Drummond's death (1897) it had sold 120,000 copies and had been translated into four languages. It earned more than twenty rebuttals by scientists and theologians alike, several being book-length treatments: none affected its popularity. Herbert Spencer later incited Eliza Lynn Linton to write a ferocious review of another Drummond book for the *Fortnightly Review*. But even she was finally unable to explain away Drummond. 'That he has succeeded in his aim is proved by the enormous success of his book – mere hash of other men's labour as it is – a plagiarism from first to last. It is a thing of this kind which makes one despair of one's generation. The greedy haste to swallow any form of quackery while neglecting truer and better-grounded expositions – the seeming inability to grasp first principles or to go to the fountain-head for knowledge . . . all point to a state of mental confusion': 'Professor Henry Drummond's discovery', *Fortnightly Review*, n.s., vol. 56 (September 1894), p. 457. In reality, of course, the flow from all the fountains was polluted to some lesser degree. If Drummond's way of arguing had really been *sui generis*, it would be of no interest today. Being highly symptomatic, however, of a very widespread application of scientific data, it is possible to see that most of the contemporary criticism of Drummond was beside the point. In *Pseudo-Philosophy at the End of the Nineteenth Century* . . . by 'Hugh Mortimer Cecil' (i.e. Ernest Newman) the author's taunt was that the jargon is the jargon of pseudo-science, but the ideas are the ideas of the curate at the YMCA (p. 119). Undeniably true; but Linton's and Newman's energies would have been better expended questioning the social function of this type of reasoning and why even the best minds succumbed to it. If biological discourse had atrophied to the point where the ideals of the YMCA were its highest level of aspiration, Drummond could hardly be blamed for capitalising on the fact. Quotations in the text are from the 16th edn, 1885.

18 Michael T. Ghiselin, *The Triumph of the Darwinian Method*, p. 71.

19 I. F. Clarke, 'The nineteenth-century Utopia', *Quarterly Review*, vol. 296 (1958), p. 82.

20 I. F. Clarke, *The Tale of the Future: From the Beginning to the Present Day. A Checklist . . . between 1644 and 1960*, p. 12.

21 Frederick L. Polak, *The Image of the Future: Enlightening the Past, Orientating the Present, Forecasting the Future*, trans. Elise Boulding, Vol. 1, p. 302.

22 These terms are used, respectively, by Richard Gerber, *Utopian Fantasy: A Study of English Utopian Fiction since the End of the Nineteenth Century*, and Robert DeMaria, 'From Bulwer-Lytton to George Orwell: the Utopian novel in England, 1870–1950', unpublished PhD thesis, Columbia University. DeMaria further distinguishes between physiological and orthodox eutopias, and other dystopias which he lumps together in one class.

23 Frank E. Manuel oversimplifies the biological influence in this fashion in his essay 'Toward a psychological history of Utopias', in his *Utopias and Utopian Thought*, p. 86.

24 In the preface to *By and By: An Historical Romance of the Future*, new edn (London, 1875), p. vi, Edward Maitland ventures no examples of such literature. Possibly, but hardly probably, he had in mind Edward Bulwer-Lytton's *The Coming Race* (1871), but the tone there is hardly one of despair. More likely he is referring to such gloomy forecasts as W. R. Greg's speculative *Enigmas of Life* of 1872. The same comment appears in the first edition (1873) of *By and By*.

25 This phrase is P. N. Furbank's, used in a discussion of *Erewhon* in his *Samuel Butler*, p. 90.

26 [Ellis James Davis], *Pyrna: a Commune; or, Under the Ice*, p. 58.

27 Norman and Jeanne Mackenzie, *The Time Traveller: The Life of H. G. Wells*, pp. 57–9.

28 H. G. Wells, *Experiment in Autobiography: Discoveries and Conclusions of a Very Ordinary Brain – since 1866*, Vol. 1, p. 204.

29 Quoted by 'Geoffrey West' (Geoffrey Harry Wells; no relation), *H. G. Wells: A Sketch for a Portrait*, p. 62. The odd punctuation is as given by G. H. Wells.

30 Or, if not from Huxley's lectures, then definitely from his published essays, which Wells read assiduously. 'We read his speeches, we borrowed the books he wrote, we clubbed . . . to buy the *Nineteenth Century*' (quoted in ibid., p. 49).

31 H. G. Wells, *Text-Book of Biology*, with an introduction by G. B. Howes, 2 vols (London,

1893). There was a revised edition of part I in 1894. This work, despite being Wells's first book, is very little known. It is not mentioned, for instance, in Ingvald Raknem's meticulous bibliography *H. G. Wells and His Critics*. Indeed, the only published reference to it by Wells himself seems to have been in a chatty letter to Grant Richards of 6 November 1895, where he calls it 'a cram book – and pure hackwork . . . facts imagined' (from the unpublished letter as quoted by Bernard Bergonzi, *The Early H. G. Wells: A Study of the Scientific Romances*, p. 24). These contemptuous remarks – which are untrue – may have been occasioned by Wells not, in his own opinion, having been given sufficient credit for some revision undertaken by himself and a friend. There are some relevant letters in the Wells Archive, University of Illinois. The *Text-Book* was reissued in 1898, fully revised by A. M. Davies.

32 H. G. Wells, Julian Huxley and G. P. Wells, *The Science of Life* (London, 1931), pp. 262, 265.

33 H. G. Wells, 'Human evolution, an artificial process', *Fortnightly Review*, n.s., vol. 60 (October 1896), p. 594.

34 The six surviving much-revised drafts of *The Time Machine* were written between 1887 and 1895. The two serialised versions, in the *National Observer* (1894) and the *New Review* (1895), differ – in the first case radically, and in the second slightly – from the Heinemann first edition of May 1895. West, *H. G. Wells*, pp. 289–95, supplies the basic facts of its composition, and Bernard Bergonzi in *Review of English Studies*, vol. 11 (1960), pp. 42–51, a detailed discussion of the various versions. We are only concerned here with the final form. Quotations in the text are from *The Time Machine: An Invention* (London, 1969) throughout.

35 Norman Nicholson, *H. G. Wells*, p. 26.

36 Quoted by Vincent Brome, *H. G. Wells: A Biography*, p. 64.

37 A. L. Morton, *The English Utopia*, p. 187.

38 *The Short Stories of H. G. Wells*, p. 753. 'A Story' first appeared in the *Pall Mall Gazette* in 1897.

39 Bergonzi, *Early H. G. Wells*, p. 49; though earlier Bergonzi defines the mythic qualities of the novel as operating in such socio-literary terms on one level only. He concedes that 'its further significance is biological and even cosmological' (p. 42). The same limitation may be found in William Bellamy's attempt, in *The Novels of Wells, Bennett and Galsworthy: 1890–1910*, to find *The Time Machine*'s theme comparable with other 1890s novels: 'the indirect transcription of "abnormal" mental phenomena into literary terms' (p. 70).

40 H. G. Wells, 'Zoological retrogression', *Gentleman's Magazine*, vol. 271 (September 1891), p. 253. Robert M. Philmus, '*The Time Machine*; or, the fourth dimension as prophecy', *PMLA*, vol. 84 (May 1969), refers to this knowledge of Wells's of degeneration 'at this early date' as certainly eliminating the possibility of the influence of Max Nordau, which 'Bergonzi adduces as a source for the vision of the future' (p. 530). But Bergonzi does no such thing. He carefully notes that '*The Time Machine* was already completed' when *Degeneration* appeared in English in March 1895: see his '*The Time Machine*: an ironic myth', *Critical Quarterly*, vol. 2 (Winter 1960), p. 297. Further, Wells's 'early' knowledge of degeneration dates to a good deal earlier than 1891 – in fact to 1884–5, if Wells's own testimony is to be trusted (see n. 29, above). As shown earlier in this chapter, degeneration was in the air among biologists and other interested parties from 1880 or earlier: long before any of the literary manifestations.

41 Darko Suvin, '*The Time Machine* versus *Utopia* as a structural model for science fiction', *Comparative Literature Studies*, vol. 10 (December 1973), p. 338.

42 Richard Jefferies, *The Story of My Heart: My Autobiography*, pp. 152, 158. Jefferies's few other recorded comments on biological theory show him to have been, like most field naturalists of the period, a rather vague Lamarckian. His diaries display an acute lack of sympathy with Darwinism: see, for instance, the entries for February and March 1887 of *The Nature Diaries and Note-Books* ed. Samuel J. Looker, pp. 249–50.

43 William Winwood Reade, *The Martyrdom of Man* (London, 1925), p. 423. First published in 1872.

44 H. G. Wells, 'The Man of the year million: a scientific forecast', *Pall Mall Budget*, 16 November 1893, p. 1796. The anonymous parody of this article, '1,000,000 AD' in *Punch*, vol. 105, 25 November 1893, p. 250, consisting of verses and an illustration, testifies to the vividness of Wells's imagination. The poem follows the argument closely, and the drawing shows ovoid heads with attached hands, with one bathing in a crystalline bath of pepsin. The reference to Wells's article is incorrect.

45 Wells's bold assumption that evolution on Mars may have paralleled that on earth may be thought to have been the wildest fantasy considering the radically different and hostile environment on that planet. Right through our period, however, Victorian astronomers with their relatively crude instruments were of the opinion, in the words of the astronomer-journalist Richard Proctor (1837–88), that 'in all essential habitudes the planet Mars resembles our own earth'. See his 'Lands and seas of another World', *Fraser's Magazine*, vol. 78 (August 1868), p. 256. Imaginary 'seas' had been detected and named; the atmosphere was believed to have similar constituents; the temperature range to be comparable, and so on. These misconceptions lasted for an extraordinarily long time. Wells's own two columns on the same topic twenty-eight years later echo Proctor: 'there is no doubt that Mars is very like the earth . . . It has lands and oceans, continents and islands, mountain ranges and inland seas.' See his 'Intelligence on Mars', *Saturday Review*, vol. 81 (4 April 1896), p. 345. It was not until 1907 that a race of canal-building Martians, as proposed by Percival Lowell, at last became too much for the biologists, as represented by Alfred Russel Wallace, to swallow. He answered his own question *Is Mars Inhabitable?* in the negative. So there was nothing intrinsically absurd in Wells offering his readers Martians as *Homo sapiens* at the far end of evolution.

46 H. G. Wells, *The War of the Worlds*, p. 4.

47 However, we know that Wells was very assiduous in his preparatory reading: before working on the later *A Modern Utopia* (1905) he steeped himself in Utopian theories as diverse as More, Campanella, Cabet, Howells, Bellamy, Morris and Hertzka (MacKenzie, (*Time Traveller*, p. 189).

48 To give just one example, the parasitical model also forms the basis, in Davis' *Pyrna*, of the relations between the people of the Mune and a subservient race dwellling below them whom they regard 'with the sort of feeling with which we regard animals – as an inferior creation, made to be of service' (p. 67). This race, or species, of domestics seems to exist primarily to deal with the communards' sewage.

49 T. H. Huxley, 'The struggle for existence in human society', in his *Collected Essays*, Vol. 9, p. 199.

50 The phrase is Mark Hillegas' in *The Future as Nightmare: H. G. Wells and the Anti-Utopians*, p. 18.

51 'The discovery of the future: a discourse delivered at the Royal Institution on Friday, January 24 by Mr H. G. Wells', *Nature*, vol. 65 (6 February 1902), pp. 326–31. Thirty-two years later, Wells spoke of this lecture as marking his 'steady invasion of the world of influential and authorititive people'. His own evaluation of its contents was: 'vague, inexact and rhetorical, but that is the measure of the progress in definition that has been going on in the intervening third of a century. When it was read, that lecture was well abreast of its time.' See Wells, *Experiment in Autobiography*, Vol. 2, pp. 636, 649. Certainly it drew out in detail certain threads of thought common to all the romances.

52 T. H. Huxley, 'Evolution and ethics', *Collected Essays*, Vol. 9, p. 86.

5

Remember, Beethoven's Father Was a Drunkard:

The Dubious Appeal of Eugenics

> If some fancier with the catholicity of Shakespeare would take us in hand, well and good; but I would not trust even Shakespeares, meeting as a committee. Let us remember that Beethoven's father was an habitual drunkard and that his mother died of consumption.
>
> WILLIAM BATESON (1914)

It is a recognition almost as old as civilisation itself that, given the determination, a free people can selectively control their own breeding to make the succeeding generations approach more and more closely to some ideal of physical, mental and moral excellence. In seventh-century Lacedaemon the statesman-hero Lycurgus (if he was in fact a real person) began the deliberate and highly successful conversion of a city-state into an army camp, and the population into professional soldiers. In Sparta private family life was dissolved. Advantageous matings were encouraged and sponsored by the state, and infanticide practised regularly as part of a ritual which saw to it that only the physically hardiest babies could survive. Three centuries later in *The Republic*, which is Sparta elaborated, formalised and given a strong class-consciousness, Plato can be found advocating a similar scheme for the propagation of the Guardians, his ruling class. In Plato's Utopia there are legal, in fact compulsory, infanticide and abortion; there are marriage festivals with unlimited opportunities for copulation for those men who have distinguished themselves in war; and there are faked lotteries to quieten the resentment of the majority not selected for breeding. (Plato's exact words justifying this fraud are: 'we shall have to arrange an ingenious system of drawing lots, so that our inferior Guardians can. . . blame the lot and not the Rulers'[1] – eugenics has continued to bring out the worst in thinkers for two and a half millennia.) Plato, like all eugenists from his day to ours, faced the perpetual question: What qualities shall we breed for? The ancient polities answered it by taking their cue direct from nature. In nature the strongest, the fittest, the most çunning survive. Virtues good enough for nature were deemed good enough for Sparta and Plato's idealised Athens. But for the post-Darwinian eugenists the virtues of the most prolific survivors were deemed not good enough at all.

For the Greeks' eugenic schemes to succeed, no more understanding of the machinery of heredity and evolution was demanded than the rule-of-thumb expertise of the stock-breeder. Plato introduces his prescripts for the conception and rearing of children with allusions to the most advantageous ways of breeding hunting-dogs, gamebirds and racehorses, plainly regarding the problem and the technique as being much the same in both cases. We notice, too, that the Greek ideal was to produce more of the same, or a constant supply of an already known and tested item. The Spartans selected for military prowess on the pattern of the heroes of the past, and Plato's imaginary Athenians for more wise and Socratic Guardians. There was little or no awareness among the Greeks that improvement, or at least change, might be continuous and the base stock moulded beyond recognition. With very few exceptions, this notion of human eugenics being no more than a complicated form of animal husbandry (complicated not biologically but through the obstinacy of the subjects) was the one that persisted right up to the 1860s. It is true that by the later eighteenth century a few rationalists like Condorcet and Godwin were already giving the doctrine of perfectibility a biological dimension; and at least two of the pre-Darwinian evolutionists mentioned earlier, William Lawrence and William Wells, had noticed the great power of artificial selection. They perceived that new varieties of fruits and vegetables, obtained by nothing more complicated (in principle) than the continued selection of the desired quality, seemed to move away without obvious bounds from the character of the parent stock. They, and certain other perfectionists of the Enlightenment, suspected that such a process of variation and selection under pressure might proceed endlessly. It was left to Malthus, in the course of his *Essay on Population* at the end of the century (1798), to do his best to scotch such metaphysical and certainly doctrinally dangerous speculation. He used a very homely example in the course of a short digression from the main argument of the *Essay*. Suppose we compare in imagination a brilliant garden flower with the drab wild ancestor from which it has been bred. As we do so we may well be impressed by the apparent degree of malleability that exists in natural forms. Can any limit be set on this malleability? In Malthus' opinion, there can. He appeals to good sense. Whatever breeding techniques are applied, Malthus is clear that 'no carnation or anemone could ever by cultivation be increased to the size of a large cabbage . . . a careful distinction should be made, between an unlimited progress, and a progress where the limit is merely undefined'.[2] Therefore, Malthus concludes, every species has a built-in resilience. Forced selection may for some generations pull a species out of the shape which the Creator has ordained that it should have. But when left alone such deviant forms will spring back: the phenomena of reversion prove as much. If we assume that what is true of carnations is true of men, then this line of argument has an important consequence. There is no advantage in condemning the 'unfit' to celibacy. Men also bring forth after their own kind. Reversion to type – the type of Adam – will quickly obliterate any temporarily damaging influence of hereditary disease or asocial behaviour. Since for Malthus humanity had not evolved, he could see no dangers in devolution and therefore no need to combat it with eugenic measures.

Malthus' sober argument recommends itself to biblical scholar and grower of prizewinning marrows alike. It sounds right and it appeals to one's feeling for natural harmony and regulation; but it is, of course, wrong. The evolutionist reading the *Essay on Population* at any date after 1860 could at once see the force of taking Malthus' 'careful distinction', excising the word 'progress' and substituting 'change' of a Darwinian kind. If change is endless, if there are no limits whether defined or not to the plasticity of the stock, then the distinction Malthus insisted upon fades away. The Darwinian evolutionist was required to swallow the transformation of a bear-like mammal, swimming with its mouth open to gather food, into a whale.[3] He was not likely to strain at the improbability of Malthus' anemone attaining the dimensions of a cabbage. Darwin himself, while absorbing from Malthus the vital concept of a population pressing always against the barrier of limited resources, also saw in the nearly magical improvements wrought by professional breeders like Robert Bakewell (1725–95) something more than an analogy between selections artificial and natural. Rather, it was *because* 'we know what artificial selection can do'[4] that Darwin was led to place the significance he did in natural selection's power to shift successive generations so far away from a root stock that they might eventually cease to be identifiable with it. These two concepts go hand in hand so fittingly that it required only one further argumentative step by those Darwinians who were interested in eugenics. If natural selection was the true cause of evolution, then it must be possible for it to induce limitless, consistent and persistent change. But what natural selection does blindly and slowly, man can do consciously and very rapidly. He can initiate limitless, consistent and persistent *improvements*. Post-Darwinian (or, as we should more correctly call it, Galtonian[5]) eugenics therefore is fundamentally different from the classical version because it rests on a new appreciation of nature's flexibility. With a few changes of emphasis it lasted from the mid-1860s to just beyond the end of our period. As late as 1909 it was still just within possibility that Darwinism might yet form the sub-structure to an applied science of human breeding. But even before the First World War a reaction was setting in, and G. K. Chesterton's castigation of all advocacy of eugenics as a 'crown of crime and folly'[6] was certainly the most appropriate comment on brutal sterilisation programmes in America before, and in Europe during, the Second World War.

Late-Victorian Eugenics: An Overview

In his careful history of the involvement of eugenics in Utopian speculation, J. M. Smith distinguishes three phases of complexity in such proposals, which he calls 'selectionist eugenics', 'transformationist eugenics' and 'biological engineering'. In the first category he places such measures as the destruction or segregation of undesired variations from the base stock, the storage and later use of reproductive materials, and the imposition of various social and legal controls over mating; in the second, the direct manipulation of genetic material by tampering with the inheritance code or with the embryo; and, in the third, surgical interference with individuals, fitting artificial organs, transplantation and the like.[7] Smith's phases correspond roughly to the skills of medical technology available or imaginable at a particular time. It is true that some Victorian imaginations reached even as far as biological engineering. Samuel Butler made a leap in this direction in his essay 'The mechanical creation' of 1865, where he suggests quite casually and humorously that different varieties of men will develop (more, be developed deliberately) to cope with different kinds of machines. But the main concern was with theoretical selectionist eugenics only. 'Theoretical' because in England, unlike America, eugenics remained a matter for discussion only. There was, as far as is known, nothing to correspond to J. H. Noyes's voluntary programme within the Oneida Community of 'stirpiculture' which resulted in the birth of fifty-eight children from carefully selected matings – the most determined and best-planned attempt in a free group in modern times.[8]

In England, then, discussion ranged very widely but inconclusively over the virtues and limitations of, at one extreme, the direct and allegedly positive procedures at Oneida and, at the other, the negative ones of sterilisation and restraint legislated for and practised in several American states. This distinction between 'positive' and 'negative' eugenics was one the English theorists were very fond of. In their writings, however, a case was usually made for working at both ends; and since they were more impelled by the fear of social degeneration than by any genuine hope of improvement the most common mood was one of despondency appropriate to those who believe themselves to be fighting a rearguard action. At the same time they were united by the rather paradoxical optimism that man, that product of blind forces, has been supplied by those forces with enough intelligent insight into his condition to remedy or at least to reduce the consequences of his own deficiencies. This was certainly not a conclusion that every Darwinian was brought to by the inner logic of his position. For a purist among the Darwinians like Hardy, who apparently believed that all prospects of human felicity were lost for ever about the time that nature crossed the border from invertebrates to vertebrates, and that misery and an advanced nervous system are simply conterminous, eugenics, either positive or negative, held nothing whatever. Even T. H. Huxley, very well aware of what degeneration could do to a species, would have no truck with the crudities and vulgarities of eugenics. At the end of his Romanes Lecture of 1893 there is indeed a brief attempt to moderate the argument by the expression of the flimsy hope that some such transformation as has been worked on the dog, 'the brother of the wolf', to convert him into 'the faithful guardian of the flock' might be worked, too, on man, to make him into a truly social being.[9] In the more considered Prolegomena, however, even this thin canine analogy is forgotten as Huxley penetrates deeper into his distaste for what he mocks as the 'pigeon-fancier's polity' – a kingdom where, by some miraculous attainment of wisdom, the pigeons can get to be 'their own Sir John Sebright'.[10] (Sebright published *The Art of Improving the Breed of Domestic Animals* in 1809.) The breeder, Huxley claims on two rather muddled grounds, should play no important part in the evolution of society. The social and biologic worlds must stay autonomous because we are simply too ignorant to merge them. We can, with some difficulty and disagreement, discern the good points in a puppy, but how are we to tell the good points of a new born citizen? This is a task of quite a different order of complexity, and not an order where utilitarian and aesthetic considerations can be dovetailed neatly with physical appearance. It may be noticed that Huxley does not say eugenic breeding will not work, given the right technique and a general agreement about values and qualities. On the contrary, he is well aware that any real objection to eugenics must be on ethical rather than on biological grounds. It emerges that Huxley most greatly fears that inflammatory climate of opinion where there exists the importunate urge to do something dramatic simply because a powerful tool is available. The ones who most attract his ire are those who 'rank medicine among the black arts', who 'contemplate the active or passive extirpation of the weak, the unfortunate, and the superfluous' and who would refer sexual relationships to 'the principles of the stud'.[11] Huxley's concern does not centre on the biology of eugenics at all; nor, despite surface appearances, on eugenics as a potent instrument of practical totalitarianism. He is most fearful of eugenics as a disruptive social force which must (most offensively) loosen the bonds of the family. For him and for those who shared the sympathies of his class and the prejudices of his politics, eugenics by the 1890s had become simply a synonym for sexual radicalism.

If Huxley detested the main trend of eugenic speculation – what might be called the neo-Darwinian, Galtonian trend – he equally detested the teachings of the schismatic, social reformist eugenicists. He casts both aside with their cases almost unheard. And yet that schism is crucial: it is only possible to understand how eugenic ideals were assimilated into literary culture by taking note of the division which first became manifest in the middle 1880s and was completed by the end of the century. Let us first pursue the main stream of eugenic controversy, which flows smoothly through the decades from its spring in the rock of natural selection as this was understood in the early 1860s. We will then return to pick up the branching tributary which is formed by the thought of Allen, Wallace and Wells.

Long before Galton coined the word 'eugenics' and put it into general circulation, it had become plain to some of those who accepted the Darwinian principles as the sole, or main, agency which has raised our species to its present eminence that those civilisations highly valuing their individual members are the ones most prone to internal decay because they remove the selectionist pressure, letting the 'unfit' multiply. Many orthodox Darwinians after Darwin – they included Francis Galton and Karl Pearson – therefore saw the imposition of selective breeding as civilisation's remedy to civilisation's own problems. As early as 1868 the oligarchist writer W. R. Greg (1801–81) brought together into a single polemical essay what he conceived to be the antitheses of desirable social progress and humanitarianism. In dilating 'On the failure of "natural selection" in the case of man',[12] a short but rhetorical piece which circulated very widely, Greg supplies not a whit of really new information, but his attitude is the very quintessence of those fears which later, ordered and refined, produced what its apologists thought to be the regrettable necessity of introducing negative eugenics.

Greg entertains no doubts about 'the great wise, righteous and beneficent principle' (p. 356) of natural selection, which Western society had recently thrown so troublesomely out of gear. For individuals and races alike, the best are those who have survived the brutal conflicts of population pressure and food shortages: Greg trots out the ubiquitous examples of Roman succeeding Greek; of barbarian succeeding Roman. Curiously, he claims to have no illusions about the excesses of British colonialism. He takes it for granted that the imperial pageant conceals naked power and even corruption and injustice. But to place against that he has the comfortable assurance that 'in the counsels of the Most High' (a fine orotund phrase for what Greg takes to be biological necessity) 'a commanding will, at least in this stage of our progress, [is] a more essential endowment than an amiable temper or a good heart' (p. 358). Having thus trumpeted the cause of the STRONGEST (that superlative is always worth capitalisation for Greg), and having well flexed his simple Darwinian values, he is ready to confront his own effete civilisation. Again, targets which became irritatingly familiar later on in the century are picked out. The hereditary carriers of disease; the subnormal; the delinquent; the merely incapable and improvident: all these types which a truly kind nature would want to see extinguished are, thanks to the almost criminal liberalism of the social reformists, allowed to pass on their taint. Although he does not lay the charge specifically, the reforms of Edward Chadwick would appear for Greg to be hardly preferable to the depredations of Attila the Hun, the Scourge of God. Greg's not inconsiderable pride in sanitary conveniences and epidemological death control is quite offset by his vision of pullulating millions with 'subtle and hereditary mischief in their recesses' (p. 359) breeding, not with the blind vigour of rabbits, but with the slack irresponsibility of *men*.

It is interesting to notice that Greg's alarm at dysgenic disorder cuts squarely across the social fabric. He is as ready to curse the inheriting rich who 'hand down their vapid incapacities to numerous offspring' (p. 359) as the poor with 'tainted constitutions' (p. 358). The offence of the former is, if anything, the greater: flushed with the insolence of wealth and rank they carry away healthy brides and reduce them to 'mothers of a degenerating, instead of an ever improving race' (p. 361). He is abler, however, at apportioning blame than he is at suggesting remedies. Possibly given a government of unremitting paternal despotism, the present stunted and dissolute population might be turned into (Greg does not question the advisability of doing so) 'one glorious congregation of saints, sages, and athletes' (p. 361). But, miserably, a country gets the government it deserves, and Greg knows that his eugenic paradise with its collection of such oddly assorted vocations will never escape from the pages of *Fraser's*. Mournfully he lists the factors which together must help to speed us on the way to biological perdition: the growth and expectations of individual freedom; our mad determination to thwart the wise economy of nature in letting the unfit die; the advancement of democracy, which is to say the mob rule of precisely those least acquainted with the implacably rigid laws of hereditary transmission. Despairingly, Greg ends with a token gesture to the need for more education in the principles he has drawn out. He sounds less than confident; and the title of one of his last books, *Rocks Ahead, or The Warnings of Cassandra* (1874), stands as testimony to the deep pessimism so often induced in those who set out on the quest for eugenic perfection armed only with the inadequate, and even fallacious, formulations of later Victorian inheritance theory. The further crucial significance of Greg's essay is not only that it is quoted approvingly in *The Descent of Man*, but also that the actual course of its argument is followed unremittingly, as though Darwin had nothing to add to Greg's case. Certainly Greg's deep pessimism was found congenial enough; not only in 1871 but right to the end of Darwin's life, if Wallace's evidence is to be trusted. Writing in 1890, Wallace claimed that eight years before Darwin had given vent to the darkest admonitions about those societies where the Darwinian law had ceased to run.[13]

It is usually taken for granted that Herbert Spencer's socio-political theories best define the amalgamation of Darwinian biology and the social sciences. Spencer certainly clung to the principle that whatever increases a conflict of interest helps to resolve the struggle for existence and therefore to maintain the selectionist pressure at a maximum. Open warfare between the classes and sectional interests are, for Spencer, the most effective way of diminishing the dysgenic effects of civilisation. In this reading Social Darwinism means, simply, the carrying-over of Darwinism into social life;

and it is the offensive implications of this that Darwin's defendants have in mind when they assert that Social Darwinism is 'no *necessary* inference from any principle of Darwin's' or, more strongly, that it is 'an utterly erroneous concept . . . which he would have been the first to disavow'.[14] There is indeed much evidence that the humanitarian Darwin was repelled by the spasmodic identification of might with right made in his day. Suppose, however, that we reject this anarchic, freewheeling version of Social Darwinism and substitute a version that insists on more social controls, not less; that sees a carefully managed eugenic programme as a saner way of benefiting from the casual violence of nature?[15] Certain passages in the *Descent* which adopt a strong tone towards the 'inferior stock' thereby become clearer in their implications, as does the fact that the eugenics movement in this period was truly derived from a reading of Darwinian biology – a crude reading, to be sure, and one full of internal inconsistencies, but one to which Darwin himself, despite some protestations, gave the support of his immense reputation. He may, as we have seen in the previous chapter, have been confused about the precise connections between evolutionary change and progress, about the semantics of 'higher' and 'lower' species, but his few remarks in the *Descent* about altruistic behaviour in primitive communities are swamped by the constant reiteration that struggle lies behind social advancement, and that struggle does and should squeeze out the weak, the poor, the malformed. At a key point, having spoken of the great effectiveness of the social controls over the natural eliminative checks, he continues, exactly in Greg's vein:

> Thus the weaker members of civilised societies propagate their kind. No one who has attended to the breeding of domestic animals will doubt that this must be highly injurious to the race of man. It is surprising how soon a want of care, or care wrongly directed, leads to the degeneration of a domestic race; but excepting in the case of man himself, hardly any one is so ignorant as to allow his worst animals to breed.

And a little later, so that no reader could miss the implications of the phrase 'a want of care, or care wrongly directed', in words that are practically a paraphrase of Greg's:

> [if we] do not prevent the reckless, the vicious and otherwise inferior members of society from increasing at a quicker rate than the better class of men, the nation will retrograde, as has too often occurred in the history of the world. We must remember that progress is no inevitable rule.[16]

The final correspondence with Greg (and this more than anything else captures Darwin's reliance on and sensitivity to social attitudes in general circulation, right down to his reading the standard lesson from the fate of the Hellenic and Roman civilisations) is the realistic caution he maintains with respect to the imposition of any eugenic solution. As a biologist Darwin entertained no doubt that the techniques of the stud could be just as effectively applied in the marriage bed, to the intellectual and moral advantage of our race. The first passage quoted demands such an interpretation. So does his refusal to give any support to the defence of Charles Bradlaugh, who had been arrested for publishing a handbook on contraception. Darwin was against this on the grounds that it would encourage profligacy among women and because more people were needed to fill up the Empire; but chiefly because he felt that a vigorous multiplication of the human stock should be encouraged in order to permit the fittest to be selected out. As a social philosopher, however, Darwin at the same time recognised that the chances of any number of people refraining from marriage and children because they knew themselves to be unfit were zero, or, as he cannily put it, 'more to be hoped for than expected'.[17] And, though he concludes his discussion in the second edition of the *Descent* (1874) with an impressive tribute to his cousin Galton's work on inheritance as bringing forward the day when eugenics might be practised (or imposed?) with real wisdom, it may be doubted whether this really reflected his private opinion. Only four years later he is found being as contemptuous as it was in his nature to be about Galton's scheme for issuing breeding certificates.[18] (Unfortunately Darwin's exact opinion of this cannot be known because of his editors' excisions in the letter in question. Probably they wished to prevent Galton's learning his real attitude to this absurd scheme.)

If, leaping for the time being over Galton's practically single-handed creation of Darwinian eugenics, we move on to the end of our period and the philosophy of his disciple and biographer Karl Pearson (1857–1936), who originally trained as a statistician, we may observe the extreme inflexibility of Darwinian eugenics. Nothing was learnt in forty years. The targets of hatred, the neuroses, the naïve reliance on natural selection are all just the same. For Pearson, who was one of the sternest of all the eugenists, it was indeed the slow grinding sureness of selection that made him especially zealous in offering this method of improvement to the public. As he told a Newcastle audience in 1900: 'I want to justify natural selection to you. I want you to see selection as something which renders the inexorable law of heredity a source of progress which produces the good through suffering, an infinitely greater good which far out balances the very obvious pain and evil.'[19] This is familiar enough, for it is only the creed of *Self-Help* (and it is curious that Samuel Smiles's famous handbook appeared in 1859) dressed up as biological edict. Selection is being offered as the fiery crucible out of which can come only a finer metal. It is the same imagery of stoicism with which we have already become acquainted, calling for the joyful embracement of suffering – especially for the non-European races. Here there is the paradox of Darwinian eugenics that, first presented as a human alternative to blind force, it eventually itself lapsed into mindless cruelty. In the ideal community, according to Pearson, the public weal would best be served by statesmen capable of guaranteeing and insisting that 'the fertility of the inferior stocks is checked, and that of the superior stocks encouraged'; statesmen who, as eugenic busybodies, would 'regard with suspicion anything that tempts the physically and mentally fitter men and women to remain childless'.[20] While they remain on the rarefied heights of theory these prescriptions of Pearson's for the ideal eugenic society may sound innocuous enough. Only when they are brought down to the plains of social practice, of humans living together within a code of justice, can we understand the true import of this area of biological speculation. In reality it is only the briefest step from Pearson to the Third Reich's *Lebensborn* foundation, which promised any unwed German girl a child in financial security and which employed coyly named 'procreation helpers' (*Zeugungshelfer*), usually SS officers, to act as studs. The literary imagination, rooted as it is in the concrete personal situation, can help to make this clear.

Literature and Galtonian Eugenics

The eugenic beliefs of Francis Galton (1822–1911) comprised more than a philosophy: they were the dogma of a religion which he spent the greater part of his immensely productive mature years evolving and which claimed his emotional being totally. On much the same grounds as Greg and Darwin, he took the need to introduce a eugenic programme into society to be self-evident. For Galton was a good Darwinian. The impact of his half first cousin's book was, as he recognised at the time, 'a real crisis in my life'.[21] Later he tested with an elegant experiment Darwin's theory of heredity – pangenesis – and his negative results made him yet more resistant to any environmentally modifiable theory of inheritance. His admiration for the principle of natural selection, and for what a simulated version of it might accomplish for humanity, was practically unbounded. It was for him practically a moral obligation to seize hold of the evolutionary tiller and make that which had been for infinite generations semi-random, egotistical and mostly ineffective, deliberate and systematic and altruistic. Altruistic, because Galton for all his life was that rather rare creature, an optimistic, hard-line Darwinian. He fervently believed that artificial selection, aping natural selection, could generate the highest moral values as well as the mere instruments of survival. Further than this, Galton actually aligned the social virtues with sharp teeth and high intelligence, and claimed that, because it guaranteed a deep unity of purpose and allowed some sympathy with deviant behaviour, altruism had always been a powerful factor in human evolution. By thus giving natural selection such a high moral tone, Galton sidestepped the eugenists' usual difficulty of first having to define the desired characteristics of the 'fit' who are going to be bred for. Breed for what the highest societies most revere, said Galton, and we will therefore be following the logic of evolution, only we will be substituting a 'more merciful and not less effective'[22] technique for something mostly destructive and always wasteful.

What is basically a religious tenor in Galton's eugenic proposals is obvious at an early stage – long before, in fact, he had come to propose any special avenue of reform. In the course of a long article on inherited abilities, we find him manoeuvring into a position where he can offer, almost casually, a biological substitute for the doctrine of original sin. The image of man in the *Descent* and in Lyell's *Antiquity of Man* is of a creature not fallen from high estate, but rising up from a lowly one of barbarity and, earlier, brutishness; a creature whose moral nature has yet to catch up with his material prosperity. Not too much should be expected, Galton tells us with his curious and rather typical mixture of complacency and humility, of a recently uplifted savage. His immediate argument, and the normative anthropology of the 1860s, leads him on from there to organise a ranking of racial abilities, according to each race's distance from its bestial origins: from the 'jabbering, quarrelling, tom-tom-ing, or dancing' negroids, to the 'self-complacency of a steady-going Chinaman'.[23] But, anticipating the later trend of his thought, Galton is already betraying a certain impatience with the natural forces which after long years may, optimistically, shift the qualities of the human races in his spectrum from the black towards the yellow end. He is already hankering after a bolder vision: of a race to whom our present peoples will seem to be 'what the pariah dogs of the streets of an Eastern town are to our own highly-bred varieties'.[24] Galton, as early as 1865, is already turning from the ideal of a Victorian anthropologist to that of a dog-fancier. He has already caught on to the frightening yet piquant notion that within a reasonable, human time-scale man can remake himself in his own image. It is in this respect that Galton's eugenics was unique, though in other respects the ways Galton proposed for reaching his grand objectives were of no greater refinement than the Spartans'. He, like them, believed implicitly that by blending 'advantageous' characteristics of the parents the children must exhibit them doubly enforced. Even after the Mendelian, particulate theory of inheritance moved into the region of repeatable, experimental fact after 1900, Galton took no special notice. When he delivered the Second T. H. Huxley Lecture in October 1901, for instance, on the topic of 'The possible improvement of the human breed, under the existing conditions of law and sentiment', Galton showed himself to be much more diverted with the problems of the second half of his subject than with those of the first. In his last decade he refused to acknowledge that the biology of eugenics had any further mysteries to explore. He felt that the implementation of a eugenics programme was simply a question of effective public relations work. He was finally persuaded that the information already available to the biometricians about the inheritance of wanted and unwanted traits was already complete enough to pursue what he calls 'the grandest of all objects'.[25] In this lecture his confidence reached so far as a proposal for the immediate and resolute segregation of habitual criminals under whatever conditions of surveillance are necessary absolutely to prevent them from reproducing. Galton leaves it vague whether this virtual blanket sentence of life imprisonment (he does not mention sterilisation) for every recidivist should appeal more to the nation's parsimonious instincts or its concern for moral improvement: he carefully mentions both.

To express his eugenic convictions Galton turned, at the end of his life, almost instinctively to the Utopian romance; for as a medium of propaganda for lofty but impracticable causes nothing better had yet been devised for reaching the audience most responsible for the 'existing conditions of law and sentiment'. Eugenics was indeed a component in most of the Utopian writing after 1870, though the concern with raising a better stock only rarely went so far as to consider what we would now call 'genetic engineering': that is, the direct interference in the process of inheritance. Those writers who espoused eugenics normally added little to the classical arguments for the state's involvement in, and supervision of, the selection of parents for the next generation. In a typical example, *A Thousand Years Hence* (1882) by 'Nunsowe Green', the narrator describes his long prophetic dream reaching that far forward into the future. Even before the dream begins Green looks forward to a time when the state will participate directly in the rearing of children, particularly street orphans. At a time three hundred years in the future, the benevolent government makes no secret that it is 'primarily interested in suitable marriages, and in the highest health, moral as well as physical, of family life'[26] – an interest not diminished by the expressed belief (perhaps derived direct from Galton?) that moral health is no less inheritable, and therefore modifiable, than physical. No restraints, however, are placed on free unions, other than those vetoed by public opinion. National selections are made: and, if mating follows by free will of the parties, the children become the special beneficiaries of the state. Another hundred years farther on, and Green's dream of 'social resanitation' is seen as coming true. In a kind of eugenic Dr Barnardo's Home orphans are to be raised up as a new aristocracy. 'They were to be specially brought up in separate institutions, where they might be duly educated so as to complete all the rudimentary advantages nature had given them, and thus be sent forth into the world as a kind of superior race – a national nobility – to take, by force of pure personal quality, their natural lead in society' (p. 225). It will readily be seen that in this transformation of the orphanage and workhouse into a greenhouse where *Homo superior* might be forced like exotic orchids there is more assuagement of social guilt going on than any pursuance of the hard-headed methods of the stockman. There is also a rather pathetic belief in the power of education to correct innate defects, and the usual anti-hereditarian bias which, a century ago as today, went along with environmentalism and behaviourism. The message of *A Thousand Years Hence*, then, is that among novelists as among theorists eugenic ideals could exist without reference to considerations brought into notice by the new biology. In Galton's Utopia *Kantsaywhere* of twenty years later, by contrast, these considerations are paramount. The Darwinian facts of variation and selection (positive selection to deflect the current of variation into desirable channels; negative selection to prevent degenerative collapse) push the didactic part of Galton's fiction right out of the tradition of Plato and More.

Galton began to write *Kantsaywhere* in May or June 1910, in his 88th year. By 23 October he had it finished in rough draft, at which time it was evidently the length of a long short story.[27] By 29 October it had been fully revised and typed and, on 4 December, Galton offered it to Algernon Methuen for submission to his reader. Galton was not sanguine about its chances of acceptance for, as he wrote, 'at first sight he [Methuen] was very dubious. He takes no interest in Eugenics.'[28] Methuen did decline to take it, and Galton, the most sensitive of authors, cast it aside. '*Kantsaywhere* must be smothered or be superseded. It has been an amusement and has cleared my thoughts to write it. So now let it go "Wont-say-where".'[29] After Galton's death the next month one of his nieces, finding the plot absurd, destroyed a large part of the typescript including all the fictional episodes. The fragments which survived this rather arbitrary and ruthless critical judgement are no more than flat descriptive pieces of the hero's experiences in the imaginary eugenic state. They passed eventually into the hands of Karl Pearson, who printed some (or possibly all: he is not clear on this point) in his biography of Galton.[30] This partial destruction of *Kantsaywhere* makes it impossible to pass any comment on the literary merits of the Utopia, but the surviving sections are the clear lineal descendants of Galton's earlier essays both in clarity of exposition and in content. *Kantsaywhere* was not the product of any new line of thought, but rather that of leisure and mature reflection; it is practically the summary of nearly half a century's pondering on the structure of a workable genetic eutopia. Even in the 'Hereditary talent' essay, from which we have already quoted, Galton muses (in the protective first-person plural) that we should 'give reins to our fancy, and imagine a Utopia – or a Laputa, if you will – in which a system of competitive examination for girls, as well as for youths, had been so developed as to embrace every important quality of mind and body'.[31] In every important sense, then, the fragmented *Kantsaywhere* is a product of the biology of our period and will be considered in that light.

Pearson's comments, which were based in turn on the memory of one who read the complete story, suggest that Galton adhered carefully to the traditional formula for his plot. To the shores of Kantsaywhere he brings an intelligent and curious visitor; he makes that visitor, Professor I. Donoghue, fall in love with a girl of the community, Augusta Allfancy;

and Galton enlightens the reader and his hero simultaneously by having the latter guided about the Utopia by voluble cicerones. Other complexities, which certainly existed, have vanished altogether. Donoghue finds that the colony of Kantsaywhere is ruled by the Trustees of the College of Eugenics, the College having been founded in 1820 thanks to the munificence of a citizen who left money for granting diplomas to those future citizens who were judged to possess valuable inheritable qualities, and for the nurture and education of children born to highly diploma'd parents. Galton tells with obvious approval of the benefactor's firm condition that under no circumstances should any part of his gift go to charity, to the poor, or to the naturally feeble. His interest reached only to sponsoring the multiplication and nourishment of those 'who were strong by nature' (p. 414). Kantsaywhere at the time of writing is held to contain 10,000 inhabitants, acquired by natural increase and immigration; the College has become the supreme arbiter without the bother of any democratic paraphernalia, and it levies a rate to finance its selectionist activities.

As an immigrant Donoghue is required to take the Pass Examination to become a citizen: a Pass in Genetics gives him the right, depending on the combined marks of himself and his marriage partner, to produce a certain number of children. The caste system is unbreakable. At the bottom are the Untouchables, here defined as the eugenically Unclassed. A special Bureau tries to make the lot of these Unclassed 'as pleasant as might be, so long as they propagated no children'; but the chances of happiness seem to be reduced by the imposed need to 'work hard and live in celibacy' and by the fact that in the event of any resistance 'kindness was changed into *sharp severity*' (p. 416: Galton's italics). In Kantsaywhere the ends justify the means, as they have a habit of doing in other Utopias; and for Galton the end is the creation of overlords racially and perhaps even specifically distinct from ordinary humanity. A tentative move towards this still-distant end is another hurdle which Donoghue tries to leap and which he describes extensively: the Honours examination, taken by only eighty candidates a year. Success here bestows special privileges if the graduates marry within their own ranks. Any resultant children are adopted by the College and so perpetuate an elite.

The detailed series of tests which form the eugenic heart of Kantsaywhere are of quite extraordinary crudity. The first group of tests ascertain the candidate's physical prowess, by ranking on a simple scale the acuity of his senses, his stature, power of grip, and so on. In the second group he is tested for aesthetic sensibility by being required to read poetry and prose, to sing, and finally 'a few athletic poses were gone through as well as some marching past' (p. 417). In the third group he is medically examined with a simple external inspection, and in the fourth information is gathered about his heredity. By aggregating the scores thus obtained, so the authorities of the Eugenic College believe, everything necessary to establish human superiority might be deduced. Their views definitely reflect Galton's own:

as Pearson himself attests, the procedures described very closely parallel the ones used in Galton's first eugenics laboratory in South Kensington.

In reading through the remnants of *Kantsaywhere* it is difficult to know whether to be amused by Galton's arrogance or alarmed by his dangerous naïveté. We are told that in this eugenic paradise 'the 'arry and 'arriet class is wholly unknown' (p. 417), and that the Honours examination is at some point divided to accommodate the people 'intended for the more cultural class' on the one hand and 'the hard workers, whether on farms or in town', on the other (p. 418), so even eugenics quails before the class structure; but these are merely the habits of thought of the English upper classes of the time and specifically of an influential segment of the intelligentsia, and as such excusable. More frightening is Galton's readiness to view the breeding of the unfit person as a crime to be punished by compulsory emigration (with 'surveillance and annoyance' ordered for those who defiantly remain) or, in extreme cases, by imprisoning the transgressor in 'a Labour Colony with hard work and hard fare until it is considered that he has purged his debt' (p. 420). How this Draconian legislation to thwart the reproductive impulse can be squared with Galton's reputation as a humane and compassionate man, it is difficult to say.[32] Certainly C. P. Blacker, an apologist for Galton and for eugenics generally, does not impress when he defines the political structure of this outrageous 'pigeon-fancier's polity' as 'amiably paternalistic'.[33] Galton, he insists, disliked tyranny: he quotes a passage where Galton denies the value of allowing a Spartan dictator to exercise absolute eugenic power. But the polity of *Kantsaywhere* is, if not a tyranny, an oligarchy which is supposed capable of making fearful decisions – decisions which Galton himself must have been ready to endorse, or why base a Utopian fiction on them? Of the eugenic examination system Blacker comments: 'the man who is ploughed may be disappointed; he may even feel bitterly aggrieved; but he does not, if he is sane, regard himself as a victim of fascism' (p. 123). Discounting the point about the mental condition of the Unfit (which is irrelevant), we may protest on their behalf that their attitude will be formed by their perception of the fairness and the reliability of tests – including 'marching past' – which according to Galton's own dispensation of eugenic justice may be about to condemn them to permanent involuntary celibacy. It is a measure of the absurd ingenuousness of the Victorian eugenists that they could seriously hope to dam off the sexual instinct by exhortation coupled with a little genteel force. By 1905 eugenics, resting on a base of unproven but tenaciously held assumptions, was the sector of biology most heavily charged with blind emotion. For this condition the biometricians were solely responsible; and *Kantsaywhere*, most ignominious of Utopias, stands testament to it. As for the gifted Galton himself, if he were indeed the epitome of the ideal eugenic type (he learnt his capital letters at 1 year old; could read at 2½; could read fluently in English, some French and Latin by the age of 5), then he is the best possible argument against a eugenic policy!

A disgust of and a disbelief in Galtonian eugenics began to set in by the early 1880s and flourished as the appeal of 'hard' Darwinism waned. By the later 1890s the schismatic social reformist eugenic party led by Wells, Wallace and Grant Allen had become strong enough to dominate the thinking of the Fabians and other socialist theoreticians on eugenics generally, and therefore to produce the decisive rejection of real-life eugenic legislation in favour of liberal reform. Wells's imaginative and didactic prose exhibits this shift most interestingly.

At the end of the last chapter we saw how Wells's first decade of literary activity pointed more and more conclusively to a programme of eugenics to solve the difficulties which, he believed, would quickly obliterate the human species if its evolution were not taken in hand. Wells's successive models for the future of humanity left to the mercy of a Darwinian universe – the Man of the Year Million, the Morlocks, the Martians – culminate in the grotesque Grand Lunar of his last romance *The First Men in the Moon* (1901). Significantly the society of the Selenites, being carefully modelled on that of the anthill, is a far cry indeed from the brutal free-for-all of the Martian society or that of the future earth revealed to the Time Traveller. Instead the Selenites have a rigid caste system with every worker knowing and being fitted for his place. But, again significantly, Wells imagines a caste system which is not a product of natural evolution like the caste system in *The Time Machine* but one that is the result of conscious Galtonian experiment and enforcement by the Selenites themselves. They have chosen to live this way. They have chosen to break down their once homomorphous species into a multitude of specialised anatomical forms, all ideally suited to their various tasks. 'Quite recently,' the stranded Cavor informs his audience on earth by radio,

> I came upon a number of young Selenites confined in jars from which only the fore-limbs protruded, who were being compressed to become machine-minders of a special sort. The extended 'hand' in this highly developed system of technical education is stimulated by irritants and nourished by injection, while the rest of the body is starved . . . It is quite unreasonable, I know, but such glimpses of the educational methods of these beings affect me disagreeably. I hope, however, that may pass off, and I may be able to see more of this aspect of their wonderful social order. That wretched-looking hand-tentacle sticking out of its jar seemed to have a sort of limp appeal for lost possibilities; it haunts me still, although of course, it is really in the end a far more humane proceeding than our earthly method of leaving children to grow into human beings, and then making machines of them.[34]

The charged phrases – 'highly developed system', 'quite unreasonable', 'lost possibilities', 'humane proceeding' – in this quite remarkable passage must serve to reinforce our uncertainty about the exact limits to Wells's irony.

On a first reading it is a sick jest based on Cavor's difficulty in accepting the reality of a really thorough eugenic programme. But on looking over it again one's second thoughts are that the reader is being directed away from the ant-like Selenites and towards the human advantages of such an educative system. Covertly Wells seems to be asking us to try the effect of regarding that helpless hand sticking out of its constraining vessel as a *vain* symbolic appeal and one to which we might feel justified in hardening our hearts. Certainly the Selenite mathematician, with his attenuated frame, his voice a mere squeak for stating formulae and his mind furnished with no ideas beyond those required by his discipline, is a bizarre advertisement for the eugenic society. The Selenites are far from being an ideal; but somewhere in the strange *mélange* of surgery and brainwashing which his imagination has conjured up, Wells seemingly suspected that there might lie the only solution to human decay.

For what, after all, is the alternative? Wells's fears, as expressed with painful clarity in a collection of forecasts of social change which won him the respect of the Fabians, *Anticipations* (1902), align him on much the same axis as Greg, Galton and Pearson. Like them, Wells places the causes of degeneration firmly in the differing reproductive rates of the social classes. Like them, he reacts almost with nausea to the dysgenic spectacle of 'a mean-spirited, under-sized, diseased little man, quite incapable of earning a decent living even for himself, married to some underfed, ignorant, ill-shaped, plain and diseased little woman, and guilty of the lives of ten or twelve ugly ailing children'.[35] We notice that the vocabulary of distaste here ('mean-spirited', 'under-sized', 'ignorant', 'plain', 'ugly') has completely lost connection with the Darwinian definition of the Unfit in terms of one particular environment. Wells's Unfit, those whom he believed had already spawned so unremittingly that 'in the more civilised states of the world, the average of humanity has positively fallen' (p. 307), are so only in respect of some nebulously conceived Efficient State run by eugenically superior men belonging to what he is pleased to call the New Republic. Wisely, Wells attempts no ostensive definition of the 'average' which has fallen, and it would hardly be appropriate to expect him to do so in this kind of discourse. Without pause, then, Wells moves forward to the expected prescription: the euthanasia of the weak, silly and fecund, coupled with 'good scientifically caused pain that will leave nothing but a memory' (p. 301) in order to quieten (one supposes) any recalcitrant survivors of the eugenic purge. Certainly it is surprising to find this kind of rhetoric spilling out of high-minded Fabianism. It is something which the political historians have been happy to gloss over. Yet, as in the case of the biometricians, it is not difficult to appreciate why, in laying down eugenics as a stout bridge between what is and what ought to be, Wells's political liberality should suffer. For when one is forced by the pressure of accepted inheritance theory to view the spring of human potentiality for excellence – excellence expressed more and more in terms of hard genetic laws – as being constantly unwound by random mutations and weakened by the undesirables' procreation the shift to authoritarianism and to talk about the need for 'social surgery' becomes almost irresistible.

Nevertheless, *The First Men in the Moon*, as well as being his last romance, effectively marks the conclusion to Wells's fifteen-year-long meditation on the human meaning of Darwinian biology. As we concluded after surveying his response to the powerful anti-Darwinian coterie, Wells found nothing crucial to quarrel with in natural selection. Most probably his ambivalent feelings about repressive eugenics owed something to that capable Darwinian publicist Grant Allen, to whom he gave generous praise.[36] But Allen died in 1899, and soon afterwards Wells was obliged to recognise that Darwin's equation of the fecund and the fit simply would not do. Early in the new century, then, in his heavily didactic 'novel' *A Modern Utopia* (1905), Wells begins the attempt to put his public on its guard against 'all nonsense of the sort one hears in certain quarters about the human stud farm . . . from anyone in the days after Darwin, it is preposterous'.[37] The nonsensicality he alludes to is not that of oversimplifying the Mendelian laws in applying them so crudely to breeding superior men. His outburst is more the cry of a man finding his principles clashing with natural law. Struggle is good: Wells was enough of a Victorian to credit that as axiomatic. But, though he saw no place in any desirable human future for the ventral dream of peace, Wells shrank from the Meredithian approval of permanent and unresting combat. Caught in this dilemma Wells and some other eugenists began to turn to such nostrums as the emancipation of women, the separation of sexuality and reproduction, sanitary reforms and other general euthenic measures. Among these thinkers socially undistorted biological process gradually became emptied of all the rich human significance which the Darwinian moralists had once found in it, and did indeed become no more than a backdrop to mundane concerns.

Reformist Eugenics: Wallace and Grant Allen

Eugenics by social reform, though it gained a certain political currency that was never granted to the biometricians, was predominantly the creation of the brisk and practically minded biologist with a social conscience. Such men – Alfred Russel Wallace and Grant Allen may serve as worthy representatives of the type – were in one sense or another Darwinians, believing in the efficacy of Darwin's mechanisms. What they could not stomach was the inevitable cruelty and illiberality that must result from allowing nature free play. They were united in being disdainfully impatient of the old romantic stoical pessimism. They were inclined to turn their backs on a far-distant New Jerusalem which was going to be brought into being by means neither speedy, sure, nor flexible enough for their taste. And yet the Utopian sentiment was as strong in them as in the Galtonians.

Their response was to turn, not to the *Origin*, but to the *Descent*: not to natural selection, but to Darwin's subsidiary hypothesis of sexual selection, first offered in detail in 1871.

To see how this worked in practice, and to sample the arguments of the reformist eugenists, we shall conflate two essays by Wallace, 'Human selection' and 'Human progress', published in 1890 and 1892 respectively. The opening premiss of the second paper in fact is antecedent to everything in the first, because it supposes that, in respect of intellectual and moral advance, the 'high water level' of ability (Wallace has in mind the few supreme donors – Jesus, Buddha, Confucius – of complete ethical systems) has subsided over the last two millennia. The *mean* level of ethical sensibility, however, Wallace takes to have been continually elevated, thanks to the purely Darwinian factor of intraracial conflict – conflict which, Wallace rather curiously assumes, has tended to favour the survival of the offspring of the sober, virtuous and healthy sectors of a populace. Wallace was no retrogressionist, and he abominated the kind of mentality which was nostalgic for more harshly natural days. He opts without hesitation for the 'decided gain' of an increased sympathy with suffering as 'the most marked and most cheering of the characteristics of our age'.[38] Yet Wallace still exercises himself over the problem of increasing the proportion of the most capable intelligences and the most subtle moral natures – and, what is more, within the Darwinian limits. (For Wallace spends time in both essays disabusing the reader about Lamarckian philosophies of education. He takes all talent, even up to the level of genius, to be a spontaneous genetic endowment; successive generations of a capable family do not rise 'progressively to loftier heights of genius and fame': p. 155.) He shares Galton's biological presuppositions but is highly critical of his eugenics, which he correctly asserts to be elitist in conception and effect. He will have nothing to do with compulsion.

Wallace's grand solution for increasing the selectionist pressures in a well-regulated but open society is, quite simply, to improve the freedom of choice for women by removing the economic incentives to marriage. When women are no longer legalised prostitutes but independent, the number unmarried will rapidly increase since relatively few seek marriage for 'personal affection or sexual emotion'. Since these clear-eyed business-girls will be trained from their cradles to look with contempt on 'all men who in any way wilfully fail in their duty to society – on idlers and malingerers, on drunkards and liars, on the selfish, the cruel, or the vicious',[39] and since they will be able to decline, and certainly will not tolerate, any unions other than marriage, all will receive many offers which they can afford to reject. These latter-day Lysistratas will blackmail some of the sexually impetuous males into good conduct, and the residue they will scornfully leave to celibacy and oblivion:

> In such a reformed society the vicious man, the man of degraded taste or of feeble intellect, will have little chance of finding a wife, and his bad qualities will die out with himself.[40]

So there will be no need after all for Galton's eugenic concentration-camps. It is women who will act as eugenic police, and their soft yea or nay will be much more effective and civilised than any sterilising scalpel. As the agents of sexual selection, Wallace's women are supposed to be able to choose the 'best' males by unalloyed reason; they are of course conformist enough to equate society's 'best' with their own tastes, and apparently are so devoid of viciousness, unseemly appetites, or stupidity themselves that they needs must love the highest when they see it. The point is hardly worth making that the Victorian stereotypes of women are here operating in a pseudo-biological context. So manifestly blind is Wallace's idealisation that he can conclude, without a trace of irony, that 'we may safely leave the far greater and deeper problem of the improvement of the race to the cultivated minds and pure instincts of the Woman of the Future' ('Human selection', 337).

The last catchphrase is an echo of the title of a series of essays in the *Universal Review*, where Grant Allen's contribution to this engaging debate appeared only four months before 'Human selection'. Allen had his own distinctive line on reformist eugenics, and a more radical one. The mature life of Charles Grant Blairfindie Allen (1848–99) covers almost the whole of this study, and his hybrid career as a writer is almost a stereotype of those literary interests that we are most concerned to trace. For Allen was a man of very many parts – a journalist in the style of Jasper Milvain in Gissing's *New Grub Street*; a poetaster; a scientific populariser in many diverse fields; a social commentator; a controversial novelist – but his most lasting interests revolved round the triple poles of speculative biology, natural history and middlebrow fiction. There are two reasons why his attitude to eugenics is important here. Although a great deal of his writing was done under financial pressure and displays the weaknesses of a fertile but undisciplined mind, this very fact of his being by necessity a thinker with no time to glance up from the immediacies of the hour gives his ideas (and his writings always illustrate some scientific, often biological, or else moral, speculation) a peculiar relevance. A second reason is Allen's presence within both the literary and scientific fraternities of the 1880s and 1890s and the extraordinary freedom with which he was able to move from one to the other without invoking antagonism. How he managed this is something of a mystery. He had none of Thomas Hardy's deference to scientists; none of his patience for picking up stray crumbs of information tossed his way at polite evening parties. And he was quite unlike W. H. Hudson, content to be an honest hodman for the theoreticians. His approach to delicate points at issue was not dissimilar to Samuel Butler's, except that unlike Butler he retained a measure of respect, or at least tolerance, from most quarters. This was made easier, too, by his ability (which Butler signally lacked) for picking up a heresy at just the right point along its journey into orthodoxy. This gift may be seen at work no less in his biology than in his handling of tricky social issues, and it certainly helped in preserving his image as a daring but never an anti-Establishment writer. The forbearance he enjoyed of scientists and literary men both cleverer and more successful than himself also derived from, we must recognise, his reputation as a maker or breaker of writers. Richard Le Gallienne, writing after Allen's death, alludes to that uncomfortable talent in his ironically patronising way:

> Grant Allen always had this happy knack, by the sheer innocence of his almost childlike sincerity, of attracting, or shall I say, repelling, immediate attention for any cause he cared to espouse . . . He could not avoid making the poster phrase, the poster word.[41]

And we may well suppose he was feared for it. Again, his almost precognitive capacity for spotting and vigorously promoting the next cultural trend worked over an astonishingly wide spectrum of intellectual and creative activity: all the way from Weismann's important essays on inheritance theory (which he boosted) to William Watson's first significant volume of verse, *Wordsworth's Grave*, the review of which practically handed over to Watson his volatile fame as the last traditional Victorian poet.[42] Grant Allen's own more substantial work rather modifies any impression of a readiness to compromise to any degree with the market. His bibliography shows very well that he expanded himself over a number of genres for no other reason than that his interests were wide-ranging. It was his good fortune that, whether composing light verse (*The Lower Slopes*, 1894) or dilating on *The Evolution of the Idea of God* (1897) from the 'shocking' anthropological standpoint made fashionable by Spencer and the *Descent*, he managed to appeal to readers who must otherwise have been miles apart in taste and sensibility.

Allen has been the subject of no important biography since Edward Clodd's *Grant Allen: A Memoir* (1900) and therefore the shifts in his involvement with the life sciences must be extracted entirely from his published work and from outside references to him. From these sources a coherent picture emerges which is strikingly in tune with the introductory remarks on the biologico-literary interaction with which this study opened. Allen's career is, so to speak, a lodestone against which the deviations from the main trends of thought of other and greater creative writers may be measured.

Allen's important intellectual life began in the late 1860s at an Oxford still reverberating with the Owen-Huxley confrontation, and the residuum of the Spencerian evolutionism which he imbibed there is apparent in all his purely scientific books, especially in *Physiological Aesthetics* (1877), which won more than a token approval from both Spencer and Wallace, and also in its later expansion *The Colour Sense: Its Origin and Development*

(1879). The latter, since it traces the evolution of colour vision over geological aeons, naturally met with approval from the evolutionists. (It contradicted those philologists and critics who believed that there had been an actual growth in the eye's sensitivity to colour 'within a period of which we possess literary memorials'[43] because of the alleged lack of colour words in ancient Greek poetry.) So did Allen's two later series of essays on natural history topics, *The Evolutionist at Large* (1881) and *Vignettes from Nature* (1881). Of the latter T. H. Huxley wrote that he had no fault to find on the score of accuracy; that 'on the contrary, I find much to admire in the way you conjoin precision with popularity – a very difficult art'.[44] So far as one can gather, Allen gave complete satisfaction to the inner Darwinian circle for the rest of his productive life, attracted by neither the excesses of neo-Lamarckism nor those of neo-Darwinism; and, as the pressure against Darwinism mounted, Allen was there with facts gleaned from the field naturalists like W. H. Hudson in support of such subsidiary explanations as sexual selection. He was such an excellent propagandist that this alone might account for his high regard among the Darwinists. Huxley, at least, was astute enough to realise that Allen could command the ear of a wide, profoundly non-intellectual, fickle, middle-class audience: that stratum of English society which was, and remains, most impermeable to abstract theories. Huxley in his own lectures to working men was preaching to the converted; to a group of radical and lively intelligences who were hungry for facts and speculation and had no time for spoon-feeding. But spoon-feeding was Allen's forte. Even Darwin recognised it in his pleasantly enthusiastic way; in an undated letter he told Allen: 'who can tell how many young persons your chapters may bring up to be good working evolutionists!' (quoted in Clodd, *Grant Allen*, 111).

Since we are most interested in the position of equilibrium which Grant Allen finally attained, we will sketch in only lightly the modulations in his biological opinions. We have already mentioned that Allen's biology typifies the period, and it is therefore only natural that he should have been attracted to purposive, goal-directed evolutionary systems. This is clear from his review of Butler's *Evolution, Old and New*,[45] and may profitably be contrasted with George Romanes' scarifying review of another Butler book nearly two years later,[46] and Allen's own views four years later still in his *Charles Darwin* (1885). Both Allen and Romanes consider Butler's brand of evolutionism to be important as a monitor of quasi-scientific trends, but they differ absolutely as to their evaluation of Butlerism itself. For Romanes, speaking *ex cathedra* from the heart of Victorian scientific professionalism, Butler is quite simply a charlatan – an arrogant, vain, stupid and even mildly insane charlatan. While Butler restricted himself to 'entertainments' (so Romanes contemptuously terms *Life and Habit*) he may have been occasionally amusing; but Romanes' response to his daring to enter 'the arena of philosophical discussion' is one of simple outrage. Grant Allen takes a notably more pacific line. He prefers to talk about the confusions inherent in Butler's brand of irony than to confront his teleology head-on. On the whole he reacts mildly to Butler's aggressive relocation of Darwin himself as merely the last in a long line of evolutionists. While he cannot agree about the otiosity of Darwinism itself, he is nevertheless guardedly interested. Six years later his attitude is slightly more positive. The preface to his short critical biography of Darwin pays Butler a generous compliment: of Allen's having extracted from *Evolution, Old and New* 'many pregnant suggestions with regard to the true position and meaning of Buffon, Erasmus Darwin, and the early essentially teleological evolutionists' (*Charles Darwin*, v). But he still dissents from Butler's acrimonious and ill-judged verdict on Darwin himself. So far as one can discover, Allen expressed no more positive interest in the teleological case than this. But, equally, he objected to any theoretical extremism of an opposing brand. His presentation of Weismann's *Essays upon Heredity* another five years on in 1890 is a model of defensiveness. He discerns the 'almost classical authority' of the *Essays*, but shies at their insistence that we are to 'give up at once the whole remaining Lamarckian element in evolutionary biology, and to hand over everything to the arbitrament of natural selection . . . If Weismann is right, we shall have to begin all over again.'[47] These are the words of no neo-Darwinian, but of a publicist protective of the revised and augmented 'soft' Darwinism of the decades after Darwin's death. Allen takes what comfort he can in noting that just one single case of somatogenic inheritance would be instantly fatal to Weismannism, and he speaks vaguely of psychology providing that case.

Grant Allen's intellectual career, then, is not a complicated one and his attitude to Galtonian eugenics is just what might be expected. As a Darwinian body, Allen moved in a highly predictable orbit. In the 1860s and early 1870s he was pulled irresistibly into the powerful gravitational field of those central luminaries the *Origin* and the *Descent of Man*. In the early 1880s he came like many others under the influence of a rival field of force, that of teleological evolutionism, as his attitude to Butlerism suggests; but this was amply balanced out by Weismannism at the end of the decade. Before he died, Allen achieved some sort of equilibrium by publishing his reformist eugenic programme which, like Wallace, he hoped to see realised by emancipating women. But, as his 1890 essay and still more his *succès de scandale, The Woman Who Did* (1895) make amply clear, his definition of emancipation is a good deal more permissive than Wallace's. Before thrusting the whole burden of eugenic responsibility on to the one sex Allen wants to empower women with real control over their own fertility as well as untying them financially; he would have maternity made a responsibility of the state and (most daringly) allow the choice of a mating partner to be unhampered by any consideration other than that of fitness for paternity. At the heart of Allen's eugenic millennium, in short, would rest an Amazonian cult among whose members motherhood 'would soon crystallize into a religious duty . . . The Free woman will choose which lord she shall serve. And do you think her choice will be for the colonial broker?'[48] His fantasy, obnoxious for obvious reasons to Huxley and Wallace alike, is when considered coolly no less absurdly sentimental than Wallace's modification of it. Biologically speaking, both proposals rest on the transposition into civilised life of the theory that in nature all sexual distinctions have arisen from an attempt to attract, or fight for, the opposite sex – that 'the most vigorous males, those which are best fitted for their places *in nature*, will leave most progeny' (*Origin*, IV:67–71; my italics). Since Darwin had here no more sophisticated distinctions in mind than the stag's antlers, cock's spurs, or beetle's mandibles both proposals are open to precisely the same accusations we have already levelled against the Galtonian eugenists. Such devices evolved for survival or ostentation relate in no meaningful way to human goals. One might as well expatiate on the valuable eugenic potential of fancy dress for men, on the evidence of the attraction of the peacock's plumage to the peahen – providing one is content with a peahen's instinctive values. Equally, only someone already predisposed to prize women's powers of selection could expect them to serve as eugenic referees.

Blind though Allen may have been to the blemishes of his own eugenic proposals, no one made crueller fun of the Galtonians than he; for no one else, it seems, bothered to inquire imaginatively into what a practical scheme of negative eugenics might signify at the level of personal relationships. Ten years before 'The girl of the future' Allen had first turned his hand to the short story, allowing his gift for scientific exposition to spill over freely into the new medium. One of the most successful results was 'The Child of the Phalanstery', which Allen thought highly enough of to include in an anthology nearly twenty years later.[49] According to its headnote, the story was built around the experience of seeing a child with a club foot and listening to a friend advocate socially licensed infanticide for such cases, to curtail the transmission of inheritable defects. 'The Child' is set in some indeterminate future when the population has been regrouped into Fourierian phalanxes. In Charles Fourier's (1772–1837) scheme, the phalanxes were designed to guarantee harmony: each one was to occupy three square miles, made up of the main building of the phalanstery itself housing 160 people and a surrounding park. Allen strips away all this detail to leave the pleasantly vague outline of a pastoral community ordering itself according to the unwritten rules of an anonymous Founder – who, one suspects, is probably meant to be Galton himself. The plot is elementary: Olive and Clarence, two young lovers, wish to marry despite Olive's slight physical weakness. They put this proposal as custom dictates to the hierarch and council of the phalanstery. Since there is no real impediment on eugenic grounds the marriage is permitted; but a child is born slightly deformed. After a statutory period the child is killed, again according to custom, whereupon the mother dies heartbroken. The real appeal of the story lies in the irony with which Allen develops his friend's serious proposal of eugenic infanticide, not in the trivial plot. It may possibly be objected that in any Western community, no matter how Utopian, the taboos against infanticide are so strong that no ironically rational defence of it can possibly rise above the level of farce – at least, in any less masterly hands than Swift's. But this is not so in the Utopian fiction of this time, where quite serious defences of infanticide and abortion on biological grounds are commonplace (as of course both were in real life, and were even covertly licensed in the institution of baby farms). Galton in *Kantsaywhere* stops short of it, probably believing that his elaborate Pass and Honours examinations for parental fitness could eliminate nearly all genetic abnormalities. Begetting an abnormal child in Kantsaywhere does not attract disapproval for the parents, but the unfortunate offspring is of course prohibited from mating, classified as Unfit, and removed to a Labour Camp if he dares to disobey.[50]

Elsewhere, though, there was no flinching from making the ultimate alliance with the observable law of nature. Among the icy, passionless people of Davis's *Pyrna* (1875) arrangements are Spartan but within their limits effective. Each baby is examined at birth and at once killed 'if not perfectly sound and healthy, or if it was monstrous or deformed in any way'. This procedure is held to be automatically beneficial – just as in Kantsaywhere or in Allen's phalanstery, the consequence is that no inhabitant grows up with 'a shattered frame, with a feeble pulse, with an aching head, or with that worst of all evils that is bred in a decaying civilisation, a nervously diseased constitution'. The anonymous visitor to Pyrna has, to be sure, a few residual moral qualms; but on hearing that infants have in fact to be put down in a ratio of less than one per thousand he is persuaded that 'like most terrors, this disappeared when it was approached'. In case we fail to appreciate the elevated principles governing the communards, the guide offers an elaborate appeal to natural law. Davis has nothing new to say about this flexible doctrine, but his vigorous defence of it binds him closely to the compact body of thought of the hard Darwinians:

> Everything that hath existence obeys one law – a law which has no limit, and knows no mercy. While nature lasts, while the world exists as it does now, that law will bind in its rigid clasp all created things . . . What has the neglect of obedience to this law entailed on you and your race over and over again? The decay of every system of civilisation . . . a system that preserves and breeds what nature requires should be exterminated.[51]

Those unimpressed by this fervent appeal could be solicited from another angle. The author of the anonymous *Etymonia* (1875), in describing life on a North Sea island – an idealised England – considers, only boldly to dismiss, Malthusian restraint as an answer to the twin evils of overpopulation and degenerating stock. Only selective child murder will work; and man's law is so much more humane than nature's:

> The sacrifice [of children], if but dispassionately, philosophically, aye, and charitably considered, would, in this form, be far less painful than hunger, – far more advantageous than a universal and constant struggle, – far less injurious than vice, – far more effective than chance, – far more limited in its operations than the innumerable aggregated evils attending upon the haphazard system of checks now in force.[52]

These attitudes, whose connection with those of Greg, Galton and even Darwin are self-evident, were certainly current; and there was no extravagance in Allen's satirising them. By making the people of his phalanstery disciples of Congreve and Harrison, like his young heroine Olive who tries to feel 'first and foremost of the progressive evolution of universal humanity' (p. 168), Allen is able to attack both the camp of the appealers to nature and of the 'humanitarians'. While no one in the community believes that the birth of a baby with distorted feet is anything other than a random stroke of fate and meaningless in itself, two sets of responses are possible, and Allen dramatises both of them. For Eustace, the phalanstery's physiologist, the pointless accident serves to reinforce his determination to follow through the methods of Darwinian nature to the end. As a stoic and spokesman for natural law, he welcomes the internal struggles which must go on for generations 'before the higher and more abstract pity conquers the lower and more concrete one' (p. 168). Like Davis's hero surveying Pyrna's society, Eustace sees the feelings of distaste for infanticide as nothing more than atavistic scruples outdated by a sounder biological ethic. Every victory of rationality in the community paves the way to the time when it will react against weakness and disease in the newborn as instinctively as a sow devours her runt. Cyriac the old hierarch, of more thoughtful temper, confesses to there being 'a great deal of the old Adam lingering wrongfully in me yet' (p. 168) – he knows he will never consent without inner conflict to the extinction of unhappy and imperfect lives. Between Eustace and Cyriac, and to a lesser degree between the earnest Clarence and Olive with her stirrings of idolatrous love, Allen builds up the same tensions which he also projects on to the world beyond the phalanstery. The mother's love for the afflicted child versus the higher ethics of the Founder are reflected in the eugenism of 'the highly civilised Euramerican countries' versus 'the old colonial societies' where cripples are suffered to live; and most concretely in 'the hearty, healthy, sound-limbed useful persons' (p. 169) versus the useless cripples, twisted also by the awareness of their misfortune. In the final scene, however, Allen relies completely on satire. After eighty days, after the opinions of three other physiologists have been secured and all hope of surgery abandoned, the child Rosebud is sorrowfully put to death – on the day renamed Darwin! – with the final invocation 'I release you, Rosebud, from a life for which you are naturally unfitted' (p. 175). After chloroform has come to the aid of nature, the drama ends with the death of Olive and ritual chanting to the name of Supreme Humanity.

'The Child of the Phalanstery' exists to show the true colours of biological utilitarianism by inverting the present conventions of behaviour towards the handicapped. We can only release the spring that Allen's story has wound up by reaching after the 'common sense' interpretation: that regenerate humanity will continue to cherish the Unfit even when applying the felicific calculus leaves no doubt as to where the greater happiness lies. Unfortunately Allen does not make his behavioural norm, 'the old antisocial prephalansteric days' (p. 163), sound sufficiently attractive to *be* a norm. To set up a clash between maternal emotion and the principles of a rational society is too easy; it does not help towards resolving the problems that crop up once eugenics is taken seriously. Allen wants to take eugenics very seriously, and thus we have the wavering of his ironic tone and the eventual collapse into melodrama.

Eugenics and Prevailing Theories of Inheritance

Allen's most inclusive target in 'The Child of the Phalanstery' is that most vulnerable part of the Galtonian eugenics, on which it was founded: the conviction that the mysteries of heredity had been thoroughly sounded. The birth of a pathetic cripple among a people who forbid marriage on such tenuous grounds as one of the partners having 'a distinctly bad or insubordinate temper' (p. 165) is the greatest ironic gloss of all on the footling activities of the biometric school. Yet when he writes positively Allen is prone, like the other reformist eugenists, to many of the Galtonians' misconceptions. He takes for granted, because Darwin supported it, the dysgenic effect of different reproductive rates in the different social classes. He assumes that recklessness and improvidence are inheritable. He is prepared to fling away as useless the operational definition of the 'Fit' but he is less than honest in refusing to redefine the 'Fit' clearly enough to permit debate about it. The Galtonians did at least see dimly that there was a teasing value judgement to be made; and according to taste they asserted the socially fittest to be the best intelligences, or the physically perfect, or those of the most approved beliefs, or those who measured up to any of a dozen other parameters. Though they did not openly call for clones of Lycurgus, Socrates or Galton, they could put out the call at once for the segregation and the purification of the model Fit, whatever the model was. Allen and Wallace shorted out the whole issue by reinstating sexual selection as the omnipotent eugenic agent. Hence the Fit became the descendants of those males taken as sexual partners by the mysterious Free Womankind. The whole subject thereby becomes esoteric.

The reformists and the Galtonians may also be linked by their totally incorrect theory of heredity. Properly, this is a matter for the following chapter; but one may anticipate just a little by noting that the very superficial view of heredity current at the end of the 1890s was far from being a minor academic confusion restricted to the lecture-hall and the technical journals. The rediscovery of Mendel's particulate theory still lay a decade in the future; and meanwhile prevailing opinion had it that, genetically, the child was a smoothly blended mixture of the inheritable characteristics of his parents. The appeal of this ancient idea was strong enough to secure the unquestioning acceptance of nearly all biologists, not just eugenists, from 1859 onwards. The mistake that heredity is essentially qualitative surely arose from the custom of describing the various observable degrees of inheritance in terms of blood-lines, consanguinity and such related concepts. Since blood is a fluid, and given its primitive status as the very essence of life, it becomes a reasonable proposition that 'blood' may be diluted or blended, fortified or polluted. When Galton, in assessing the dysgenic effect of centuries of sacerdotal celibacy on the intellectual community, speaks of a diminished stream of virtue 'flowing in the veins of Europeans',[53] his metaphor of intermingling streams of good and evil, capability and stupidity, reveals how effectively language had him in thrall. But the reformists, as good applied Darwinians, had no more adequate a basis to their daring proposals; and they, too, made some wholly unrealistic statements about society's failure to preserve its pool of excellence. Allen, bewailing the restrictions imposed on the endowed individual by the need to conform to monogamy, finds that in consequence his offspring must be 'a single stream of *diluted* and homogenous children, in whom his native excellencies are *masked and marred*' ('Girl of the future', 53; my italics). We are supposed, in our mind's eye, to see the various colours of inheritable skills mingled together on some microscopic artist's palette, with the distinct bright hues of talent muddied together into some grey mediocrity. In this respect as in others, Victorian eugenics as a whole may have been post-Darwinian in origin but it was pre-scientific in its conceptual equipment. Neither biologist nor man of letters realised this until well into the present century.

Notes: Chapter 5

1 Plato, *The Republic*, trans. H. D. P. Lee, p. 215.

2 Thomas Robert Malthus, *An Essay on the Principle of Population*, 6th edn, p. 269.

3 *Origin*, VI:98: 'I can see no difficulty in a race of bears being rendered, by natural selection, more and more aquatic in their structure and habits, with larger and larger mouths, till a creature was produced as monstrous as a whale.' Most people could see many difficulties, and in the face of great ridicule Darwin dropped this example after the first edition.

4 Letter to an unknown American correspondent, 14 March(?) 1861. So dated by George Sarton, 'Darwin's conception of the theory of natural selection', *Isis*, vol. 26 (March 1937), p. 340.

5 Francis Galton coined *eugenics* in his *Inquiries into Human Faculty and Its Development* (1883). His fullest definition is the one adopted in this chapter: 'the study of agencies under social control which may improve or impair the racial qualities of future generations either physically or mentally'.

6 G. K. Chesterton, *The Victorian Age in Literature* p. 94.
7 John Maynard Smith, 'Eugenics and Utopia', *Daedalus*, vol. 94 (Spring 1965), pp. 487–505.
8 Raymond Lee Muncy gives a full account of Oneida's eugenic experiment in *Sex and Marriage in Utopian Communities*, pp. 186–92.
9 T. H. Huxley, 'Evolution and ethics', in his *Collected Essays*, Vol. 9, p. 85.
10 T. H. Huxley, 'Evolution and ethics: prolegomena', in his *Collected Essays*, Vol. 9, p. 22.
11 ibid., pp. 36–7.
12 [William Rathbone Greg] , 'On the failure of "natural selection" in the case of man', *Fraser's Magazine*, vol. 78 (September 1868), pp. 353–62.
13 Alfred Russel Wallace, 'Human selection', *Fortnightly Review*, n.s., vol. 48 (September 1890), p. 325. Wallace records this as the gist of his last conversation with Darwin.
14 These two defences of Darwin from the taint of Social Darwinism are separated by more than thirty years, but the indignation remains much the same. They are by A. O. Lovejoy in 'Some aspects of Darwin's influence upon modern thought', in Franklin L. Baumer (ed.), *Intellectual Movements in Modern European History*, p. 126 (first published 1909); and by M. F. Ashley Montagu, 'Further comment [on Social Darwinism]', *Science and Society*, vol. 6 (Winter 1942), p. 74.
15 It is beyond our scope to enlarge on the reasons for making such a redefinition. R. J. Halliday offers several good ones in his 'Social Darwinism: a definition', *Victorian Studies*, vol. 14 (June 1971), pp. 389–405. The new definition has the advantage that a long line of opinion can then be traced all the way from the *Origin* to the American eugenists of extreme opinions, like Charles B. Davenport (1866–1944).
16 Charles Darwin, *The Descent of Man, and Selection in Relation to Sex*, 2nd edn, pp. 206, 216.
17 ibid., p. 206.
18 In a letter to G. A. Gaskell of 15 November 1878: *More Letters of Charles Darwin*, ed. Francis Darwin and A. C. Seward, Vol. 2, p. 50.
19 Karl Pearson, *National Life from the Standpoint of Science*, 2nd edn, p. 23. The lecture from which this book was developed was given on 19 November 1900.
20 ibid., p. 61.
21 Letter to Darwin, 24 December 1859; printed in Carlos Paton Blacker, *Eugenics: Galton and After*, p. 83.
22 Francis Galton, *Memories of My Life*, p. 323. The final pages of Galton's autobiography give his eugenic testament in its maturest form.
23 Francis Galton, 'Hereditary talent and character', *Macmillan's Magazine*, vol. 12 (August 1865), pp. 321, 327.
24 Francis Galton, 'Hereditary talent and character', *Macmillan's Magazine*, vol. 12 (June 1865). Certainly Galton did not overestimate the efficacy of the technique he is advocating. J. B. S. Haldane once calculated that the rate of change of a characteristic under continuous artificial selection might exceed the natural rate of change by up to 300,000 times! See his 'Natural selection', in P. R. Bell (ed.), *Darwin's Biological Work: Some Aspects Reconsidered.*
25 Francis Galton, *Essays in Eugenics*, p. 2. This lecture was read before the Anthropological Institute on 29 October 1901.
26 'Nunsowe Green', *A Thousand Years Hence: Being Personal Experiences as Narrated by Nunsowe Green*, p. 206.
27 According to Galton in a letter of 23 October 1910, the completed manuscript 'would now fill about 20 to 30 pages of *Nineteenth Century* size'.: Karl Pearson, *Life, Letters and Labours of Francis Galton* (hereafter *Life*), Vol. 3B, p. 613.
28 Letter of 6 December 1910: *Life*, Vol. 3B, p. 615.
29 Letter of 28 December 1910: ibid., Vol. 3B, p. 616.
30 The fragments are published with some critical commentary in ibid., Vol. 3A, pp. 414–24. Further references are in the text.
31 Galton gives as areas for the girls' examination 'grace, beauty, wealth, good temper, accomplished housewifery, and disengaged affections'. It is not clear if these are in order of priority: 'Hereditary talent and character' (August 1865), p. 165.

32 This evaluation of Galton is evident throughout Blacker's *Eugenics*. L. S. Penrose calls him 'a humanitarian' in 'Ethics and eugenics', in Watson Fuller (ed.), *Social Impact of Modern Biology*, p. 94.
33 Blacker, *Eugenics*, p. 122.
34 H. G. Wells, *The First Men in the Moon*, pp. 229–30.
35 H. G. Wells, *Anticipations of the Reaction of Mechanical and Scientific Progress upon Human Life and Thought*, new edn, p. 306.
36 In his *Experiment in Autobiography* (1934), Wells speaks of Allen's 'very pronounced streak of speculative originality' (Vol. 2, p. 546) but he does not elaborate further on the degree of creative affinity between them. Certainly they shared the freedom (and the isolation) of existing on the fringes of a discipline, and the published work of both is also evidence of their joint desire to subjugate social theory to biology. Both wrote for the *Pall Mall Gazette* in the 1890s and presumably met through their common literary agent, J. B. Pinker.
37 H. G. Wells, *A Modern Utopia* (London, 1905), p. 163.
38 Alfred Russel Wallace, 'Human progress: past and future', *Arena*, vol. 5 (January 1892), p. 148.
39 Wallace, 'Human selection', p. 333.
40 Wallace, 'Human progress', p. 157. The 'reformed society' to which he refers is modelled on Edward Bellamy's future Boston in *Looking Backward* (1888). The eugenic effect Wallace describes, however, is purely his own invention. W. Warren Wagar, in his large study *Good Tidings: The Belief in Progress from Darwin to Marcuse*, mentions Wallace's eugenic proposals but incorrectly dates them 1913 instead of (at the latest) 1892. The point is of some importance because the earlier date disconnects Wallace from the political struggle for emancipation, for which he probably had little sympathy, especially in its more activist form.
41 Richard Le Gallienne, 'Grant Allen', in his *Attitudes and Avowals: with Some Retrospective Reviews*, p. 190.
42 Allen's important review of Weismann was for the *Academy*, vol. 37 (1 February 1890), pp. 83–4, and is considered further below. There is a full account of the part played by Allen in bringing Watson into the public gaze in James G. Nelson, *William Watson*, pp. 19 ff.
43 Edward Dowden, 'The scientific movement and literature', in his *Studies in Literature, 1789–1877*, p. 90. On this assumption Dowden proceeds to build a case for the evolving structure of literature having a partly biological basis. This issue of the development of colour vision was given a generous airing in the Victorian periodicals, and the literary evidence was well mustered by the orientalist Max Mueller. The whole thoery is now generally considered to be exploded. Allen's *The Colour Sense* (1879) was accepted by Trübner's on Samuel Butler's recommendation, which may in part explain Allen's fairly magnanimous treatment of his 'evolution' books (see below).
44 Letter from Huxley to Allen, 2 May 1882. Quoted by Edward Clodd, *Grant Allen: A Memoir . . .*, p. 112.
45 Grant Allen, '*Evolution, old and new . . .*', *Academy*, vol. 15 (17 May 1879), pp. 426–7.
46 George John Romanes, 'Unconscious memory . . .', *Nature*, vol. 23 (27 January 1881), pp. 285–7.
47 Grant Allen, '*Essays upon Heredity*' (review of Weismann's *Essays*), *Academy*, vol. 37 (1 February 1890), pp. 83, 84.
48 Grant Allen, 'The girl of the future', *Universal Review*, vol. 7 (May 1890), p. 61.
49 Grant Allen, 'The child of the phalanstery', first appeared in *Belgravia*, vol. 54 (August 1884), pp. 163–7, under the pseudonym 'J. Arbuthnot Wilson'. It was reprinted with slight revisions in *Twelve Tales*, pp. 45–64. Quotations in the text are from the earlier version.
50 Galton does not allow for any exceptions to these general edicts. Yet, incongruously, in the next paragraph he piously maintains that 'the child should be in no way discouraged on account of its natural defect'! *Life*, Vol. 3A, p. 416.
51 [Ellis James Davis], *Pyrna: a Commune; or, Under the Ice*, pp. 105–12.

52 *Etymonia*, p. 199. It is not, perhaps, entirely redundant to confirm that both this author and Davis are perfectly serious in their proposals.

53 Francis Galton, *Hereditary Genius: An Inquiry into Its Laws and Consequences*, 2nd edn, p. 411. The phrase is in both the first (1869) and second (1892) editions.

6

Nemesis without Her Mask:

Heredity before Mendel

> By revealing to us the absolute mechanism of all action, and so freeing us from the self-imposed and trammelling burden of moral responsibility, the scientific principle of Heredity has become, as it were, the warrant for the contemplative life. It has shown us that we are never less free to act than when we try to act. It has hemmed us round with the nets of the hunter, and written upon the wall the prophecy of our doom. We may not watch it, for it is within us. We may not see it, save in a mirror that mirrors the soul. It is Nemesis without her mask. It is the last of the Fates, and the most terrible. It is the only one of the Gods whose real name we know.
>
> OSCAR WILDE (1891)

Oscar Wilde, in the dialogue on art criticism from which the epigraph to this chapter is drawn, attempts to make the fearful facts of human heredity as the late nineteenth century revealed them a subject of aesthetic contemplation; to turn them into art.[1] Despite his confident definitions we may well suppose that in reality he knew little and cared less for the specific difficulties which biology found itself in by the early 1890s; yet shorn of its flamboyances his strongly emotional attitude was shared by his more sober and better-informed peers. Like Wilde, all manner of folk from T. H. Huxley to Zola found an almost hypnotic appeal in the prospect of reducing to simple laws even the most complicated patterns of inherited behaviour. A few found this a cheering prospect. Huxley claimed to find nothing repugnant in the new biology reducing him to an organic automaton, just as long as the cellular computer could be programmed so that he would automatically do good. Wilde himself was eager to be released from the moral responsibility of his actions. But many more were genuinely afraid of learning too much, about things with which the human spirit might not be able to cope. Taine asserted that vice and virtue are just as much the products of organic chemistry as vitriol and sugar. Where might this lead? At any moment the social pyramid might be undermined by some ultimately persuasive demonstration that the fault did, after all, lie in the protoplasm and not in ourselves; that it might really make as much or as little sense to do as the Erewhonians did, and punish people for being tubercular and to sympathise with their criminal acts. Every one of the significant Victorian advances in the life sciences struck a fresh balance between freedom and necessity, but over the burning issue of the mechanism of heredity the scales of informed opinion swung more violently and took longer to reach equilibrium than over any previous biological hypothesis since that of evolution itself. And what is also true, and most relevant to our present concern, is that the debate on heredity reached out and engaged the literary sensibility more radically than ever before. The engagement was, however, just one particular outcrop of a broad public interest in heredity, which was in turn the consequence of the peculiar status which the topic occupied in turn-of-the-century biology.

The problem of how human characteristics are transmitted from one generation to the next has always been given an answer of one sort or another; but as a scientific problem it was not faced squarely before the perfecting of microscopic techniques and the rise of embryological studies with the work of K. F. Wolff about 1760. Only then did it become clear that, however the information of inheritance is packed, the spermatozoon is observably not just a minute version of the adult. For the most popular theory of the seventeenth and early eighteenth centuries had been that the total number of individuals who were ever going to be born from the Creation to Armageddon were packed into the primordial egg like a series of Chinese dolls. 'Dans cette hypothèse,' one summarist of this strange theory tells us, 'il n'y a plus création, ni formation, mais Evolution de germes préexistants où toutes les parties de l'être futur sont indiquées par avance et n'ont qu'à grandir et à se développer'.[2] The proponents of *emboîtement* were not of course troubled by any considerations of limits on the divisibility of matter; and this makes Delage's prints of early drawings (allegedly taken from life) of spermatozoa which illustrate this layered arrangement of mannikins specially quaint, as well as incidentally supplying good examples of preconceptions controlling observation.

Pari passu, the same strictures apply to the theories of inheritance current at the end of the nineteenth century. Not only were these numerous and short-lived – Delage lists no less than thirty different systems current in 1896, few of which survived far into the next century – but also their advocates changed allegiances with bewildering rapidity. A biologist-philosopher like Hans Driesch (1867–1941), for example, began as a pupil of Ernst Haeckel, a fierce evolutionary materialist who had an explanation for heredity based on the mechanical vibration of the muscles; later he worked with August Weismann, the founder of the neo-Darwinian school, in 1886; yet he lived to become an aggressive vitalist, to whom Darwinism was 'a discarded theory . . . an insult to the reader'.[3] As was the case with several other of the theoreticians in this field, Driesch's early training among the Darwinians made him a formidable opponent and his criticism very highly pointed. He handled Darwin with particular roughness over his treatment of the problem of variation. It was a keystone in the original natural selection hypothesis, of course, that no saltatory factors like mutations were of any significance, and Darwin was thus forced back to rely on the exclusive action of individual differences. Darwin could not satisfactorily explain why these occurred, either; but he also knew as an empiricist that the only real scientifically persuasive knowledge is that of cause and effect. Convinced that the enormous effects of variation aided by population pressure could account for all evolution, Darwin at first simply assigned the cause of variation to chance. In most contexts this is a disagreeably ambiguous term, and even after Darwin had tried to refine it two meanings can still be distinguished in his writings. Sometimes Darwin obviously wished it to be translated as 'ignorance', and therefore not to be taken as a *vera causa*. This is how T. H. Huxley always took it: 'because no law has yet been made out, Darwin is obliged to speak of variation as if it were spontaneous'.[4] But at other times Darwin and especially some of his followers went farther to speak of variation as a truly random process, conforming to no discernible law at all. They spoke of it as unwilled, unplanned, as not being 'called out' by any specific circumstances, as not moving forward to any observable end. For this second assertion the Darwinians had no proof. As the negative phrasing in their definition suggests, the only value of the assertion of randomness was simply in its lying at the most distant possible remove from teleological explanation. Placing on one side the actual truth in this matter, the ambiguity in the Darwinian view of heredity is still of great interest because while it existed it deflected many theorists into unverifiable speculation against which no observable facts might be set.

Darwin himself, having laid down the cause of variation to chance or to unknown factors, then did his best to fill the gap by presenting a theory, pangenesis, which essentially is no more than the reworking of an ancient Greek idea. Pangenesis represents Darwin's final movement into a position where he accepted the workings of 'soft' factors in heredity, though it came surprisingly early in his career – it may even pre-date the *Origin*. His first public account of it forms the twenty-seventh chapter of *The Variation of Animals and Plants under Domestication* (1868).[5] According to this hypothesis, all parts of the body – or the cells which constitute it – manufacture and then throw off particles (the so-called 'pangenes' which obviously owe something to Spencer's 'physiological units') which move through the bloodstream and conglomerate in the reproductive organs. Here they become the components of heredity in egg and sperm. Darwin did not imagine his pangenes to be themselves cells, but only to be capable of constructing or turning into cells. He further imagined that the actual quantity of pangenes produced by each organ of the body was controlled by the activity of that organ or by its development, or by the intensity of the environmental pressures upon it. To explain the transport of pangenes to the reproductive organs, Darwin spoke of a mysterious 'mutual affinity' between them; possibly he supposed that the ova and spermatozoa are themselves just aggregated packets of pangenes, for he certainly shared in the common mistake of his time about the physiology of reproduction in believing that the whole mass of sperm is the fertilising agent and that the sex of the embryo and the resemblance to its father will depend on the amount of sperm released. One immediate difficulty with pangenesis, which Darwin was very much alive to and tried to deal with at once, is that it would in action be too grossly Lamarckian to square with common observation. It is not *only* Lamarckian, for Darwin supposed that the rearrangement and redistribution of pangenes could explain the blending of characteristics, along the lines of modern genetic theory; but certainly, if the theory were true, one could expect that the inherited acquired skills and other modifications passed from parent to child would swamp all others. To resolve this difficulty, Darwin's further supposition was that most pangenes remain dormant and exercise no influence on inheritance, the ones that do being only those from the most mature parent cells. This, though ingenious, does not disguise the fact that natural selection and pangenesis do tend in opposite directions conceptually; nor that Darwin's theory, like Samuel Butler's, was an armchair theory without any physiological backing. The merit of pangenesis derived solely from the reputation of its originator and in itself had no firmer empirical basis than any of the dozens of other theories popular at the time. The only experimental test done for pangenetic effects – by Francis Galton, who tried transfusing blood (*ex hypothesi* containing pangenes) from one variety of rabbit to another – naturally yielded negative results.

Francis Galton, whose whole career as a eugenist was shadowed by his inability to resolve the issue of particulate versus blended heredity in favour of the former, well conveys the general imbroglio in this part of biology in the 1880s when, looking back in his autobiography from 1908, he recalls how

> the subject of human heredity had never been squarely faced, and opinions were lax and contradictory. It seems hardly credible now that even the word heredity was then considered fanciful and unusual. I was chaffed by a cultured friend for adopting it from the French.[6]

And if that were already true by 1908 how much more difficult it is now to reconstruct the clashings of those ignorant armies! There is, after all, nothing quite so dead as dead science; nor anything quite so unappealing as quarrels over what were, as it turned out, inadequately substantiated facts. And yet to understand some of the literary activity which was set in motion by ideas of inheritance we must think ourselves into that group state of mind which appears even to the most capable scientific historian of the period to be one of 'documented qualification and nagging doubt'.[7]

To try to capture something of that atmosphere we cannot do better than refer to the retrospective account of one who lived through those days in active engagement but who never occupied a position of extreme partisanship. The career of William Bateson (1861–1926) fully spans the relevant period, and in the later 1880s and the 1890s he was a prime mover in establishing the exact limits of the variation problem even though he made no real contribution to its solution. Bateson began as a Lamarckian; but not (unusually) as a Lamarckian content to argue his case without extensive fieldwork. He abandoned it with some reluctance only after he had convinced himself that there was no evidence for it in the wild. Later on he was instantly receptive to the rediscovery of Mendelism – there is an anecdote of him completely rewriting a lecture in Mendelian terms while in a train *en route* to deliver it, after he read of that rediscovery in a journal – and in his spirited defence of Mendelism (1902) he refers to the occasion as 'a moment of rejoicing'. Though much the larger part of this defence turns on highly technical questions and the need for it has long since passed by, the preface, where Bateson looks back on the frustrating pre-Mendelian days, is particularly evocative. Bateson's usually spare prose takes a metaphorical turn as he describes how progress in theoretical biology had ceased and how 'the more prudent' researchers left for fields 'where the harvest is less precarious or the yield more immediate'. We get a strong impression of the intellectual sterility surrounding the active young biologist in the 1880s, and the impression is reinforced as Bateson goes on (probably with one of his own arduous and disappointing field expeditions in mind) to speak of those admirable souls who remained to push through 'the jungle of phenomena' in contrast to those who were, he concludes bitterly, 'content supinely to rest on the great clearing Darwin had made long since'[8] – an image combining imputations of laziness, arrogance and a slack readiness to confuse partial formulations with definite solutions. Nor did the passage of several more years exhaust Bateson's lode of scorn. In an article commemorating Darwin's birth he returned to the attack, now abandoning figurative expression for the round assertion that one of the 'melancholy sequela' of Darwinism was 'the process of attrition' in the stock of evidence about variation and heredity. Should you be interested in the prehistory of genetics, warns Bateson, you will have to do more than thumb through the back copies of *Nature*: there is 'little to reward [your] quest in the leading scientific Journals of the Darwinian epoch'.[9] Now, the charge is a good deal more serious than of idleness among Darwin's apprentices. It is the one which we have already several times mentioned: that Darwinism with its appeals to some nebulous 'chance' provoked a great deal of woolly theory but little experimental work. It is no wonder that Mendelism, once it had become common currency after 1900, produced a great sense of intellectual release in those like Bateson who were trained in the biometric school of Galton, Pearson and Weldon. Ernest Jones, Freud's biographer, reflecting on his early life at the turn of the century, has spoken of the 'leap from the sombre Galton–Pearson period into the bright day of Mendelism'.[10] The oddity is that the 'leap' had in reality been made thirty-five years before.

Gregor Mendel's experiments showing particulate inheritance in peas (begun 1856; first published 1865) effectively superannuated all previous speculation about the mechanism of inheritance, by dispensing with what Garrett Hardin has called the 'paint-pot' or blending theory[11] which Darwin and the eugenists had taken for granted, and thereby placing natural selection on a firm statistical basis. Mendelism and Darwinism, the two prime axioms of modern biology which perfectly interlock with each counteracting the deficiencies and limitations of the other, came to fruition in the same decade. The baffling problem of why Mendel's important paper 'Versuche über Pflanzenhybriden' should have been neglected for so long, and the equally baffling fact of three independent researchers duplicating his results in the single year 1900, continue to haunt the historian of ideas and to intrigue those desiring to integrate scientific discovery into one prevailing cultural *Zeitgeist*. Fancifully, one might almost suppose that the pressure to make some sense out of the confusion in biology grew so intense that the enigma suddenly capitulated on three fronts simultaneously. Many more prosaic solutions have been proposed, ranging from the obscurity of the original paper in the *Transactions of the Brunn Natural Science Society* to the blunt accusation that Mendel or an assistant faked certain critical ratios to accord with a predicted distribution of characteristics, which made strict repeatability impossible.[12] None of these really satisfies. After a most careful review of the whole problem, Elizabeth Gasking arrives back at a simple generalisation with which we, holding in mind the literary trends of this period, can wholeheartedly concur: 'one is forced to conclude that Mendel was ignored because his whole way of looking at the phenomena of inheritance was foreign to the scientific thought of his time . . . [while in 1900] his problems were the problems of the time'.[13] The first half of this statement can in fact be sustained purely in terms of biological history. First, one of the few contacts Mendel had in the larger scientific world, Karl Nägeli of Munich, let him down badly by condescendingly laying aside his statistical treatment not as false but as simply trivial; and in fact Nägeli's massive synthetic volume of 1884, devoted entirely to problems of inheritance, does not mention Mendel's name once.[14] Second, all the evidence suggests that even had Darwin had Mendel's laws before him as he prepared the fourth edition of the *Origin* he would not have altered his conclusions one jot and certainly would not have retreated from that part of his explanation for evolution based on the use and disuse of organic function. This is not unsupported speculation.[15] Some experiments on plant hybridisation other than Mendel's were carried out in the early 1860s by Charles Naudin (1815–99) and, though he obtained no results consistent enough to be dignified as 'laws' he discovered enough to publish a testable hypothesis of particulate inheritance. Darwin rejected it in a letter of 13 September 1864.[16]

We do not have to search for the reason. Darwin was firmly caught in a cleft stick. One the one hand, any theory of blending inheritance was, after Fleeming Jenkin's critique, fatally damaging to natural selection

(Mendel was no evolutionist, but he did note that his laws solved the Darwinians' 'paint-pot' problem); on the other, Darwin had already committed himself irretrievably with pangenesis to a highly fluid view of variation as a phenomenon modifiable in many ways – and Naudin's findings hinted at indivisible units, 'atomic' in the original sense. Darwin, then, was in the extraordinary position of being unable even to consider the explanation which removed a major criticism because it flew in the face of another, supplementary, hypothesis. The Darwinian school took its cue from him, and the stage was set for a whole generation of hotchpotch conjecture.

The reasons why Mendel was unacceptable in the 1860s therefore accord well enough generally with our knowledge of the selective blindnesses of that time. The second half of Gasking's résumé, the 'rightness' of the *fin de siècle* as the incubator of more empirical studies of heredity, offers a lot more resistance to being pinned down by a few extracts from scientific history. Even to attempt a gloss would mean pushing well on into the present century to a time when Mendelism had undergone a mathematical transformation and its relevance to evolutionary theory had been fully delineated. Here we must content ourselves with the observation that, quite apart from Mendel's work, the end of the nineteenth century marks another transformation in ways of thinking about heredity. In 1903, Theodor Boveri and Walter Sutton, working on the recently discovered chromosomes, proved that these do indeed carry information in some inconceivably condensed form between parent and child. They went on to show, in a whole series of exceedingly delicate operations on sea-urchin eggs, that the movement of chromosomes during fertilisation actually accords with the statistical abstractions of Mendel; and in the same year Sutton postulated a series of sub-microscopic 'genes' strung out along each chromosome. At first glance, there may seem to be only trifling differences between Sutton's new gene and the earlier pangene, id or plastidule. Was this not just another ghostly concept? It was not, because attitudes had changed. Sutton, just like his Victorian colleagues, knew that to get as far as the facts one has to step beyond the facts. But, unlike them, he extrapolated only one step at a time and waited for observation to catch up. He was not interested in armchair experiments, and in that of course he represents the modern mood. The opposing theories of inheritance of Samuel Butler and August Weismann, to which we now turn, represent the former tradition of reaching for the truth – a manner Platonic rather than Aristotelian. Just as Socrates, in his famous demonstration, dipped into the mind of an illiterate slave and drew forth the fundamental manipulations of geometry and arithmetic, so Butler and Weismann (to be fair, the first much more than the second) believed that they could reach within themselves for *a priori* explanations and expect nature to conform with them.

Butler on Heredity

> [*The Origin* is] a book for which I can never be sufficiently grateful, though I am well aware how utterly uncapable I am of forming any opinion on a scientific subject which is worth a moment's consideration.[17]

Samuel Butler wrote this sentence in 1872: today it must strike us as one of the most unconsciously ironic that that master of irony ever penned. Perhaps in 1872 he did believe its sentiments to be true, for in that year he was known not as a scientist but as the writer of occasional essays, of a book on being a colonist in New Zealand, and of a popular and clever satirical romance, *Erewhon*. Certainly there was on this past form every reason for his correspondent, who was the ageing Darwin himself, to take the above statement at face value. Yet within two years of making it Butler had become fascinated by the most profound puzzle of his age; within six years, obsessed; within eight, had set his face against every competent biologist alive, had abused the revered Darwin personally, and had accused him wildly of the most wretched chicanery, plagiarism and fumbled attempts at fraud. By the year 1890 he had conceived and presented the original ideas of his first treatise, *Life and Habit* (1877), had gone on to write three further works on related topics, and had finally published a connected group of essays under the title of 'The deadlock in Darwinism'.[18] The man of letters incapable, according to his own words, 'of forming any opinion . . . worth a moment's consideration' had transformed himself into a self-taught scientific controversialist for whom speculative biology was the dominant interest of life throughout his most energetic years. We notice, however, that it was never the sole interest and was not one that lasted till death. Butler's 'evolution' writings (that term is much too limiting, but serves as shorthand) fill only four and part of a fifth of the twenty volumes of his *Collected Works*. In his last decade Butler turned his attention and his well-oiled argumentative powers to quite different problems which obliterated any residual interest he may have retained in biology. There may even be a touch of pettishness in a remark to Marcus Hertog in November 1891 that he felt himself to be 'too busy on other matters to meddle with biology just now'.[19] The dates just quoted are critical ones. They at once limit what we can expect of Butler's theories; that is to say, his 'evolution' work fits in its entirety into Peter Vorzimmer's long period of 'nagging doubt'. Butler, by withdrawing his interest at just the time he did, isolated himself by accident or design from Bateson, from Weismann, from the reappraisal of Mendel. While these men were publishing and drifting towards the bright centre of disputation Butler was off hunting that Snark of his later years, the Authoress of the *Odyssey*.

When considered as what in fact they are, a single extended essay in controversy, the evolution books have no even texture. The best-written is undoubtedly the first, *Life and Habit*, where the thought is freshest, Butler's odd wit thoroughly meshed with his material, and the entertaining illustrations are carried over from his earlier exploitation of topsy-turveydom for its own sake. Most of his exegetes quote heavily from it and ignore that most mature and deliberate piece of argument *Luck, or Cunning?* (1886), which is at least as good. The two intermediate books were written hastily and are little more than rag-bags stuffed with replies, denunciations, oddments of translation from sympathetic German biologists (Butler had a much better press there than in England), and even fragments of autobiography. They supply nothing that is not available in livelier form elsewhere. But to open any single essay or volume by Butler with the expectation of being able to extract therefrom a considered and stable point of view is to be disappointed. We are ushered instead into a wildly untidy storeroom full of fragmentary notions which have a treacherous solidity but which disintegrate as soon as they are grasped. The reader's perceptiveness is not at fault. Rather, it is Butler conducting his own education in public by a curious pedagogical technique which might be described thus: he does not, as he adds to his stock of knowledge, gradually extend outwards an area of what is for him certainty. Instead, having selected a topic intuitively, he rotates it steadily before his mind contemplating each facet in turn – even incompatible facets – as though it alone expressed the truth. Then, over the years, we find him stripping away the unacceptable and the inconsistent until his truth emerges bare and clean. And it does so emerge, given time: 'I was not sure that my *Life and Habit* theory was more than ingenious paradox,' wrote Butler years later, recollecting his first days incubating that book, 'but by February 14th [1876] I had gripped my meaning and knew it to be sound.'[20] Often moving from paradox to assurance took a lot longer than that; and only when we range over his whole career, recognising the excrescences for what they are, does a simple and consistent pattern emerge.

It has not been observed before that Butler's so-called 'theory of evolution' is in fact made up of two strands of ideas. One of these had imaginative potential for Butler himself; the other not, although effectively made use of by others. The strands are, first, the theory of human inheritance proper which we will take up in a moment; the second a speculation about future human evolution, which must be dealt with summarily here. The two strands were never woven into any intellectually satisfying union, though both persist (rather bewilderingly) into Butler's last novel.

Nothing tells more about Butler's method of first twisting about all the possibilities latent in a notion (to the confusion of his public) and then pursuing only the useful ones than his prognosis of humanity's future. Yet the sequence is really easy enough to trace, and permits us to see in detail the gradual solidification of Butler's real, serious, view. His two articles in the Christchurch *Press* (1863–5) comprise the first stage. In the first, which is disguised as a letter to the editor and replete with bogus concern, the pseudonymous correspondent proposes machines as the natural inheritors of the planet: just as vegetable life developed from minerals, so is the mechanical kingdom beginning to lord it over the organic. Humans, by placing more power in machines and relying more upon them, are conniving at their own obsolescence: 'in the course of ages we shall find ourselves the inferior race'.[21] Here Butler has certainly not escaped from the spell of the *Origin* (he may have read it only the year before) and, rather than trying to establish whether he is 'serious' or not, it suffices to recognise that the logic of the piece is exclusively Darwinian. Beleaguered man is offered only two choices: capitulation to the machine or war to the death. There is nothing obviously parodic about the essay at all, any more than there is about the following piece written in London but also published in the *Press*, or the third and final piece, 'The mechanical creation', which supplements it. The second *Press* contribution, couched as a reply to the previously distraught correspondent, puts forward machinery not as the natural enemy of humanity but as humanity's symbiotic partner. Instances of the coming 'machinate' phase of evolution are already about us: a man's watch is more use to him than most of his hair; a notebook in the pocket substitutes for an infallible memory. We are already witnessing in the gulf that is opening between rich and poor a division into two distinct sub-species of man. The rich man 'can tack a portion of one of the P. and O. boats on to his identity' and make the poor regard him with the 'affectionate reverence' a dog feels for man.[22] Butler claims to be passing no moral judgement ('we do not say that the thousand-horse man is better than a one-horse man, we only say that he is more highly organized and should be recognized as being so': p. 220) but this is disingenuous, for in reality he is already having as much trouble as the Darwinians in keeping evolution and ethics apart. Later on, when, paradoxically, he had thrown over Darwinism, he succumbed entirely and spoke freely of the rich as the chosen medium of expression of the evolutionary Purpose. For, if his vision of man's prosthetic future was ever a joke at all, it very rapidly ceased to be one. The final vision is specially elaborated in a perfectly serious chapter of *Luck, or Cunning?* entitled 'Property, common sense and protoplasm'. Among many similar examples he suggests that we see our stomachs as purses for storing nutriment and, by the same token, our purse as 'a kind of abridged extra-corporeal stomach wherein we keep the money which we convert by purchase into food'.[23] The inversions are reminiscent of similar ones in *Erewhon*, but the tone is more like that of a textbook. Butler has found his truth, has 'gripped his meaning'.

Butler's theory of inheritance pursues quite a different line and ramifies outwards far more thoughtfully than the unsophisticated, mildly comic sequence of ideas which we have just traced. There, Butler has tried to give us a series of snapshots illustrating some ways in which the destiny of man might be involved with the machine. It is accomplished journalism, but no more than that. But the inheritance theory analyses in quite exact detail a troublesome problem in theoretical biology. It takes rational notice of rational objection and is offered, as we shall see, as a real contribution to group scientific enterprise. To be sure, both speculative sequences of ideas have their origins among Butler's first literary work. There are quirky ideas about heredity even in *Erewhon*, in the chapters 'Birth formulae' and 'The world of the unborn'. Earlier than that, indeed; for Butler tells us in his preface to the revised edition (1901) of *Erewhon* that these two chapters were written, and may have been obscurely published, as early as 1865.[24] But the theory that is expressed in *Life and Habit* already has a firmness of outline, an attention to detail and to potential criticism which together bespeak hard thought and constant rewriting. Common sense tells us that such a theory did not spring from Butler's forehead fully armed, but the only indisputably accurate description of its writing – by Butler himself in a long letter to Francis Darwin (Darwin's third son, who was for a short time an acquaintance of Butler's) on the eve of the publication of *Life and Habit* – is not only vague about sources but also poses some problems about the exact nature of Butler's disagreement with Darwin. As this letter is germane to the present line of argument, and as its significance has not been generally noticed, a few essentials will be extracted from it.[25] At the time it was written Butler's main concern was the offence which Darwin might find in the newly completed *Life and Habit*. He begins by asking Francis Darwin to evaluate it before passing on a copy. He then continues:

> it has resolved itself into a downright attack upon your father's view of evolution, and a defence of what I conceive to be Lamarck's. I neither intended nor wished this, but was simply driven into it. (*Memoir*, 1:257)

He goes on to tell how the impelling force came about two months earlier (from internal evidence, this must have been on 23 or 24 September) when *Life and Habit* was already 'three parts done'. On looking into the latest edition of the *Origin* – this was the sixth, published five years before – to check what Darwin had to say on the causes of variation, Butler was confronted with Darwin's maintaining it to be a 'serious error' to believe that very many habits are directly transmitted (he does not, of course, say that *no* habits are transmissible). Butler describes his feeling of shock and how he was inclined to burn the manuscript of his book at once. Very soon, though, he saw his way afresh. He would have to break with Darwinism altogether. Therefore he simply 'cut out all support of natural selection and made it square with a teleological view' (1:258). All of these readjustments must have occupied less than two months.

What do these remarks teach us about the successive stages in the construction and application of Butler's theory? First and most obviously, they modify the usual impression of Butler as the fierce and highly original anti-Darwinian theorist who positively enjoyed setting about the biological establishment. On the evidence of this letter Butler first conceived his theory of inheritance as a supplement to the theory of natural selection. In offering such a supplement, in trying to find a true cause of variations, Butler was doing no more than tackling the same problem which was fully occupying Romanes, Spencer and of course Darwin himself. In this sense, then, he is in the main line of theoretical development and in no sense a renegade thinker. Second, Butler, despite his early lucid précis of the *Origin*'s argument, had not kept up with the gradual drift of the *Origin* back to use-inheritance and related spurious explanations for variation. Third, practically all the really substantial objections to Darwinism were taken with a single grab from Mivart's *Genesis of Species* (published six years before), which Butler had rapidly decided was on certain points unanswerable. Fourth and most relevantly, the date of the letter offers the intriguing possibility that Butler, in the light of the new neo-Lamarckian principles to which he had exposed himself, changed his plans for his new novel which was to become *The Way of All Flesh*.

Butler finds it convenient in *Life and Habit* to direct the reader first to the three levels of human attention and to make him see these as habits (that is, as 'learned' involuntary behaviour patterns) which have been relegated with greater or lesser success to the unconscious. At the deepest level below the threshold of consciousness proceed such autonomous habits as the cellular defence mechanisms or blood circulation which cannot be brought under voluntary control even if we wish. A little higher in the level of attainment exist such subconscious habits as seeing, hearing, walking, piano-playing and so on – skills which at some early stage had to be deliberately acquired and practised before they could be 'forgotten'. Finally, on the surface for most people, are such deliberate activities as doing mathematics, social routines and decision-making which demand a continuous input of conscious mental energy. That these levels exist and are not clearly demarcated Butler thought conclusively proven by the capacities of child prodigies and calculating wizards: such people can draw on, as low-level habits, skills which for the majority have to be painfully deliberate.

Butler quickly noticed that these psychological facts about habits (which he did not of course discover) were comparable with other facts about heredity; and he quickly went on to suppose that the relationship was not in fact a mere analogy at all, but a true affinity or even identity. These points of resemblance which especially struck Butler may be conveniently summarised in a table which tries to synthesise statements from the four evolution books stripped of supportive examples:

	Memory	*Heredity*
1	The most recent impressions are recalled most readily.	Offspring resemble most closely their immediate forebears.
2	Memory works by associating items of information which, once fixed, may only be recalled in complete trains (e.g. whistling a tune).	*In embryo*, offspring recapitulate the entire history of organic development.
3	Deeply engrained habits (not low-level habits like breathing, but fixed patterns of activity) need the occasional shock of a new experience to prevent their becoming obsessive.	Most species (but not humans?) benefit from hybridisation.
4	But a wholly unfamiliar situation, for which no habits are appropriate, can be disruptive or even fatal.	Inter-species sterility.
5	A perfected habit is totally unconscious and not subject to rational appraisal.	Some animal instincts are inappropriate to new conditions, or even lethal (e.g. migration of *Lemmus lemmus*).
6	Some disorders of memory like fixations and obsessions and phobias are characterised by the upheaval of old, forgotten and obsolete memories.	Atavism.
7	Creativity is the recombination of memories in new patterns until a satisfactory or effective one emerges.	Variation.

These parallels, weak though some of them are, would be entirely meaningless if Butler had not made two further assertions, one of which he defended vigorously (the issue threatens, in fact, to become the sole issue), the other of which he never questioned the truth. In order to offer heredity persuasively as inherited habit, Butler's theory required him to see all life as continuous; as being chopped into arbitrary units called individuals to suit our immediate convenience alone. It must at once be stressed that this insistence on the essential identity of all life (he traces the sequence back from octogenarian adult to newborn child, to embryo, to gamete and so to that primordial cell 'which has differentiated itself into the life of the world, all living things whatever being one with it, and members one of another'[26]) is in no sense a mystical or pantheistic insistence. The linkage he has in mind has nothing to do with reincarnation: it is purely physiological. All bodies, he is saying, are of one flesh; for how else could memory persist? Given this, each life-cycle is but a repetition of a habit built up by this same 'individual' over every preceding cycle; and heredity is but what he can recall, or have others point out to him, of his experience in his earlier manifestations. Should new and unforeseen circumstances arise in which memory cannot be trusted, the organism must turn to its own resources, must act with whatever consciousness it can muster, and must either adapt or be obliterated. Butler gives us an amusing but highly fanciful example of the second contingency: that of a corn kernel picked up by a hen. At first the kernel thinks it has been sown and tries to germinate. Then it becomes aware of the grossly unfamiliar conditions, 'gets frightened, loses its head, is carried into the gizzard, and comminuted among the gizzard stones'.[27] The hen's habits have mastered its own vegetable habits, so it is digested, forgets its own nature and partakes of the hen's past instead. From now on it will attack its brother kernels with ferocity. This is excellent fun if we allow for the not very helpful anthropomorphism, but behind the humour is the implicit assertion that experience is transmissible directly through the generations, whether or not these are 'really' one creature. Butler never saw the slightest reason for doubt here: and with the voluminous work of Darwin, Spencer and Romanes plus their massy reputations behind him why should he have doubted? As late as 1886 he was still referring to his system as being an easy corollary on Lamarck's – not, we notice, a substantiation of Lamarck for which he obviously saw no need, but just a humble adjunct. To his mind, he could supply himself with no better credentials.

Butlerism and Mainstream Biology

Whether we look at it from the standpoint of a well-informed biologist of, say, 1903 or as modern historians of science tracing the mainstream of advance in theories of heredity, the whole corpus of Butler's 'evolutionary' writings are, as serious contributions to the sum total of knowledge, absolutely spurious. This is a harsh judgement and not a popular one; it is one that almost all of Butler's admirers have tried to evade. Even in his own century when his personal abuse of Darwin made him an Ishmael among nearly the whole of the scientific fraternity, there were still a few independent souls willing to praise his possession of ' "scientific imagination" and logical consistency very rarely found among scientific men';[28] and in the euphoric, immediately post-Mendelian days there was sober talk of a Butler revival: his theoretical work was hopefully seen to be 'emerging from oblivion',[29] and he himself was hailed as a depolariser of the extreme opinions of his own era. Butler's more recent critics have had a purely literary orientation and this has certainly made it easier for grossly inflated claims for his significance beyond that of satirist and novelist to get a footing. Clara Stillman, writing at a time when Lamarckism still retained a precarious respectability, proudly affixed Butler's name to the roll of early twentieth-century neo-Lamarckians – a tribute which now reads flatly since all their reputations have sunk to zero.[30] The collapse of the neo-Lamarckian school has meant that two further roles have been invented for Butler: the clear-eyed demolisher of Darwinian cant, or else the visionary who forestalled discoveries in other sciences altogether. So we have the Butler who gave 'a salutary check to the uncritical acceptance of Darwinism'; the one who wrote 'pioneer works in psychology' having 'much in common with Jung's theory of the racial unconscious'; and yet another who allegedly anticipated William James, August Weismann, Freud, and the entire science of genetics.[31]

The truth is that Butler's greatest selective blindness was his profound lack of comprehension of Darwin and the experimental technique, and his gross insensitivity to the kinds of question science can hope to bring into its purlieu. Not an incomprehension of Darwinism itself – he came to understand natural selection better than most – but of the hypothetico-deductive system which is the core of science itself: the amassing of facts, the disinterested testing of potentially disprovable hypotheses, and the expunging of personal beliefs. No piece of hard evidence cited anywhere in Butler's work is first-hand, and his examples of the genetic transmission of habits are dubious indeed. Even in the summarising 'Deadlock in Darwinism' essays we are offered nothing more convincing than a child masking his eyes with one hand and later passing on the same habit to his children, or the kids of a goat which had been kept chained acquiring the same struggling motions from their parent. It is quite true and should always be kept in mind that such cases were the common currency of the biological speculation of the day, with even the most eminent accepting these standards of evidence; it is also true that laboratory experiments in biology were in this period a lot rarer than the armchair variety. But Butler's thought is demonstrably eclectic to an extent to which, as a self-styled renegade philosopher, he was not willing to admit. Time and again he takes facts from Darwin even while abusing Darwin for failing adequately to acknowledge his sources. Elsewhere he blends together with scant respect such mutual adversaries as Erasmus Darwin, Weismann and Ernst Haeckel. Haeckel's 'molecular' theory of inheritance, which Butler spoke of enthusiastically, helps to place Butler's speculations within the context of the general crudity of thought at the time. Haeckel taught that as, for example, a bird's wing is strengthened by use the protoplasm of the wing muscle is altered in that its constituent molecules take up a different frequency of vibration. He supposed that these were transmitted through the bird mechanically and caused the reproductive cells somehow to resonate in harmony![32]

But even by the undemanding standards of the day suggested by this example Butler's argumentative strategy is slack. Often he relies on metaphor and even in places on the pun. We are told solemnly that a baby 'knows' hydrodynamics as it pumps blood in its body; that a sodium atom 'knows' how to combine with chlorine to form salt; that a pianist 'knows' a work by heart. Butler makes no distinction (seems indeed unaware that there is any to be made) between these different senses of *know*. His standard retort to charges of inconsistency in reasoning is that nature is inconsistent, too. Typically he treats scientific evidence as a lawyer treats the evidence of his client. He measures his success in reasoning not by directing a question at nature and cajoling from her a wholly unambiguous reply, but rather by his ability to persuade enough men off the street that what he has to say expresses the whole truth. 'What nonsense!' Butler is reported as having briskly remarked when told of T. H. Huxley's blistering opinion of *Life and Habit*. 'The matter is one which any barrister or business man can judge of just as well as Huxley himself.'[33] For Butler the story of the facts of heredity was a philosophical problem and one where all the facts are in. He saw scientific controversy simply in terms of getting up the best possible defence – in which logical tricks like *ad hominem* appeals are just part of the routine – and then waiting for a verdict from a jury of laymen. This legal habit of mind also expresses itself as an implicit contempt for inconvenient fact, as a belief that a hired advocate has no moral responsibility to present a fully rounded case. Allied with this is another form of contempt, directed at the narrow specialist who is incapable of drawing back his gaze long enough to capture the total implication of his labours. The specialist may be used to build up a case, certainly; but he must not be allowed to fill in a pristine theory with a mass of superfluous and pettifogging detail. Butler saw no shame in professing an ignorance of science: 'I know nothing about science . . . I neither know, nor want to know, more detail than is necessary.' Rather, this was a boast, as was the extraordinary admission that 'I have left my scientific inaccuracies uncorrected even when aware of them'.[34] This is partly romantic defiance of professional scientism; the pawky cheek of the amateur who has bumbled his way to great discovery. It is a potent myth in English scientific history. Even Francis Darwin, publishing his reminiscences of his father in 1887, dwelt lovingly on the slovenliness of Darwin's practical work – the inaccurate micrometer, the use of the household's ruler, the unreadable gradations on the dirty measuring-glass – and rightly assumed that these details could only enhance his portrait of the greatest biologist of the day. In Butler's case, though, there is also something rather less pleasant detectable. There is arrogance coupled with a wish to hedge, to keep open an escape route back to the relativistic land of Erewhon. Simultaneously he wants 'simply to entertain and interest' without having anyone pin him down on particulars; yet he does not, presumably, want to be called a liar when avowing 'I am in very serious earnest . . . from the first page of my book to the last'.[35] Even Butler's virtually unchallenged status as 'the most dogged, perceptive, and forceful opponent of Darwinism in the nineteenth century'[36] will not withstand a moment's investigation. Compare his performance with any of the teleologically minded capable biologists of the 1880s and 1890s, compare him even with Herbert Spencer, who like him relied on others to do the experimental spadework, and Butler at once stands revealed as the lightweight theoretician he is. His wholly undeserved reputation in this regard is the fabrication of his too reverent interpreters.

What value, then, does remain in Butlerism? Is there anything to be said for it? To the scientific historian it does retain the negative virtue of indicating unmistakably the essential vacuity of all the hereditarian ideas current in the closing decades of the century, even as it loses the positive virtue of being a set of notions that might have filled that void, could Butler only have been granted a hearing. He himself of course inclined towards a conspiracy explanation which sometimes approaches a condition of mild paranoia: 'why have so many of our leaders shown such a strong hankering after the theory, if there is nothing in it?'[37] he asks plaintively as almost his final utterance, apparently oblivious of the strong Lamarckian currents swirling all about him. But, in truth, Butler cared little for the outcome of the battle between the two main biological factions; cared little, even, that as he was writing Weismannism was deliquescing before his eyes. What he wanted in his embittered middle-age was far simpler: public recognition that equating memory and habit had been his contribution alone. But in fairness we must add that these considerations were not yet in the foreground in 1877, when it is Butler's remarkable prescience expressed in *Life and Habit* that 'heredity of any kind is little more than a term for something which one does not understand' that commands our admiration. His polemic, though it eventually relapsed into mere diatribe, has at its best moments a certain crude vigour. Even if the matter was lifted from Mivart, the expression is his own; and as a destructive critique of Darwin's evasions in the successive editions of the *Origin* his work is a mine still not quite exhausted. His misfortune lay in his trying to mould at that early date a few wretched straws of fact into a solid conceptual brick. In the end he paid the penalty of the innovator, as did the originators of all those other too hasty coinages: 'gemmules' and 'plastidules', 'ids' and 'physiological units'. Such another was Weismann's concept of 'the germ plasm', whose fame in its heyday far transcended these others.

Weismann and the Germ Plasm

August Friedrich Leopold Weismann was born in 1834 and, as he did not die until 1914, had a professional life which more than covers our entire period. In the first part of his career he did important work in zoology, but his theory of heredity which alone concerns us here was worked out in the 1880s after bad eyesight had brought his laboratory work to a halt. The essence of Weismannism in its pure, early form is to be found in his treatise of 1885, *Die Continuität des Keimplasma's als Grundlage einer Theorie der Vererbung*, although because this was not translated into English until 1893 by Parker and Rönnfeldt and probably also because of its limited appeal, Weismann did not achieve widespread recognition until 1889 when his *Essays upon Heredity* were collected and translated.

Weismann begins, in the earlier *Continuity of the Germ Plasm*, with the prosaic observation that at the unicellular level life is continuous and immortal because the mode of reproduction – simple division – makes death impossible except for accidents. He goes on to assume that this continuity and immortality is essentially no less true for multicellular organisms, including those as complex as man himself. For each organism may be traced back to a sperm and an egg cell which were in turn part of an organism; and so the line may be traced back to the (speculative) genesis of life from organic matter. In this conception of heredity the individual organism is only of secondary importance, and ought to be viewed as a kind of 'housing' generated by the parasitic germ plasm within as a form of protection. The ribbon of inheritance is continuous and unbreakable, and the genetic endowment alone is transmitted down the ages, absolutely unaffected by anything happening to its transient carriers short of death before the reproductive age. At first glance the persistence of the plasm may not seem to be wholly distinct from Butler's idea of persisting habits; and indeed at a later date Bernard Shaw was content to draw upon both without noticing that there are in reality absolute dissimilarities. We may express these simply in this way. Butler's argument at core is that a psychological phenomenon, habit, may be forced to bear the weight of a biological phenomenon, heredity. We have noticed that he argues as though in a courtroom and that the few facts he musters are wholly derivative. Further, the validity of his whole case may be fined down to the simple Lamarckian postulate. Demonstrate that no habits are inheritable (or, better, put the onus of proof on Butler) and his case falls to dust. Weismann believed that he could make that demonstration, and on his belief erected his concept of the inviolability of the germ plasm. Unlike Butler he had the resources of the laboratory at his back. When he declared war on the neo-Lamarckians, he did so on the basis of a heroic series of experiments in which he amputated the tails of no less than 1,592 mice of twenty-two consecutive generations. When he found that even the very last mouse in that long line continued to bear young with full length tails, Weismann concluded that acquired characters are not transmissible.

The later history of Weismannism – and by extension of the whole neo-Darwinian movement of which Weismann was the chief figure – repeated uncannily though more speedily the history of Darwinism itself thirty years earlier. The parallelism with the initial fortunes of Darwinism is explicable enough in so far as the most serious deficiency of Darwin's natural selection – its failure to explain variation – is repeated, in another form, in Weismann's continuity theory. One curious feature of that theory is its similarity to the old semi-metaphysical one of *emboîtement*. To suppose that the germ plasm is continuous and deathless is to imply that cached within it are all the possibilities, all the metamorphoses of structure through which life will pass into the indefinite future. Besides being deterministic at both the personal and racial level to an absurd degree, this implication seems to deny adaptive evolution its raw material of a constant supply of wide-ranging variation, however caused. Weismann did not of course deny the existence of variations, but he asserted that they are due to amphimixis alone: that is to say, arising from the blending of the characteristics of a bisexual parentage; these characters being divided differently, perhaps at random, in each separate fertilisation. Weismann here came very close to the idea of the Mendelian lottery, with the serious reservations that he had no sense of discrete particles of inheritance, no supportive experimental work, and nothing more definite than the logic of his position to make him adhere to it. Within his lifetime, Hugo de Vries proved that radical mutations do occur in nature; and after his death Hermann Muller (b.1890) gave some reasons why: the ribbon of inheritance, far from being immutable inside its body-case, can be disrupted fairly easily by crude environmental changes in heat and radiation levels. By 1895 at the latest Weismannism had effectively surrendered all its uniqueness by adding on something called 'germinal selection', which supposedly worked in a Darwinian way on changes induced by the environment.[38] Even Weismann, arch-opponent of vitalism in all its many forms, could not avoid making this final truce with his enemies.

The best contemporary account of the rise and collapse of Weismann's reputation is to be found in G. J. Romanes' three critical analyses of 1890, 1893 and 1894–5,[39] which taken together illustrate *in extenso* the defects of Weismannism already noted. They are particularly interesting in that Romanes, a subtle but occasionally devious and often too vehement critic, saw more clearly than any other of Darwin's first followers the massive problems posed by variation. His interest in Weismann's solution was first sparked by the publication in English of Weismann's influential *Essays upon Heredity and Kindred Biological Problems* (1889–92). The eight essays it contains, all closely connected in theme, were written in German over a period of seven years but not collected until, the importance of his work having come to be recognised in England, three admirers collaborated in producing an edition in translation together with a preface by the author. On publication the *Essays* were reviewed with almost lavish praise and soon attracted more attention outside the narrow confines of their speciality than any strictly biological text since the *Descent*. The young H. G. Wells made some play with Weismannian panmixia in an early version of *The Time Machine*, and the consequences of Hardy's reading the *Essays* will be taken up in the next chapter. But Romanes in his review early the following year was notably more cautious in his praise than certain other Darwinians who saw Weismann as the bearer of the master's words purified of all dross. Romanes saw that on some cruces Weismann and Darwin were at loggerheads. It would be impossible, he thought, to find any common ground between the germ plasm theory and Darwin's pangenesis, which he accurately described as the former's 'logical antipodes'. In his youth Romanes had been one of Darwin's most fervent and uncritical admirers, and he was forever afterwards eager to turn his bellicose pen to the defence even of his teacher's inconsistencies and delusions.[40] Romanes came to value cases of allegedly Lamarckian transmission because they supplied him with a stout stick with which to belabour Weismannism. We find him, after giving some second-hand facts about tailless cats producing tailless kittens, concluding triumphantly, 'it would be a grand thing to knock down Weismann's whole edifice with a cat's tail'.[41] This is Romanes in an off-duty mood, but it is suggestive that Romanes really thought Weismann could be upset that easily. A year later his feelings, now shared more generally by the professional world, have hardened considerably. He accuses Weismann of asking for evidence of the germ plasm's manipulation by outside factors, and then ignoring what Romanes took to be the quite conclusive work of Charles Brown-Séquard on inheritable mutations. Further – this is a petty touch – Weismann no longer deserves the honourable title of 'neo-Darwinist'; let him be relabelled an 'ultra-Darwinist'.[42] So the wheel of fortune inexorably bore Weismann down. As Romanes' final attitude shows, his offence had been to be more royalist than the king. In the end he acquired the scientifically dangerous reputation of extremism, of being a man of one idea at a time when biology was necessarily eclectic and fond of invoking multiple causes when no testable hypothesis offered itself. Not natural selection *or* use-inheritance, not saltation *or* smooth variation, but both together; and often other factors thrown in for good measure. In the last analysis neither Weismann nor Butler did justice to the facts, particularly when one bears in mind that the experimental foundations of the germ plasm theory were hardly less shaky than those of Butlerism. Weismannism never really squared up to the objection that in most organisms, once one gets above the level of quite primitive plants, the reproductive cells form directly out of the somatic cells and hence preserve no simple continuity at all. The plasm tends to become something invisible and undetectable, and Weismann's best reason for promoting it a feeling that it ought to exist, like phlogiston before Lavoisier or the ether before the Michelson–Morley experiment. Why, then, was Weismannism so popular? Certainly it was later partly vindicated by the discovery of DNA, which does have some of the plasm's characteristics. But we still have to account for the adulation of the time; still have to explain, as Herbert Cannon incisively puts it, 'how any serious-minded person would come to believe in Weismann's fantasies'.[43]

As Cannon is obliged to concede, such persons did believe for a while, and in large numbers. The evidence for why they did so is still there, buried in those ruminative essays by biologists with the popular touch. For Alfred Russel Wallace, for Ray Lankester (who completely reorganised his own book on inheritance of 1876 after reading the *Essays* in 1889–90), as well as for others lacking formal biological training like Hardy, Weismannism excited attention because rather than qualifying the Darwinian struggle for survival it intensified and confirmed it, thereby awakening all those demons who had been lulled to sleep by the gradual thirty-year softening of that bleak scenario of the past. Even the harshest Darwinians had not been opposed, or not uniformly opposed, to theories of social progress set upon humaner principles than killing or being killed. Now, a decade after Darwin's death, the human condition looked as gloomy as it had done for a brief spell in the early 1860s. Wallace, in his essay 'Human progress' to which we have already referred for its contribution to the eugenic debate, makes much more of the main social difficulties posed by Weismannism. Since the germ plasm cannot be touched, the line of human inheritance cannot be improved by any education to which the somatotype is exposed. Social selection, imitating natural selection, can be the only method of improvement; and yet this kind of control, Wallace argues, can only be to our moral detriment. But Wallace, we recall, was in a difficult intellectual position. Despite his antipathy to the depressing quality of Weismannism, Wallace was at the same time eager to defend a theoretician who was, after all, the first important voice for some decades to support Wallace's lonely insistence on the efficacy of hard Darwinism. He does his best therefore to load the other side of the scales as heavily as possible: we are told comfortingly that at least Weismannian heredity, if true, cannot transmit the *bad* effects of either education or environment. The mistakes and vices of the past cannot directly infect the present. There is no biological original sin: 'street arabs of our great cities, when brought up under healthy and elevating conditions in the colonies, usually improve both [sic] physically, intellectually and morally, so as to be fully equal to the average of their fellow-countrymen'.[44] Thus every generation starts afresh with an unsullied inheritance and the opportunities for social improvement rest squarely in the hands of the reformers. Wallace's attempts to lighten the picture in this fashion could not really be successful, however, because of the undeniable stress placed by Weismann on the necessity of not tampering with the selection mechanism. He was sure that retrogression and a host of other organic evils would follow rapidly upon any attempt to mitigate the operations of nature. Weismann pointed frequently to the same potent example often chosen by the degenerationists. In the dark and placid waters of certain subterranean lakes in Kentucky live fish who have no eyes. Sight for these creatures after they had become trapped in permanent darkness no longer had any selective advantages over blindness, so that the hereditary factors producing blindness, and those mutations born without any eyes altogether, were not suppressed but spread inexorably through the population. There could be no cheating nature in Weismann's account: the species which is not evolving through natural selection is devolving instead. Others, like Benjamin Kidd in his *Social Evolution* (1894), were not slow in making the extrapolations to human society. In doing so they undercut the rationale of those static Utopias of the previous two decades which had relied on circumventing nature by opting for zero growth not merely of their economies but of human potentiality generally. Earlier we examined the shift in emphasis from the static Utopia to the ever-blossoming, ever-dynamic Utopia which dare not slacken its rush into the future. It is the shift implicit in Wells's dystopian stories and made explicit in his polemics of the late 1890s. Here we need further remark only that this transition, along with the hyper-Darwinian sociology of Kidd, originated obscurely in that biological school of which Weismann was the most active and learned proponent.

In conclusion, then, one might say that Weismannism lost ground just as rapidly as it gained it. But while his star was in the ascendant Weismann's influence was felt more strongly outside the narrow scientific orbit than that of any other biologist apart from the two survivors of Darwin's own group, Huxley and Wallace. And that influence was as a biologist, not as a popular social philosopher. It cannot be said, however, that Weismann permanently advanced the neo-Darwinian cause. Quite the reverse: by over-reaching itself, by pushing Darwinism to its logical conclusion and then in the end displaying a rather shamefaced willingness to toy with use-inheritance after all, Weismann furnished the vitalists with powerful ammunition. Even his simplistic experimental work was ruthlessly pilloried by the creative evolutionists: his amputated mouse-tails were thrown in his face. What possible effect, demanded Bernard Shaw splenetically, could one expect from 'injuries or accidents coming from external sources against the will of the victim'?[45] These are not habits! So up to the century's end, and for twenty years or so past it, the neo-Lamarckians pinned their hopes on the inheritance of automatic, involuntary habits which in the past some individuals had found it to their benefit consciously and painfully to acquire. For a while public opinion held that the Butlerians had triumphed over the Weismannians. In the long view, no doubt, the whole controversy shrinks to the proportions of that between the Big-Endians and the Little-Endians. Yet, like Swift's masterpiece, the finest creative works may be generated in an atmosphere of sterile and obsolete debate; and the final chapter will examine two cases proving this to be so.

Notes: Chapter 6

1 Oscar Wilde, 'The Critic as Artist', *The Works of Oscar Wilde*, ed. G. F. Maine, p. 979. This dialogue first appeared in 1891.
2 Yves Delage, *L'Hérédité et les grands problèmes de la biologie générale*, 2nd edn, p. 439.
3 Quoted in Emanuel Radl, *The History of Biological Theories*, trans. E. J. Hatfield, p. 388.
4 In a letter to Hooker of 4 September 1861: Leonard Huxley, *Life and Letters of Thomas Henry Huxley*, Vol. 1, p. 227.
5 For two discussions of the date by which Darwin had *privately* arrived at pangenesis, see 'Geoffrey West', *Charles Darwin: The Fragmentary Man*, p. 264 (which argues for

a date as early as 1840–1), and Gerald L. Geison, 'Darwin and heredity: the evolution of his hypothesis of pangenesis', *Journal of the History of Medicine and Allied Sciences*, vol. 24 (October 1969), p. 380 (which argues for May 1865).
6 Francis Galton, *Memories of My Life*, p. 288.
7 Peter J. Vorzimmer, *Charles Darwin: The Years of Controversy. 'The Origin of Species' and Its Critics, 1859–1882*, p. 20.
8 These short quotations are all taken from William Bateson's *Mendel's Principles of Heredity: A Defence*, p. v.
9 These three quotations are from 'Heredity and variation in modern lights', which was Bateson's contribution to A. C. Seward (ed.), *Darwin and Modern Science: Essays in Commemoration of the Centenary of the Birth of Charles Darwin . . .*, pp. 88–9.
10 Ernest Jones, *Free Associations: Memories of a Psycho-analyst*, p. 118.
11 Garrett Hardin uses this epithet in *Nature and Man's Fate*.
12 Conway Zirkle, *Evolution, Marxian Biology, and the Social Scene*, p. 190, speaks of the *Transactions* as 'a worthy but relatively obscure periodical'. The accusation of fraud is reported, and supported, by Arthur Koestler, *The Case of the Midwife Toad*, pp. 55 et passim.
13 Elizabeth B. Gasking, 'Why was Mendel's work ignored?', *Journal of the History of Ideas*, vol. 20 (January 1959), pp. 60, 83. She mentions also that, *pace* Zirkle, the *Transactions* circulated among 120 universities.
14 Hardin, *Nature and Man's Fate*, p. 127, expresses surprise at this, and puts it down to his condescending attitude. But as late as 1903 the second edition of Delage's *L'Hérédité* fails to mention Mendel, either. Such omissions are probably other examples of the same deep-seated incomprehension that the natural selection hypothesis had to struggle against. In both situations there was a sudden irregularity or discontinuity in the rapidly expanding mosaic of biological assumption.
15 On this same question Julian Huxley, 'The emergence of Darwinism', *Journal of the Linnean Society of London*, vol. 44 (July 1958), makes the puzzling comment that *if* Darwin had known of Mendel's work he would not have regarded that as important for evolutionary theory because 'a premature attempt at generalising Mendelian principles would merely have weakened the central Darwinian principle of gradual slow change' (p. 11). To which the sceptic might reply: it would have more than weakened; it would have made reconstruction urgent. And what is a 'premature attempt'? What significant, lasting advance was made in the study of variation between 1865 and 1900? The neglect of Mendel generated a great deal of confusion and little else. Huxley comes close to saying that whenever a discovery is made – or remade – that is the 'right' time.
16 In a letter to Lyell: Charles Darwin, *More Letters of Charles Darwin*, ed. Francis Darwin and A. C. Seward, Vol. 2, pp. 339–40.
17 Letter to Charles Darwin of 11 May 1872. Like most of the primary Butler material this is printed in Henry Festing Jones, *Samuel Butler Author of 'Erewhon' (1835–1902): A Memoir* (cited hereafter as *Memoir*), Vol. 1, p. 157.
18 *Collected Essays*, vol. 2 (XIX). Unless noted otherwise, all Butler quotations are from the Shrewsbury Edition, ed. H. F. Jones and A. T. Bartholomew. The roman numerals identify the volume number of this collected edition.
19 Letter of 6 November 1891: *Memoir*, Vol. 2, p. 116. In the same letter Butler condemns Weismannism in the tone, not of a participant in, but an onlooker at a debate. He never mentions Weismann again and, according to Jones, probably failed to carry through a resolution to read Bateson's important *Materials for the Study of Variation* (1894). See *Memoir*, Vol. 2, p. 97.
20 From a note (1901) to a letter dated 15 January 1876: ibid., Vol. 1, p. 233.
21 'Cellarius', 'Darwin among the Machines', *First Year in Canterbury Settlement* (I), p. 210. First printed in the Christchurch *Press* of 13 June 1863.
22 'Lucubratio Ebria', *First Year in Canterbury Settlement* (1), p. 219. Written in England, this was sent to the Christchurch *Press* and appeared on 29 July 1869.
23 *Luck, or Cunning?* (VIII), p. 97. Strictly, the liver, not the stomach, is the body's 'purse'

since it stores nutrients for rapid conversion into energy. Glycogenolysis was discovered much earlier in the century.

24 *Erewhon* (II), pp. xv–xvi.

25 The letter is dated 25 November 1877. *Life and Habit* was published only nine days later, which accounts for the note of urgency. Butler believed that the summary of his own development which it contained was so important that he made a special copy. See *Memoir*, Vol. 1, pp. 257–60. Certainly the letter is a psychological curiosity and has frequently been mentioned as such.

26 *Life and Habit* (IV), p. 70.

27 ibid., p. 111.

28 Alfred Russel Wallace. 'Organisation and intelligence', *Nature*, vol. 19 (27 March 1879), p. 480 (partly a review of *Life and Habit*). Wallace was the only Darwinian to give Butler any permanent credit and he attracted Romanes' ire for so doing.

29 Bateson, 'Heredity and variation in Modern Lights', p. 88 n. He goes on to call Butler 'the most brilliant and by far the most interesting of Darwin's opponents'; and, although he continues by giving the palm to Weismann for having finally disposed of Lamarckism in its crudest form, he works hard at the same time at presenting Butlerism as an impressive antecedent of Weismann's germ plasm theory.

30 Clara Stillman, *Samuel Butler: A Mid-Victorian Modern*, pp. 161–3. Her impressive list includes Vines, Henslow, Cunningham, Hertvig, Delage, Hyatt, Cope and Packard.

31 These extraordinarily excessive claims are made by: Claude Bissell, 'A study of *The Way of All Flesh*', in Herbert Davis, William C. DeVane and Robert Cecil Bald (eds), *Nineteenth Century Studies*, p. 288; by Philip Henderson, *Samuel Butler: The Incarnate Bachelor*, pp. 123, 154; and by Jacques Barzun, *Darwin, Marx, Wagner: Critique of a Heritage*, revised 2nd edn, p. 111. Barzun goes on to claim that 'Unconscious Memory, for example, was a suggestion very much akin to the renascent interest in genetics'. But genetics – whose conceptual framework did not even exist before 1905 at the earliest – far from being akin to Butlerism was deliberately built up on an empirico-mathematical basis to oppose the retreat of Edwardian biology into metaphysics. Asserting that this bifurcation existed does not of course require us to judge between the usefulness of the two methodologies: only to recognise (as Barzun does not) that they were present.

32 It is difficult to date this oddity precisely, but it was certainly current and fairly respectable by 1876. Darwin introduced it to Romanes in that year. See Ethel Romanes, *Life and Letters of George John Romanes*, new edn, pp. 49–50.

33 Edward Clodd, *Memories*, p. 260. The same legal language recurs in 'The deadlock in Darwinism' (XIX) where the sole criterion for truth presented is 'what a reasonably intelligent and disinterested jury will believe' (p. 43).

34 *Life and Habit* (IV), pp. 245, 247. However, these boasts are significantly restricted to the first book. In the later 'evolution' books, Butler took himself and his theories more seriously.

35 ibid., pp. 2, 249.

36 Lee E. Holt, *Samuel Butler*, p. 61.

37 'The deadlock in Darwinism' (XIX), p. 56.

38 Weismann's most mature handling of the inheritance problem is to be found in his *The Evolution Theory*, published in German in 1902 and translated by J. Arthur and Margaret Thomson in 1904. The 'germinal selection' hypothesis is explained at length (Vol. 2, pp. 119–22) but the many inconsistencies with his earlier position are allowed to stand, presumably because Weismann could not resolve them.

39 The Romanes analyses mentioned are: 'Weismann's theory of heredity', *Contemporary Review*, vol. 57 (May 1890), pp. 686–99; *An Examination of Weismannism*; and the posthumously published *Post-Darwinian Questions: Heredity and Utility*, which is the second volume of *Darwin, and after Darwin. An Exposition of the Darwinian Theory . . .* This treatise is a much expanded version of two courses of lectures given between 1888 and 1890. *Post-Darwinian Questions* was probably completed in Romanes' last year, 1894. The final sentence of the *Examination* (1893) runs in part: 'I intend to deal with this

question [i.e. of Lamarckian inheritance] hereafter as a question *per se*' (p. 171). And no less than five chapters out of the ten of *Post-Darwinian Questions* are devoted to this topic.

40 Their relations from Romanes' side were defined by his wife Ethel as 'unbroken friendship, marked . . . by absolute worship, reverence, and affection' (*Life and Letters*, p. 13). This may sound excessive, but not when one reads Romanes' threnody 'Charles Darwin: A Memorial Poem' of which about one-quarter is printed in the *Selection from the Poems* by T. H. Warren. It is an imitation of Tennyson's 'Ode on the Death of the Duke of Wellington'.

41 Letter to his wife of 19 March 1893: *Life and Letters*, p. 300. In the *Examination* there is the same note: 'in those passages,' comments Romanes, pointing out some inconsistencies with a certain gloomy relish, 'Weismann is annihilating his own theory, root and branch' (p. 111).

42 *Post-Darwinian Questions*, p. 12. Not to be outdone, Robert Mackintosh five years later proposed 'hyper-Darwinism' as a suitable label for Weismann's school. See his discussion in *From Comte to Benjamin Kidd: The Appeal to Biology, or Evolution for Human Guidance*, p. 238.

43 Herbert Graham Cannon, *Lamarck and Modern Genetics*, p. 38.

44 Alfred Russel Wallace, 'Human progress: past and future', *Arena*, vol. 5 (January 1892), p. 156.

45 George Bernard Shaw, 'Back to Methuselah', *Prefaces*, p. 503.

7

This Body Is an Omnibus:

The Motif of Heredity in *The Way of All Flesh* and *Tess of the d'Urbervilles*

> This body in which we journey across the isthmus between the two oceans is not a private carriage, but an omnibus.
>
> OLIVER WENDELL HOLMES (1867)

The previous chapter having supplied some background information on the state of the study of inheritance in the 1880s and 1890s, we turn finally to a fairly extended examination of Samuel Butler's *The Way of All Flesh* (written 1873–5; published 1903) and Thomas Hardy's *Tess of the d'Urbervilles* (1891), two novels which profoundly illustrate the imaginative response to particular and, it may well be thought, highly recondite issues in contemporary theoretical biology. These works will bear such analysis, first, because their density of texture is very much greater than the relatively minor works of literary art so far examined; and, second, because both Butler and Hardy explored the imaginative possibilities of the life sciences most assiduously, with the dedication of men who somehow knew intuitively that there were facts, angles of looking at the world, and a possible precision of observation in those sciences unobtainable elsewhere.

At first sight Butler and Hardy have little enough in common either as personalities or as writers, and there is certainly no evidence that either was drawn to the other's work or would have found anything congenial there had he encountered it accidentally.[1] They were quite opposed in temperament: Butler naturally acerbic, contentious, his habitual mode of response deeply satirical, contemptuous of received wisdom and even of rationality itself, yet for all that of scholarly habits, an urban creature and the product of an urban education; Hardy placid, melancholiac, a thin-skinned rebel whose instinct was to flee controversy, a rural autodidact whose over-reverence for abstractions severely lamed his art, an admirer of other men's rational creeds yet a thinker whose reflexes were basically poetic. Even the introversion which was their strongest shared feature manifested itself differently in each of them. Butler retreated into a small and rather barren private world beleaguered by largely imaginary enemies whence he emerged periodically to polemicise or shout abuse; his intercourse with other writers was practically non-existent; he was despised even by those scientists who in part shared his opinions; and he had nothing to learn from personal discussion. Hardy, by contrast, though shy, forced himself into fashionable London society and, because his role was purely that of a humble and interested layman and not therefore threatening, assimilated a good deal of biology direct. The two writers' outputs confirm these generalisations. Hardy the literary artist: fourteen novels, collections of short stories, and over 900 published poems – little else, however. Butler the man of letters: only one novel proper, a modicum of poor verse, a book on Shakespeare's sonnets, another on Homer, four on evolutionary theory, many miscellaneous essays, travel books and so on. Their relative successes in their profession is still more of a contrast. Before Butler died in 1902 he ruefully estimated his total profits from a lifetime's writing at £77 2*s* 11½*d*. By this year Hardy had already completed his famous and lucrative career as a novelist, and he had another twenty-six years of life to make another reputation as a poet before dying very old and full of honours.

There are, however, other similarities. The most noteworthy is that of simple chronology. Butler was born in 1835, the year Darwin collected his critical data in the Galapagos Islands; Hardy in 1840, two years after Malthus had provided Darwin with the vital concept of population pressures: when the *Origin* was published Butler was 24, Hardy 19. The manhoods of both belonged to the first Darwinian generation and both breathed the air of biological controversy for much of their lives; both had a firm if slightly off-centre grasp of the essentials; and for both biology fertilised their art, though the offspring were divergent enough. And below the surface there were related reasons for their equal obsession with the mechanism of inheritance and hence for their responsiveness to every twist and turn of experiment and hypothesis. Hardy was philoprogenitive but childless, and he has a deep and gloomy interest in the history of his own family which he perceived to be one of devolution and decay. Even before the neo-Darwinian movement supplied him with a theoretical sub-structure of secular Calvinism he was already set towards genetic determinism. For Butler the secrets of heredity exercised a radically different spell. His wretched childhood and despicable parents (the probable unfairness of Butler's own version of this is irrelevant); his covert homosexuality and gross fear of women allied with a pathetic fascination for fatherhood objectified notably in *Erewhon Revisited*; his odd capacity for ancestor worship and hatred: these are the contradictory ingredients which were stewed together in the four 'evolution' books which are really four magical recipes for restructuring the image of a self. Whereas for Hardy the genetic endowment is just a strait waistcoat suppressing action, for Butler a grand personal exertion of will can burst those bonds and refashion the material into a seemly garment. He is profoundly optimistic about the individual's freedom to choose a destiny and melioristic in his attitude to the family organism – an entity in which he and Hardy together, deniers of resurrection within the Christian tradition, implant the only immortality they recognise.

The Way of All Flesh and *Tess of the d'Urbervilles* have a few broad similarities despite massive differences in structure and treatment. It is true that as aesthetic products they are hardly comparable. The latter after all is the masterwork of a novelist of genius, while the former is a highly uneven *roman-à-clef* which, as we shall see, was never properly revised. As a personality Hardy merges almost completely with his creation; so much so that some critics have argued that even his authorial interpolations are not uttered in his own voice but have only a contextual relevance amounting to a sort of philosphical negative capability. Butler by contrast forces himself – and his family – wholesale into a half-fictional work which would burst at the seams were it not pinned by such a firm temporal sequence as the intimate history of several successive generations. And, far from assimilating his extra-literary preoccupations and personal neuroses, the transparently authorial persona of Edward Overton allows him to indulge endlessly his undoubted gift for wittily pursuing tangential issues well beyond the rim of the immediate action. The result is lively enough, but the few superbly balanced comic scenes are too thinly studded across the pages of essayist's prose to make *The Way* exist for the reader as more than the sum of its parts.[2]

Still, at a deeper level the novels, like their authors, can be brought into some kind of contiguity. Both were written with the awareness of biological 'facts' very much in mind, though each author according to temperament bounded his horizons with a very different selection of those facts. Idiosyncratic though each selection may be, both try to resolve the obsessional problem of inheritance, and to this end we find a centripetal movement of plot in both *Tess* and *The Way*, with all the action bearing inward on the destinies of the heroine and hero. Tess and Ernest Overton share a curious mode of existence unusual in fiction though common enough in the heroic epic; that is, their being is extended for our inspection not in three but in four dimensions. Like the epical hero whose remote ancestry was linked with the divine, the heroes of these two novels have an extra-personal augmentation. Tess's impersonal history stretches back to the Norman Conquest, and the family genotype (a concept that Hardy may have drawn directly from August Weismann) is presented as an active force within the surface 'present' of the novel; so much so, indeed, that the freedom of the protagonist is all but eroded away. Inside the frame of biological dialectics so vital to both novelists, *The Way*, though in fact written earlier, takes up *Tess*'s highly deterministic conclusions and forces past them. Five generations of Pontifexes enter the story and a sixth is mentioned; these generations form a slice out of all human history (even, given Butler's extra-fictional preoccupations, out of all *organic* history), and the decay and restitution to health of that family organism is a drama whose unities are defiantly and perfectly deliberately set against those obdurate facts of genetics which so daunted Hardy as a philosopher.

Perhaps one can best summarise the root differences between the novels in terms of their respective attitudes to the individual consciousness. In certain moods and under certain influences Hardy toyed, as he does in *Tess*, with the mournful hypothesis that human consciousness, having been spawned in a blind world of defect, could be aware of nothing but pain and therefore that a lapse back into unawareness is the best service the cosmos can render to life. In his poems more often than in the novels Hardy becomes quite maudlin on this topic:

> But the disease of feeling germed,
> And primal rightness took the tinct of wrong;
> Ere nescience shall be reaffirmed
> How long, how long?[3]

Butler also had a poor view of cognition; for him it was the response of any organism to a sufficiently complex environment which habit had not yet allowed to be relegated to the instinctive mechanisms of behaviour. Under such duress 'consciousness springs into existence, as a spark from a horse's hoof';[4] a spark, Butler taught, to be extinguished thankfully at the first opportunity, so that Ernest completes his development when he is able to *stop trying*. It may be seen that in this both Butler and Hardy reversed the ordinary conception of evolution (as we find it, for instance, in George Meredith) as life struggling into consciousness and, eventually, into self-awareness. Yet here we also have to hand a formulation of the differences between Ernest's career and Tess's. Butler has faith in the natural dispensation. For him, as for any Lamarckian, simply to be natural is to be fulfilled. Hardy, steeped in Darwinian gloom, is deeply mistrustful of nature and contemptuous of its laws. In literary terms these viewpoints result in tragedy (Tess is unable to establish the rules of her selectionist universe until it is too late) versus farce (Ernest is a fool, but his teleological universe conspires to damp down the catastrophe resulting from his folly). And in psychological terms the difference of biological perspective partly explains why Tess in the end radiates pathos and Ernest a nasty priggishness.

Butler's Roles as Biological Theorist and Novelist Compared

> Theorists may say what they like about a man's children being a continuation of his own identity, but it will be generally found that those who talk in this way have no children of their own. Practical family men know better.[5]

For 'theorists' here, we can read the childless Butler himself and the views he had expounded in the 'evolution' books; 'practical family men' can be read as a jibe against the great majority who had cast aside *Life and Habit* and its successors as bad biology. This sly remark early on in the novel by its narrator Edward Overton is one of several which forces us to notice the extraordinary ambiguity of *The Way* as a dramatic rendition of a deviant late-Victorian biological theory of inheritance. As this quotation shows, the ambiguity does not rest with the meshing of theory with narrative, for Butler rides over that by boldly allowing his scientific preoccupations to float on the surface of the novel with only the most rudimentary attempt at aesthetic distancing. At work is the same impulse to realism which led Butler to impact his own life and his art by directly copying passages from his mother's personal letters into his text. By doing so he effectively circumvented any easy reading of the novel as rancorous autobiography. The interesting problems of interpretation and tone only emerge at the point where the personal becomes universal; where the *roman-à-clef* becomes the *roman-à-thèse.* The latter genre, however, is an inherently unstable one, tending to decompose under the bright critical gaze into simpler ingredients. So the ambiguity lies first in a clash of purpose, as scenes under exact dramatic control alternate with long and embarrassingly confiding miniature essays. And there is no doubt that in the earlier chapters particularly the reader responds to the moments of high comedy and to those almost exclusively. This is quite natural, since the emotional fulcrum there is Ernest's childhood, his upbringing in a high Victorian clerical family. It was this perceivedly loathsome experience which Butler allowed to fuse inside him until, when his heated memories had just sufficiently dissociated themselves from reality, he poured them forth in a long stream of satirical invective. When it is working properly *The Way* is, quite simply, a single sustained scream of hatred; and when Butler is moulding material from his own past – material which he can work up into his personal demonology – his sense of timing and of the apposite never falters. We are rewarded with Ernest's baptism using filtered Jordan water; with Dr Skinner's dining spiritually on bread and water but in crude actuality gorging himself on oysters; and with the engaged Theobold and Christina mournfully calculating their chances of glorious martyrdom. Such vignettes need threading firmly on a string of narrative provided by viewing the entire action through the consciousness of Overton, but this expedient does not prevent the splintering off of entire segments of Overton's monologue. He is too obviously Butler's mouthpiece, with his frustrations and obsessions protruding altogether too noticeably for him to be anything like a detached satirical observer of men's follies and crimes.

The Way of All Flesh, then, is most successful as picaresque comedy. But there is a second, more radical ambiguity in the presentation of the theoretical sub-structure of the novel which derives from Butler's vacillating attitude to the craft of fiction. Although he made spasmodic attempts to write 'a novel pure and simple with little "purpose" ',[6] he probably did see himself in the true light of a novelist *manqué*, though his habitual pose of self-denigration makes it difficult to know just how seriously he felt his imaginative deficiencies. As he began to plan *The Way* he held *Middlemarch* before himself as a failure of unity and a disaster of inauthenticity, yet 'the only question is', he concludes his remarks on didacticism in the novel, 'whether, after all, that matters much'.[7] For him it did not. Consequently *The Way* constantly induces in the reader a sense of strain, for he is invited and even compelled to respond in several modes simultaneously. Butler's inability to write, so to speak, in one dimension bedevils much of his prose and expresses itself here as a tendency to turn everything into fatuity and into an object of ironical observation, up to and including (as the opening quotation shows) his own deepest convictions. Almost despite himself he forces a mannequin's role on to Overton, garbing him disconcertingly now in jester's costume, now in professorial robes. The result can be dizzying. Consider Butler's analysis of George Pontifex, the self-educated and self-conscious tourist, at loose in Europe. The portrait, while the mode is securely ironic, is delicious. George's conventional ecstasies over the Alps, lush quattrocento art and sleeping in Napoleon's bed all serve to transfix him; but Butler is not content with merely anatomising his victim. He must stretch him out on the rack of time and theorise over him and over his line of ancestry. In those terms which throughout the novel provide his most potent metaphors and most suggestive analogies, George is a sport or mutated variation:

> A very successful man, moreover, has something of the hybrid in him; he is a new animal, arising from the coming together of many unfamiliar elements and it is well known that the reproduction of abnormal growths, whether animal or vegetable, is irregular and not to be depended upon, even when they are not absolutely sterile. (p. 19)

Extracting this sentence from the elaborate context in which it appears partially masks the narrative's very noisy shift in gear at this point. When the reader is told that 'it is well known' the confidence of statement causes him to quieten his critical faculties and to take the biological fact without a quibble: that hybrids like mules are either sterile or don't breed true with successive generations. The reader is even willing, with a little prompting, to absorb the metaphorical transformation from businessman to hybrid. Given the quality of George's son, Theobald, it is not difficult to make the transference. The strain in the reader arises more from the question which insinuates itself between the words on the page and the characterisation to which they refer. How does Butler mean his bland utterances to be taken? Is it a proffered fact or is it (crudely) just a joke? And there is nothing in the novel to tell him. It is exactly this evasiveness which makes *The Way of All Flesh* so readable yet, ultimately, so profoundly unsatisfactory. We search for the high seriousness, for the solid moral concern which is so available in *Tess* at every point, but it is not to be found. Butler the anachronist is raising ingenuity, sheer playfulness, into the very subject-matter of art: he is foreshadowing Joyce and echoing Sterne while demanding from us spasmodically the response which we wish to reserve for George Eliot. There are too many clashing flavours in such a blend. To accept seriously a scientific patterning in a novel the reader first needs a reason to believe that his author believes in it.

And yet the most cogent fact on which any assessment of *The Way* as a novel of ideas must be built is that it was written in its entirety over eleven years (1873–84) which also saw the writing and publication of *Life and Habit* (1877; dated 1878), *Evolution, Old and New* (1879) and *Unconscious Memory* (1880). Since, as we have seen, the theory of heredity sketched out in those books itself suffered considerable elaboration we must have before us the interlocking pattern of Butler's scientific and literary composition.

That Butler was considering a novel as his next literary project as early as April 1873 (only a year, that is to say, after publishing *Erewhon* and well before he engaged in any anti-Darwinian controversy) is clear from the extant correspondence with Charles Darwin and Eliza Mary Ann Savage. Though at first his plans were vague, he had written five chapters of a preliminary draft of *The Way* by September of that year. Unfortunately we have no contemporary details of the genesis of Butler's theory of heredity; but what we do have is the reconstruction of his train of thought written by himself about 1880 and eventually incorporated into *Unconscious Memory* as its second chapter. From this account it is plain that both the novel and the speculative investigation which in due course became *Life and Habit* had exactly the same starting-point; that is to say, a reconsideration of the problem of heredity. According to this account of 1880 the very first paragraph of *Life and Habit* was written in Montreal in June 1874. Part of the expression here – the assertion that 'it is one against legion when a creature tries to differ from his own past selves . . . our former selves fight within us and wrangle for our possession'[8] – suggests that Butler was already searching for some sort of intellectual structure to act as truss to the imaginative presentation of Ernest's ancestry in the first chapter of *The Way* already written. Once his 'very dry but exceedingly (to me) interesting subject . . . the force of habit'[9] took a firm grip on him the novel was laid aside for several years; but when he did return to it in February 1878 he went ahead quickly in rough draft to get the ground covered, completing the second volume in some form – that is, up to Ernest's arrest for the assault on Miss Maitland – by July. Many revisions were made, very probably in the light of his adoption of Lamarckian inheritance; because if the analysis presented in the previous chapter is correct the evolutionary schema behind the first chapters of *The Way of All Flesh* must have been far less defined at first than in the finished state. The pattern of a creative *bouleversement* after a period in the front line of biological debate was repeated even more dramatically after the second and third interruptions caused by the writing of *Evolution, Old and New* (published early in 1879), and later of *Unconscious Memory* (published November 1880). When he began yet again in 1882, Claude Bissell's well-documented account tells us, 'all of the early material was rewritten, and what was intended to be merely the first draft of the concluding section was finished by the end of 1883';[10] by November, to be exact, for Miss Savage had a complete manuscript in her hands by the 17th of that month. It was not, however, a manuscript which Butler was satisfied had reached publishable form. He was particularly dissatisfied with the ending, and his own evaluation rings true after a century: 'My own idea is that the first vol. is the best of the three, the second the next best, and the third the worst . . . I must begin again with the third volume.'[11] So the relentless process went on. As late as 1884 he was talking of restructuring the final volume, and he did add a lot of material before Eliza Savage's death on 22 February 1885. She had been his sole confidante and so effective a critic that the composition at times seems almost to have been a joint enterprise. For eighteen years after her death the manuscript gathered dust. Festing Jones's contention that Butler abandoned it rather than finished it is supported by the list of addenda he prints, including yet more corrections, many to be incorporated into the last episodes.[12] The rest of his life Butler devoted to a variety of other unpopular positions in literary and artistic controversies, but his creative impulse waned along with his determination to lay siege to the Darwinian stronghold. At some deep level in his nature they were indeed twin aspects of the same impulse.

But, if his scientific heterodoxy alone enabled him to write the novel he did, Butler trimmed all the excess fat from his theory so that it is not difficult to dissect out the four main arteries which carry the main flow of events in *The Way*. First, there is the oneness of personality between the four generations of parents and their offspring. We are required to accept, in strict literality, that Ernest is the same person as his father no less than his great-grandfather, and that their personalities are fully, though latently, part of his own. Heredity is memory: 'this feature of memory is manifested in heredity by the way in which offspring (e.g. *Ernest*) commonly resembles most its nearer ancestors (*Theobald*), but sometimes reverts to earlier ones (*John Pontifex*)'.[13] But, second, if heredity is memory there must be a somatic equivalent to forgetfulness. The cumulative experience of past generations is available to the child only until he reaches the age when his parents conceived him; after that point he is thrown back on to his own resources. By the end of his life even his basal metabolism 'forgets' how to regulate itself – cellular repair fails because the body can no longer recall what healthy cells should be like. Forgetfulness, then, equals senescence; and amnesia, death. There is no more palpable evidence for the care with which Butler organised his novel than the very precise calendar that we are given to make this notion credible to the uninitiated reader – after he has done some simple calculations. John and Ruth Pontifex are both 38 in 1865 when their son George is born. Here the transmissibility of acquired characters is (within limitations) beneficial, for George has a chance in his adulthood as a successful publisher to 'recall' and apply the natural sagacity of his father and the assertiveness of his mother. George's son Theobald is born in 1802, when he is 37 and his wife 30; Ernest in turn is born in 1835 when Theobald is 33 and Christina 37. Butler devoted minute care to this chronology, altering the relevant ages many times, and this may not represent his full and final intention; but the progressive decline in male ages also marks a progressive diminution in inheritable wisdom. It puts some structure into Ernest's loss of direction, after his leaving Cambridge and his hasty ordination have thrown him so far out of equilibrium that his emergent individualism is 'nipped as though by a late frost' (p. 201) – one of several horticultural similes used in this context.

The third and fourth core biological principles may be summarily stated here since their artistic significance emerges only under closer analysis. One is that habit-memories are normally latent in each individual, and can be roused from dormancy only by environmental shock or (Butler's own idiosyncratic term for it) 'crossing'. Providing these shocks are not violent enough to damage or destroy the organism altogether, their effect can be beneficial. Presumably 'crossing' is analogous to the hardening-off process used on tomatoes to prepare them for a cool climate. The other principle is that the highest evolutionary ideal is not an intensification of consciousness but, rather, its replacement by 'grace' – a manifold virtue which encapsulates all of Butler's Hellenism, his antipathy to Pauline Christianity and his admiration for instinctive physical excellence and social poise in young men. In *The Way* the aureole of grace surrounds, of all Ernest's contemporaries, only Towneley; and in Ernest's self-lacerating description of him we note for the first time one of those odd transpositions from society to biology habitual with Butler. 'To make Towneleys,' Ernest lectures Overton after doing his six months' imprisonment with hard labour, 'there must be hewers of wood and drawers of water – men in fact through whom conscious knowledge must pass before it can reach those who can apply it gracefully and instinctively like the Towneleys can' (p. 287). Here he does not mean, as one ignorant of Butler's biology might well suppose, that Towneleys are part of an aristocratic elite and flourish only in climates of inequality. It is not a political statement at all. Ernest is a hewer and a drawer simply because the life-force must periodically become aware of itself before relapsing into blessed instinctiveness; because Ernest must struggle after virtue and invent his personality as he proceeds.

For this reason the study of Ernest as a Lamarckian hero lacks definition, as his biography continually merges with that of the Pontifexes of which he is a transient and none-too-successful component. Ernest's name itself carries some extra meaning in this context. It has a semi-allegorical significance. The family surname has often been taken as a pointer to the Catholic priesthood, though the priestly connections of the Pontifex family (limited anyway to Ernest and Theobald) makes this a feeble pun.[14] A more likely possibility, in view of Butler's interests, is that 'Pontifex' echoes biological nomenclature (cf. *spinifex, tubifex*) to remind us, by reference to such simple forms of life, of the living and environmentally sensitive family organism which transcends its members and is effectively immortal. The derivation of the '-fex' suffix then gives us Ernest 'the bridgemaker' – the means by which the Pontifex creature reaches into the future across five generations linking the two ends of the family to restore its own health in the incarnation of Ernest's children. The hero's Christian name, of course, tells his fate. As for the title, one meaning is ironic. Butler's theory repudiates the biblical admonishment that dissolution is the fate of all life. But equally the novel goes on to show the way – the method, the technique – by which life extends itself over time. If these interpretations are correct, we can make sense in terms of the novel of Butler's summary disposal of the notion of personal identity as irresolvable unity. He spoke instead of the personality as being 'a nebulous and indefinable aggregation of many component parts which war not a little among themselves . . . nothing but fusion and confusion'.[15] His description does not in fact add much to the classic demolition job by Hume, except that it substitutes a genetic flow for Hume's stream of mental impressions; but as an article of belief of a novelist it certainly does not help towards the construction of characters of Dickensian solidity. A caveat is in order here. Butler's literary perceptions are in no way similar to, say, Virginia Woolf's, and he was not interested in impressionist techniques. He does not, so to say, inspect his creations through a lens of such resolving power that the commonplace causal bonds of motivation and action fade into indeterminacy. The fuzziness of *The Way of All Flesh* arises rather from filtering a realist's perceptions through a speculative hypothesis (whose truth or falsity is not in question) which simply does not make for good novels, or at least for the good *Bildungsroman* which Butler hoped to make of *The Way*. Ernest's passivity is pressed upon him directly by his creator's vision of personal development: the vision that happiness consists, not in modifying one's surroundings to suit, but in camouflaging oneself and vanishing into them; not in discovering one's duty to the world and acting it out, but rather in seeking out an ambiguous 'inheritance'. The amoeboid Ernest's sole reflex is to shrink from hostility; or, *in extremis*, to envelop himself in a tough cyst of unearned income and social abnegation. Still, we do Butler an injustice if we blur his distinction between the illusory sense of personal identity and the person as a whole, for he is careful never to do so. Though he impresses on the reader the evanescence of every person, the logic of his position equally requires that the representatives of each generation be apotheosised. For in each Lamarckian generation each person bodies forth the total experience of his forebears and even of the totality of life itself; for he is in truth frighteningly free, almost in the existentialist sense of being forced to raise himself by his own moral bootstraps. The Lamarckian hero can choose to ignore his freedom if he wishes, or (like Theobald Pontifex) the ignorance can be his portion by sheer pressure of circumstances; but that permanent and progressive degradation of humanity, that utter impotence and futility of human striving so in the background of *Tess*, is impossible. A Lamarckian tragedy is a contradiction in terms. Though its intensity may brighten or dim as the generations pass, the course of life is always a movement into further evolutionary complexity.

The Way of All Flesh as a Lamarckian Novel of Inheritance

As soon as we move from the complexities of *Life and Habit* into *The Way of All Flesh* itself we are forcibly made aware that, from artistic necessity, the fictional events deviate a good deal from the abstractions of the 'evolution' books. To put this another way, on entering the novel we discover that Butler's real interests, partially concealed under the movement of the more rigorous argument, now become overt, displacing any material which cannot be squeezed into narrative order. Besides introducing the occasional *deus ex machina* to solve some exigency of the plot (a blatant example is the one which nullifies Ernest's bigamous marriage), Butler brings into close focus just a tiny segment of his overall account of organic development. For his biology books ostensibly cover a vast span of geological aeons no less than does the *Origin*; indeed, he goes farther than Darwin in imputing a kind of memory-consciousness and hence a potential for development even to inorganic matter. It is no part of his theory that a few generations could reveal Lamarckian, any more than Darwinian, effects. (Except, possibly, the transmission of simple injuries: Butler knew of Charles Brown-Séquard's experiments which suggested that such transmission did sometimes occur from parent to child, but he attached no special significance to them.) Yet in the novel we are required to believe that the Pontifexes in five generations show such a transparent regression and advance that a single observer can easily remark them. If Lamarckian inheritance really did improve the human stock at such a rate, we would have touched perfection long since. No; Butler like any other novelist concerns himself with psychological, not biological truths and to this end devotes the bulk of his novel to an account of Ernest's education, which is a paradigm of orthogenesis at work. In his hero's struggles to remake himself, in his blind groping towards a consummation which only gains in reality with each step on the way, we are meant to read the trend of all life. We are explicitly guided to such a reading by Overton's generalisation that 'embryo minds, like embryo bodies, pass through a number of strange metamorphoses before they adopt their final shape' (p. 207) – an analogy which, like many served up throughout the novel, sounds impressive until we reflect that the outcome of embryological recapitulation is determined and that it always moves towards a goal of structural completion which is pre-set. By contrast personal and (by definition) Lamarckian organic evolution, though goal-*seeking*, is an open-ended process without any inbuilt destination, and certainly no 'final shape'. Butler has betrayed himself with an analogy which blunts his intention. But perhaps it was intentional that, by giving Overton the role of benign and detached cartographer of his creator's aprioristic theory, some distortions had to be introduced. The point at issue is that *The Way* runs by its own logic, and that that logic quite frequently defies the simple and direct correspondence taken for granted by R. A. Streatfeild and others.[16] To read the novel this way would land us in a morass of contradictions. Some of these do exist anyway and are genuinely irresolvable; but others may be minimised by keeping Butler's biology before us as an imaginative projection, as a complicated system of tropes useful and important only for ordering the material of the novel, for defining the roles of the generations and for setting the relations between them. To do more than this means to enter the narrative itself.

The opening portrait of John Pontifex strikes a note that, by reason of its rarity among the rather limited range of effects Butler's art is capable of, sounds almost grotesque. It is that of unfeigned, wholly non-ironic approval. Old Pontifex measures up in every way to an evolutionary ideal of unconscious grace which, thanks to the intromission of his bossy wife's genetic factors, so eludes the next few generations of Pontifexes. What is more, the portrait has charm and a quiet persuasiveness, and does not alienate the reader as does that of old John's urban equivalent, Towneley. It is a portrait done from a distance in soft pastel shades, yet John is still very much the exemplar of Butler's views. He is unconsciously adapted to his narrow rural life, just as he is unconsciously an adept in his supremely practical skills of master carpenter, artist and musician. Though starting in life 'with no other capital than his good sense and good constitution' (p. 4) – and here Butler deliberately conjoins two meanings of 'capital' to show, as he does throughout, that they are identical – he has constructed a balanced and harmonious life as instinctively as he constructed his two organs (not to mention, of course, his own inner organs): without blueprint, plan or design. He rests at the centre of his Paleham house like a chambered nautilus in its complex and beautiful shell. He is unconscious even of his own reproductive capacities. Neither he nor his wife take parenthood more seriously than do the lower mammals, and indeed they do their best to emulate them by apprenticing George off to his uncle at the first opportunity, so falling, as Overton's narrative puts it with unconcealed approval, 'into their proper places, resigning all appearance of a parental control, for which, indeed, there was no kind of necessity' (p. 9). Like many more of Overton's good-humoured asides, this on parenthood prefigures Ernest's experience much later. By that time, however, all humour has been boiled off by hatred:

> A man first quarrels with his father about three-quarters of a year before he is born. It is then he insists on setting up a separate establishment. When this has been once agreed to, the more complete the separation forever after the better for both. (p. 305)

The sourness of this and its emptiness of any warmth save the heat of disgust make it difficult to appreciate that old John Pontifex has been moved by the light of nature to behave in much the same fashion. Is Butler here exploring the Oedipal situation? If we grant him so much, we should take note that he is manipulating the raw material of one science – namely developmental psychology, which did not then exist as a coherent study – within the quite misapplied conceptual framework of another, genetics. This is a recognisable quirk of Butler's reasoning, evident in many places outside the novel. For instance: 'if the mere anxiety connected with an ill-healing wound', he questions rhetorically in 'Deadlock in Darwinism', 'is sometimes found to have so impressed the germ-cells that they hand down its scars to offspring, how much more shall not *anxieties that have directed actions of all kinds* from birth to death . . . modify and indeed control, the organization of every species?' (p. 44: my italics). Here the extrapolation is an obvious one: Butler is twisting information on the alleged transmissibility of mutilations into a totally suppositious claim for the inheritance of psychic damage, simply because he has an emotional stake in the latter. Such transpositions figure very largely in *The Way of All Flesh*.

As a sophisticated product of the evolutionary process, John Pontifex is a finely balanced individual who is specially vulnerable to the grossness and fierce assertiveness in his son's nature. George Pontifex is, by comparison with his father, a failure, though he continues to be very well adapted to his new social environment. His transplantation from the rural scene to London does, it is true, reinforce his 'trifle too great readiness at book learning' (p. 8), but the disastrous consequences of this take another generation fully to materialise. George's chief claim to goodness is that he escapes poverty completely. Butler professed to be uninterested in the saintly virtues, insisting instead on the synonymity of the good of the individual and society. In this brand of utilitarianism there is to be seen yet another transition from biology into, this time, economics: if society is indeed a single organism, or one great polyp, then the reward for private virtue is going to be expressed, typically, in money terms. Poverty, then, is a natural crime – a proof of evolutionary failure. The poor are the unfit to survive. Lamarckian social morality has brought us back, by a different route, to the Darwinian imperatives. The difference is that for Butler there are winners, and the winners are those whose wills are equal to the burden of evolutionary advancement. George Pontifex is still one of the elect, as is proved by his dying rich and respected at 73.

The next incarnation of the Pontifexes, Theobald, marks the nadir in the devolution of the family organism. Instead of inheriting his father's strong camouflaging abilities, Theobald proves to be as meekly passive as old John. But, worse, we see him torn out of the Paleham setting and cursed by an education which strips away his meagre ration of instinctive wisdom even as it raises him into miserable self-awareness. The free part of him is forced into abeyance yet farther, as it is distorted more and more by the powerful wills of the Allaby family and his father until he is plunged into a hell of personal inauthenticity. Some of the scenes conveying this are richly comic ones – the passage of letters between Theobald and George before the former's ordination; the neatly timed one where the Miss Allabys settle Theobald's marital future with a hand of whist – though as usual the humour here is superficially innocent of all theory.[17] It wells up instead from Butler's only fount of literary creativity, the inexpungible hatred of his father. But we are never permitted to laugh with Theobald, only at him, and the lasting impression is one of grey despair. There is for a short while at this point in the novel a revelation of what exactly Butler meant when he said that Darwinism filled him with loathing, for Theobald has to bear the brunt of surviving in a blackly Darwinian world of random events untransfigured by the chance of personal striving to grow. The mood of *Tess* momentarily breaks through as Theobald, locked in his uncongenial life, is seen 'trudging through muddy lanes and over long sweeps of plover-haunted pastures to visit a dying cottager'. Going about his distasteful pastoral duties he sinks into a dreadful spiritual exhaustion, almost asphyxiated by being cut off from his oxygenating past identities; for 'what does a fish's best come to when the fish is out of water?' (pp. 58–9). Overton's judgement on him is stern – stern with the conviction that Theobald is no pawn of circumstances but a free moral being who has repudiated his sacred genetic obligations, who has flinched away from the stimulating 'crossings' which might have forced him to put out new shoots. Instead Theobald fails on both levels of being, cerebral and instinctual alike:

> The habit of not admitting things to himself has become a confirmed one with him. Nevertheless there haunts him an ill-defined sense that life would be pleasanter if there were no sick sinners, or if they would at any rate face an eternity of torture with more indifference. He does not feel that he is in his element. The farmers look as if they were in their element. They are full-bodied, healthy and contented; but between him and them is a great gulf fixed (p. 60)

– and that is the gulf of arduous intellectual training which has effectively destroyed Theobald as a man. His sole emotional outlet is the sadistic training he inflicts on Ernest to make him the same kind of failure, almost as if inherited decay once set in motion is self-perpetuating. Here there is neither room nor occasion to examine the account of Ernest's grim childhood in detail; it must suffice to say that the whole upbringing, though in essence autobiographical, is arranged to support *Life and Habit*'s dictum that birth 'is commonly considered as the point at which we begin to live. More truly it is the point at which we leave off knowing how to live.'[18] While in the womb the embryonic Ernest constructed a neat and efficient body-shell for himself; but the young adult Ernest, cast into the world as a London clergyman, finds his lack of real biological wisdom has at first absurd and then calamitous results. This is adumbrated in the small and light scene of Ernest as a curate preaching on the barren fig-tree, making the botanical mistake that fig-trees show their blossom (pp. 233–4); it is fully realised in the black comedy of Ernest's failure to detect the critical difference between the Misses Maitland and Snow; and it is finally pressed home in the magistrate's sentencing address – one of the most beautifully controlled examples of Butler's reliance on speech cadences to convey irony.

Ernest's fall from grace into prison is no smooth decline. For a time in his boyhood while under the careful ministration of his Aunt Alethaea, Ernest is able to turn back with relief to his latent identity as Old John, and he starts to exercise his 'muscle' of practicality by constructing in his turn another wooden organ. Personal evolution, no less than organic, *is* growth, change and development. While Butler's handling of Ernest's imprisonment must strike every reader as absurdly unrealistic,[19] this is certainly the crisis of his hero's life and the climax of the novel. It is not surprising that here, too, is revealed most distinctly Butler's deviation from orthodoxy over the evolutionary significance of the potency of the individual will. Since for a Darwinian each separate organism is entirely a product, first of a random permutation from among a near-infinite number of genetic variables, and then of an environment which moulds its behaviour, he must believe that when put under stress the organism can do nothing but die and clear the field for the better-adapted. As a neo-Darwinian by deepest conviction Hardy explores these axioms with almost frightening intensity in *Tess of the d'Urbervilles*. But *The Way of All Flesh*, as a neo-Lamarckian novel, does nothing if not extol the powers of the blindly groping will. Of course Ernest is a wretched creature with both an unstable ancestry and a miserable home life bearing down on him – things he shares with Harry's heroine – yet for all that there is also what he *chooses* to do with his nature and nurture. It naturally does not follow that Ernest's choices are all fully aware ones. One success of the novel is that it does dramatise semi-conscious impulses which help to mould a destiny. But the overall impression is of freedom, not of determinism.[20] If the game were being played out to Weismannian rules, Ernest would succumb; instead he resists and struggles to remake himself into something nearer society's desire – not his own heart's, for that is a luxury permitted only in a tragedy or a romance, and Ernest is neither a tragic nor a romantic hero.

Overton's admonition –

> if a man is to enter into the kingdom of heaven he must do so not only as a little child but as a little embryo, or rather as a little zoösperm and not only this but as one that has come of zoösperms who have entered into the kingdom of heaven before him for many generations. Accidents which occur for the first time and belong to the period since a man's last birth are not as a general rule so permanent in their effects (p. 241)

– is really an impossibility proposed for rhetoric's sake, just as its biblical source is. Ernest, no more than Nicodemus, cannot enter again into his mother's womb or reselect his genetic inheritance; so as he rises into health under the stimulus of the drastic 'cross' of prison and insolvency he chooses the next best alternative of regressing economically – digging down to social bedrock and anchoring himself to it by marrying the former scullery-maid Ellen and opening an old clothes shop in the East End. But Ernest, though a successful shopman, is not destined to remain one. 'Nature is eminently conservative,' wrote Butler in that last essay of which almost every sentence has a pointed reference to the novel, 'and fixity of type, even under considerable change of conditions, is surely more important for the well-being of any species than an over-ready power of adaptation to, it may be, passing changes' ('Deadlock in Darwinism', p. 45). Ernest's circumstances of poverty are just such 'passing changes'. Learning to be rich, to manipulate wealth, is his final adaptation, for his great expectations are his most decisive inheritance. Significantly, in view of what has been said already, the point where those expectations are fulfilled is just the point from which the novel rapidly decays as Butler's own obsessions become progressively less tinctured with wit, rapid action and sheer intellectual *élan*. At first the injury is not too serious. While Butler, via Overton, treats the financial pretensions of his hero with his customary blend of cynicism and quasi-scientific detachment ('A man can stand being told that he must submit to a severe surgical operation . . . but the strongest quail before financial ruin': p. 251), all remains well. It is only when the tone becomes heavy and preachy, redolent with confiding Butlerism, that the reader's aesthetic distancing collapses as he becomes aware that he has stumbled into quite a different book. Once again we have to rummage in *Life and Habit* to explicate the exact degree of failure. In a most ingenious paragraph, Butler there compares with deadly seriousness the slow accumulation of human capital investment over the centuries with the gradual improvement of organic function. As it is with men,

> so with the limbs and instincts of animals; these are but the things they have made or bought with their money, or with money that has been left them by their forefathers, which, though it is neither silver nor gold, but faith and protoplasm only, is good money and capital notwithstanding.[21]

About halfway through this gallimaufry, and reading back into the novel, we realise that it is really the converse proposition which interests Butler; that having money is itself a sign of evolutionary grace. To become rich, whether by personal effort or by inheritance (and, if we assume the identicality of the generations, living on inherited money, far from being parasitical, is the reward for the efforts 'you' have made in the shape of your parents or remoter ancestors), is to be signalled as naturally acceptable.

In a late essay he pushes even this oddity forward to the stage where he is saying that those people who are not highly organised enough (he means, not rich enough) to have 'grown' a solicitor or banker can generally repair any loss to their integration as freely as lizards grow new tails.[22] So Ernest, in poverty, regenerates himself as a shopkeeper; later, burdened by a large fortune, the failure of his bank, we are to suppose, would be as fatal as a coronary thrombosis! It is easy to mock this kind of absurd transposition of facts, where Butler's irony turns back and rends itself to pieces; it is easy and accurate to remark that the obsession with money and its powers revealed here is almost pathological. *The Way*, says Claude Bissell, 'is a study of the redemptive powers of the unconscious. But it might also be defined as a study of the redemptive powers of money' ('A Study', p. 166). 'Redemptive': exactly. Given the imaginative identification of financial and biological inheritance, and given that the latter is often defective and maiming (for Ernest, for Butler himself), the logic of the novel looks to the first to rectify the maladroitness of the second. And this in turn stands behind such artistically and morally obtuse scenes as Ernest's confrontation with his father as his mother lies dying. Theobald notices that Ernest looks healthy and affluent:

> here he was swaggering in grey Ulster and blue and white knecktie, and looking better than Theobald had ever seen him in his life. It was unprincipled . . .
>
> Ernest saw what was passing through his father's mind and felt that he ought to have prepared him in some way for what he now saw . . . He put out his hand and said laughingly, 'Oh it's all paid for – I am afraid you do not know that Mr. Overton has handed over to me Aunt Alethaea's money.' Theobald flushed scarlet . . . (p. 231)

Amid such scenes as these the intellectual force and delight of *The Way* are squandered to the point where Malcolm Muggeridge's castigation of it as one of the most *ignoble* novels ever written strongly advocates itself. Why are Ernest's jeering here at his ailing parents and several comparable scenes in the third volume so inexpressibly vulgar? Not inherently because of the petty-mindedness they expose beneath the carefully fashioned biological conceptions which supposedly vindicate them. The failure is not ideational, but artistic. Such scenes are etiolated because they don't work; because they are imposed on characters whose consistency of behaviour has quite suddenly become suspect. We don't believe Ernest fosters out his children to a family of bargees so they may grow uninhibitedly into fresh John Pontifexes unable to give any account of their deepest convictions. Ernest does it – Butler has him do it – for the same reason that he is laden with unearned income before the last chapter: to make him free, to deliver him from the tomfooleries of the social pecking order – because, finally, he is selfish. All these inadequacies centre most disconcertingly on Towneley, the supreme figure of instinctive *savoir-faire* who hovers elegantly on Ernest's horizons as the archetype of successful adaptation. Now, while Butler relates the changing fortunes of the Pontifexes we are temporarily inveigled into considering heredity as persisting intergenerational habits. While Ernest is exposed to the Battersby atmosphere of claustrophobic torment we welcome being told with supportive examples from natural history that filial hatred is natural. But at *no* point whatsoever is Towneley made appealing as the apex of the evolutionary pyramid, the man who has transcended the makeshift of brains to live graciously in unself-consciousness. He is incredible because the reader is being clumsily manipulated, being tugged into a web of neurosis that he cannot sympathise with because the facts, of which homosexual attraction is the key one, are not to hand.[23] Useful here is P. N. Furbank, who in what remains the sharpest short critical study of Butler offers the most valuable and succinct generalisation that 'however distant a (scientific) notion starts in this writing, it is attracted towards the emotional centre' (*Samuel Butler*, p. 17) – and in being sucked into the vortex, we might supplement Furbank, it is transmuted into imitation biology. This works no less for Butler's own contributions than for facts which he extracted from other sources; and as it is the basic process by which Butler's creative imagination fuelled itself it is worth looking at a pair of examples more closely.

One noteworthy case where Butler's grasp of his own biological notions suffers analogical transformation in the novel is his use of his own original explanation for the consequences of breeding hybrid organisms, a tactic commonly called 'crossing'. His account (*Life and Habit*, pp. 141–8) is conventional enough even in its inconclusiveness, as he wrestles with the problems of interspecies sterility (which could not be resolved without a knowledge of Mendelian genetics), and also in its prejudice, as he comments on the wickedness of half-castes. His own explanation for hybrid sterility is that it is caused by too strong a conflict between the 'memories' provided by each parent; that the hybrid is rendered sexually ineffective by being 'pulled hither and thither by two conflicting memories or advices' (p. 141). Still, provided we make full allowance for Butler's use of the term, 'crossing' conforms well enough with such standard usages as Darwin's: that is to say, the intermingling of the characteristics of two different strains. But in *The Way* we are confronted with examples of crossing which have nothing at all to do with the breeding of hybrids. By some labyrinthine analogical process the technical word apparently acquires the extra and more familiar meaning of *thwarting*, or *opposing*, or *going counter to*. Butler lets the term escape from stockbreeding into the arena of the interaction between the individual and his environment, and makes it a key to a full understanding of Ernest's slow rise to full spiritual health. We are required to see his early experiences as having been altogether too great a cross, so that it is not until his imprisonment that the native powers of resistance inherited from his great-grandfather rise from supinity; and it is not until the dissolution of his marriage that Ernest is struck down with that 'relief from tension' which only the advice of the psychiatrist (yet another Butler persona) alleviates. He tells Overton of the urgent medical need to 'cross him':

> Crossing is the great medical discovery of the age. Shake him out of himself by shaking something else into him . . .
>
> Seeing is a mode of touching, touching is a mode of feeding, feeding is a mode of assimilation, assimilation is a mode of re-creation and reproduction and this is crossing – shaking yourself into something else and something else into you.
>
> He spoke laughingly but it was plain he was serious. (p. 306)

And he goes on to make his prescription of 'a course of the larger mammals' at the zoo; Ernest is enjoined, however, to leave the *Simiae* alone: 'the monkeys are not a wide enough cross; they do not stimulate sufficiently'. A little later Ernest is seen sauntering round the elephant-house 'drinking in large draughts of their lives to the re-creation and regeneration of his own' (pp. 306–8). This little episode within eight chapters of the end is clearly to balance out the violent crosses that Ernest has earlier suffered, and again raises the perennial problem of tone. Is it anything more than diverting? To the majority of readers, innocent of Butler's theory of interspecies mnemonic continuity, probably not. Even to the rare reader with *Life and Habit* at the fingertips, such a metaphorical transition (from 'cross' equals hybrid, to 'cross' equals therapeutic opposition, almost psychotherapy), far from taking a biological theory and welding it to mundane life, rather plays fast and loose with it.

A second example of the extension of Furbank's point shows more clearly the deforming of fact under emotional pressure, since it begins with facts in the public realm. Towards the end of *Life and Habit* (ch. 11) Butler draws extensively on Théodule Ribot's *L'Hérédité* and on Sydney Smith to supply him with facts about instinctive behaviour which he can propose in turn as examples of inherited memory. For a whole page he quotes Smith on the habits of the sphex wasp which, laying its eggs alongside the bodies of previously paralysed insects, guarantees its grubs a food supply when they hatch *after* their parents' death. This is only one supportive case among many piled up in this chapter, and Butler has little difficulty in preserving his scholar's pose of detached demonstration. Early in *The Way of All Flesh*, however – it is a tempting but unprovable supposition that the two passages were written close together – exactly the same facts are placed into a context which is far more than merely descriptive.

> Why should the generations overlap one another at all? Why cannot we be buried as eggs in neat little cells with ten or twenty thousand pounds each wrapped round us in Bank of England notes, and wake

up, as the sphex wasp does, to find that its papa and mamma have not only left ample provision at its elbow but have been eaten by sparrows some weeks before. (p. 71)

The locus here is the senile grumblings of George Pontifex after his godson's christening party and, while entertaining, is an acid enough comment on filial obligation. Perhaps its saving grace resides in its hovering hesitantly between the poles of simple wit, irony and velleity without gravitating to any one of these, which would force the reader into making a decisive judgement about it. Because he can remain aloof he can simply enjoy. But Butler has not drained dry his metaphor. Moving on thirty chapters we come across this:

> He was big and very handsome . . . very good natured, singularly free from conceit, not clever, but very sensible, and lastly his father and mother had been drowned by the overturning of a boat when he was only two years old and had left him as their only child and heir to one of the finest estates in the South of England. (p. 188)

Here there can be no multivalence and no highly wrought irony. For to read *this* critical passage as ironic, where Butler is zealously soliciting our admiration, necessitates reading the whole novel as an exercise in Menippean satire or as a kaleidoscope of value patterns. And the entire didactic purpose of *The Way* militates against this; the ending, for instance, when Overton, Towneley and Ernest almost merge into a single personality, may be inept and morally repugnant, but it genuinely *is* an ending within the Victorian ethos of novel-writing. What we do have here is the familiar grasping at the reader's sleeve; we have Butler doing more than finding correlatives for the emotional wounding of his childhood; we have an elevation of this hatred into a bogus ethical deduction from biological fact: that it is natural to hate your parents and therefore that it is right. This bounding from *is* to *ought* must by this time ring familiarly. Instead of the superbly daring iconoclast pitching out all the flaccid Victorian proscriptions, we are startlingly confronted with a regression, a harking back to the Darwinian moralists of the 1860s and 1870s, who too hastily defined human conduct in terms of the survival of the fittest. Through the oddities of its composition *The Way of All Flesh* (and this goes for the evolution books as well) all leapfrog T. H. Huxley's admonition about evolutionary ethics and reveal that their fundamental suppositions are those of their originator's youth and of the culture that nourished that youth.

Butler and Heredity Summarised

The conclusion to which the foregoing reading tends is that after a little probing all of Butler's literary and scientific reflexes stand revealed as middle and late Victorian. Despite his superficial modernity he even gives *The Way* the conventional structure of three volumes and a happy ending. Ernest himself, to be sure, has attributes of the 'modern' hero in his isolation, his often pathetic yet obdurate reliance on his experience alone, his deracination both personal and cultural, even his inconclusive destiny: these things are his inheritance, just as they are for Stephen Dedalus or Camus' Meursault. It is to be observed, though, that one can make such claims for the seminality of this novel without any reference to past or present biology. The mould into which Butler poured his otherwise ungovernable feelings may at last, its purpose served, be broken and thrown away. And that is the most severe critical judgement one can pass on *The Way of All Flesh*; but it is not from the present perspective a wholly destructive judgement, for the novel still remains open to us as a supreme product of intellectual conflict. Other authors of the last century used biological notions to creative ends, but no one other than Butler first established with such relentless exactitude his position with respect to Darwinism proper. Others satisfied themselves with a little carping at the natural selection mechanism and with a refusal to believe (for not very good reasons other than those sanctioned by emotion) that this factor alone made for evolutionary advance. W. H. Hudson, Richard Jefferies and most of the other writers who looked at the matter at all took this stance. A few others – by temperament stoical – uncritically allowed Darwinism to confirm their reading of life: Meredith and Hardy. The peculiar interest of Butler is that one can see in fascinating detail the two usually opposing halves of a sensibility – the imaginative and the cognitive – here marching very closely in step, each sharing the often profound adequacies and inadequacies of the other. For, as we have seen, Butler certainly did not work out a philosophy and then write a novel around it – the sequence of events which gives Hardy's *Tess* its richness of mature and accomplished insight. Instead, a fresh idea about heredity was for Butler an artistic experience modifying his sensibility as novelist; and reading his novel we are in the presence of an almost Metaphysical mind in perpetual and energetic motion.

Tess and the Darwinian Motif

> Prof. Huxley once compared life to a game of chess played by man against an enemy, invisible, relentless, wresting every terror and every accident to his own advantage. Some such idea must have influenced Mr Hardy in his narrative of the fortunes of Tess Durbeyfield.[24]

So began an anonymous review of the newly published *Tess* in January 1892, inviting the first readers of Hardy's novel to seek for critical connections between its implicit assumptions and the world of late-Victorian speculative biology. The comment is at once perceptive and evasive; and its note of caution is premonitory, for eight decades of analysis have hardly enlarged our understanding of this aspect of the novel. We find, for example, Harvey Webster in his standard treatment of Hardy's intellectual development reporting that in *Tess* he believes 'heredity is an important and credible factor in the heroine's development'.[25] The point is not enlarged upon. Arnold Kettle handles with assurance his argument that the theme of *Tess* is fundamentally the rural decay produced by the progressive intrusion of agribusiness and that the characters, not to be seen 'merely as individuals', are largely caught up in this impersonal economic flow. Though he notes that the natural world is not a 'back-cloth' but 'the living challenging material of human existence', Kettle neglects the rich implications of this last phrase.[26] John Holloway takes the line that the plight of the rural worker is where the novel begins rather than ends; and part of his important essay moves away from Kettle towards a partially Darwinian reading. Hardy's artfully contrived selection of country matters does not, for him, plumb the depths of the novel; behind the absentee landlords, the slave-driving tenant farmers, the inhuman mechanisation stand certain solidities quite impervious to amelioration: 'the essential things which make life hard on [the land] are those which have made rural life hard since the beginning'. Gouging out frozen turnips, Holloway is contending, excoriates the spirit whether one is being driven by starvation or cruel master – or even being drawn along by a mechanical hoe. In the short space of his essay, though, Holloway merely offers a Darwinian lexicon – 'organism, environment, struggle, resistant adaptation, fertility, survival'[27] – without more than hints on how to apply it, or how those concepts are absorbed into the imagerial and metaphorical structure of the narrative.

These examples of intimation rather than exploration are quite typical of treatments of the biology, particularly the Darwinism, in Hardy's novels.[28] An important reason for this critical wariness is surely the felt need to defend him against the ready charge of risible intellectual pretensions: his readiness to lard his stories with rather too self-conscious science. It is easier and more comfortable to relegate Hardy's considerable knowledge of theoretical biology to Background Reading than to invite the comment that *Tess*, for example, is 'overloaded with ideas' which 'sometimes jut out awkwardly from the fabric of the book . . . [ideas] not so much inappropriate as imperfectly worked in'.[29] There is a genuine point here, for (to take an example from Hardy's similar use of classical mythology for a moment) the invocation of Aeschylus in the very last paragraph is such a signal failure precisely because it jars us into a different world. In the course of the novel we have detected no trace of the Greek pantheon at play, and the last page is an unconscionably delayed entrance. But the general issue cannot be burked by any such blanket dismissal, for the neo-

Darwinism in *Tess* is by no means simply a question of a few citations from the *Origin* forced into the story. Instead, Hardy's use of the very freshest data on degeneration, sexual selection and, above all, on heredity is in the strictest sense aesthetic, for these themes actually commingle with and order the narrative flow.

Hardy, *Tess* and August Weismann

That Hardy should have been a Darwinian in the 1860s is not at all surprising. That he should have continued to be one right into the 1890s is something which, in the light of the preceding pages, we can see sets him apart from almost every other writer of his generation. Few of Hardy's critics have found it exceptional that a novelist should remain firmly loyal to the profoundest synthesis of his day. The exceptionality only becomes apparent when one charts the progressive decline of Darwinism as an article of serious belief among both biologists and men of letters throughout the last half of the century.

Hardy, aged only 19 when the *Origin* appeared, probably read it almost at once; he had certainly done so by 1864. His response was immediate and vigorous. The sudden onset in him of a largely undesired agnosticism; the revulsion from nature; the temporary spiritual paralysis: all these are very well documented, both factually in his semi-autobiographical *Life* and emotionally in his earliest poems. But after a while his high seriousness of temperament, once deviated into the channels of rationalism, renewed and augmented itself. 'No intellectual influence as important as that of Darwin affected the form of Hardy's thought,' Harvey Webster has said of Hardy's formative years.[30] This judgement is equally applicable to his mature years, too, for there is no evidence that Hardy ever seriously questioned that the definitive exposition of the vicissitudes of all life was to be found between the covers of the *Origin* and the *Descent of Man*. Hardy himself was in no doubt who his intellectual masters were. On two occasions thirteen years apart, in remonstrating with critics who had tried to identify Schopenhauer as a thinker important to him, he quoted practically the same list of names: 'Darwin, Huxley, Spencer, Comte, Hume, Mill and others.'[31] Here the order of priority – four evolutionists first, two biologists and two philosophers – indicates clearly enough that Hardy was sufficiently intellectually parsimonious to spend his creative life manipulating the insights of his youth in the 1860s, just as he was sufficiently emotionally parsimonious in exhuming emotions forty years buried when he came to write the love poems of the last phase of his career. Before writing his first novel in 1871 he was conventional in sweeping away the metaphysical rubble and founding his fiction squarely on the foundation of post-Darwinian rationalism. But, more impressively than this, he at once slashed through to the very heart of the biologists' revaluation of nature and annexed natural selection as his most potent inspiration. His profound unconventionality lay in his continuing to regard Darwinism as a discontinuous mutation in the history of thought: a theory without antecedents, in need of no softening modifications, to be accepted without reservations. That is his uniqueness. One might perhaps make out a case that by the time of *The Dynasts* (1904–8) even Hardy had been affected, in his conception of the Immanent Will, by the continuing thrust away from Darwinism. There is little or no biographical evidence for this, but there is the simple fact that while the novels of the 1890s are uniformly pessimistic *The Dynasts* does end on a note of modest hope: 'Consciousness the Will informing, till It fashion all things fair!' (III:256). But the Immanent Will is itself a personification of natural selection. The metaphor Hardy most commonly employs is that of a brain, whose nerve fibres are the pattern of human events in space and time. Certainly the Will's potential growth to self-awareness is not presented as in any way automatic. Though Hardy's concept of a life-force is something like the other formulations of the time – Bergson's, Butler's or Shaw's – it lacks, on closer inspection, the irresistible drive upwards which is characteristic of those. But these caveats in any case throw little light on *Tess*: the Immanent Will is no more present there than are the Aeschylean Furies.

When, therefore, the *Athenaeum* critic in 1892 linked Hardy's and Huxley's views of nature he was, deliberately or not, isolating both outside the main theoretical trends of the late 1880s. It would be idle to maintain that Hardy could have remained ignorant of the anti-Darwinian party, despite the lack of hard evidence, for he knew many scientists and moved socially within the orbit of professional biologists: a note, for example, of May 1893 shows him taking an ingenuous pleasure in holding his own at a Royal Society social evening.[32] The *Life* records almost verbatim a conversation later the same year with Sir James Crichton-Browne, noting the latter's orthodox view that 'the doctrines of Darwin require readjusting largely; for instance, the survival of the fittest in the struggle for life. There is an altruism and coalescence between cells as well as an antagonism.'[33] Hardy restricted himself to a non-committal 'Well, I can't say', but on any reading his last two novels strongly imply that the then fashionable doctrine of Altruism appealed to him no more than did Lamarckism.[34] Sensibly, he took from biology only that which he could assimilate, justifying this procedure in one of his rare statements about his artistic strategy: 'in life the seer should watch that pattern among general things which his idiosyncrasy moves him to observe, and describe that alone'. The presence of an infinity of figures in the carpet permitted him any interpretation of nature that chimed in with his temperament, and one such interpretation happened to be Darwinism *redivivus*. But Hardy's biological interests, though genuine enough, were in no sense scientific: as this note implies, it was largely accidental that his aesthetic and emotional preoccupations happened to fit in with an 'objective' structure of thought; in different circumstances, one feels, Hardy would have extracted all he needed from a purely *a priori* system, as Blake and Yeats did. 'The result,' he then concludes his note, 'is no mere photograph, but purely the product of the writer's own mind' (*Life*, 153).

Hardy's diary shows that he had fairly complete access to Weismann's new underpinning of Darwinism when he settled down in August 1889 to write the novel which became *Tess*; but since the *Life*, our only direct source of information bearing on its composition, has been shown to be seriously defective the exact details must remain elusive. The planning of the novel apparently began as early as September 1888, with Hardy's jotting on the fortunes of his own family (*Life*, 214–15), and sixteen chapters were complete in some form one year later. There followed in October and November three rejections by various publishers on the grounds of impropriety, which led Hardy to mutilate his novel (while, presumably, in the flow of composition) so that it was at last accepted in a cut form by the *Graphic* for serialisation. Nothing appeared in print before May 1891[35] and according to Weber's account even after the third rejection (25 October 1889) Hardy was still either writing or actively revising parts of the novel. Weber claims, for instance, that it was 'in the period of enforced delay' that Hardy developed in detail the idea of great families having humble descendants; and he puts down the fully rounded character of Tess herself to the same 'prolonged delay in bringing his novel to completion'.[36] Hardy was left free to work on his nascent novel throughout 1890 and doubtless he did so, for even as late as March 1891 he was still 'putting the finishing touches to *Tess*' (*Life*, 234); and we note that this does not simply mean restoring the episodes removed for serial publication, for this final task was done 'in August and the autumn' (*Life*, 238) before the book was published in November. *Tess* from notebook to novel took three and a quarter years, including the hiatus of the whole of 1890. It was in the latter half of this year that Hardy 'dipped into' Weismann's *Essays upon Heredity* (*Life*, 230) – almost certainly while undertaking, or just after he had completed, the major revisions and reconstructions which Laird's close study of the manuscript itself has shown belonged to this year.[37] Though Laird dates the construction of the heredity theme from virtually nothing to its final vital form to approximately December 1889 – autumn 1890, it is nevertheless the case that some of the most effective passages on the heredity theme definitely did not come into the text until Hardy had finished with the manuscript altogether. The d'Urberville portraits of Tess's ancestors are to be found only in the serial *Graphic* version; the description of the psychological effect of these on Tess as she ascends the stairs with Angel to wash after the journey (a scene found first in the *Graphic*) is supplemented by two additional passages not to be found until the first edition of November 1891. It is also noteworthy that Hardy's important authorial outburst against Wordsworthian nature (to be considered later) finds its first extant treatment in the *Graphic*, for the relevant folio of the manuscript is missing.

To summarise these details of chronology, it is impossible to give an exact account of the responsiveness of Hardy's creative imagination in *Tess* to his reading of Weismann. But that biologist's work was especially well known, as we have seen, outside the ranks of his fellow-scientists; and, though the exact timing must remain vague, Hardy came across a most lucid exposition of Weismannism at a time of intense creative effort. Whether *post hoc* necessarily in this case implies *propter hoc* can never be formally proven; but an unforced careful reading of the novel itself and of a few related poems is certainly suggestive.

The *Essays upon Heredity* distinctly belong to the honeymoon phase of neo-Darwinism. The translators of them went to some pains in their eulogistic preface to give an assurance that Weismann's topics deserved a wide audience and that four of the *Essays* especially should appeal to those 'who approach the subject from its philosophical or social aspects'.[38] Hardy was himself this kind of reader, and it is easy to guess his motives for looking into a book which was well advertised in the general press: it was surely to seek confirmation of the idiosyncratic views on inheritance which he had already formed and was making use of in his growing novel. Perhaps, therefore (this must be surmise only), he turned at once to the second essay, 'On heredity', which began as a public lecture and so has the virtue of dispensing with the wealth of references and neologisms used elsewhere. This essay, like all the others, is concerned not with the reporting of new experimental work but with rapidly reviewing and marshalling the reasons for his proposing the two central ultra-Darwinian tenets: the all-sufficiency of natural selection and consequent redundancy of 'soft' inheritance-theory; and the complete immutability of the germ plasm as it creates and passes through the bodies of generation after generation of organisms. Even in this oversimplified form these tenets may seem to be poor fodder for the creative imagination, but this was not so for Hardy. For Weismann reconfirmed the pessimistic side of Darwinism, the side turned downwards since the 1860s, and that was just the side that Hardy found the most congenial. Weismann's teaching, for example, made human progress a very dubious concept, for the neo-Darwinist could no longer rely on a sort of reversed genetic Gresham's Law making the best elements in society drive out the worse. He knew that there was no particle of inheritance for social virtue; that the 'unfit' (the old bogey was raised again) had constantly and unremittingly to be extirpated. There is direct evidence, too, of Hardy's sympathy with this way of thinking. Though its inspiration cannot be traced satisfactorily to these essays specifically, the poem 'Heredity' (written about fifteen years afterwards) asserts strikingly Hardy's imaginative retention of Weismannian ideas.[39] The poet lets the genetic principle speak out directly:

> I live on,
> Projecting trait and trace

Through time to times anon,
And leaping from place to place
Over oblivion.

'Projecting' here, with its connotations of shadows thrown out and perhaps on to a screen by an inner shaping agent, is a precise description of the continuity of the germ plasm which, being the only enduring principle, produces one structure after another (the somatoplasm) to house itself; and that it has in fact been 'continuous through numberless generations of perishing organisms, from the first origin of sexual reproduction till the present time',[40] is surely a conception dramatic enough to strike even the most imaginatively flaccid. Its apostrophe at the end of the poem as

The eternal thing in man,
That heeds no call to die

deliberately awakens religious echoes supposedly suppressed by Darwin. The plasm is all that is now left to us of human spiritual essence, of the soul. Our only immortality is to live in others.

It is curious that Hardy wrote and published in 1892 a serial, 'The Pursuit of the Well-Beloved', in which the hero doggedly chases the inherited face of his loved one through three generations from mother to grand-daughter. This clumsy and tasteless romance, one of the oddest items in the Hardy canon, shows up the inherent absurdity of trying to dramatise biology directly – something which his masterpiece of only the year before resolutely avoids doing. For it cannot be stressed too strongly that *Tess* is not 'about' Darwinism, old or new, any more than it is exclusively about rural industrialisation or the abuse of land tenures or the injustice of sex-biased moral codes. Weismann, who for our purposes may be taken as representative of all those workers on the problems of heredity with which Hardy was certainly familiar, perhaps added nothing more specific to the plot of *Tess* than offering the opportunity to focus within the limits of one science his insights into time, chance and the irreconcilability of the human and natural orders.[41] Neo-Darwinism, on the other hand, does control some of the imagery and movement of a tragedy which occurs within a fictional universe operating according to wholly consistent if non-Aristotelian laws.

The structure of *Tess* conforms in broad outline with the precepts of Aristotelian tragedy in that the narrative flows smoothly up to the *peripeteia* of Alec's stabbing and afterwards ebbs rapidly away with the cathartic idyll of Phase the Seventh, 'Fulfilment'. In his presentation of *hamartia* and nemesis, however, Hardy moves outside the classical frame and draws rather on the conventions of late-nineteenth-century naturalistic tragedy. In this, the tragic flaw came to be seen as a dynamic factor existing more strongly in the external circumstances of the tragedy – e.g. in the social environment – than in the characterisation or plot; such a view is implicit in Hardy's own definition of tragedy in 1895: 'tragedy may be created by an opposing environment either of things inherent in the universe, or of human institutions' (*Life*, 274), though in *Tess* at least he is at pains to show that these form a single implacable opponent to human endeavour. Such a conception of tragedy intersected with Hardy's renewed interest in biology over the redefinition of the old concepts of fate and *hamartia* in the newer terms of instinct and inheritance. Tess's tragic flaw, her mole of nature, is her alternate impulsiveness and weak dreaminess, 'the slight incautiousness of character inherited from her race';[42] and for *hubris*, the avenging Erinys, Fortune's wheel, Hardy substitutes hereditarian determinism – innate 'wired-in' patterns of behaviour which serve, in the words of the Naturalistic dramatist Maurice Maeterlinck, to centre the interest 'solely and entirely in the individual, face to face with the universe';[43] the universe which in *Tess* is that of all organic life and its processes. As in classical tragedy, the meaning of the drama is progressively revealed to both audience and protagonist in the *anagnorisis*, the transition from ignorance to knowledge. The state of ignorance in which Tess begins is shared at various times by both Angel and Alec, for all of them foolishly believe that they are sufficiently masters of their fate to neglect the forces of heredity. Tess, for example, as she rallies after the death of Sorrow, is mistaken enough to suppose that of her ancestors 'not a thing of all that had been theirs did she retain but the old seal and spoon' (p. 132); she fails to perceive the non-material legacy bequeathed to her by the mindless forces of heredity which now control her motivation and impulses. Similarly Alec's conversion is not very convincing for the simple yet compelling reason, Hardy tells us, that his facial expression has been forced out of its natural and predetermined mould. The educative process which Tess undergoes comes to a climax in the wedding-night scene at Wellbridge where she learns, as much from the portraits of her vicious ancestors as from Angel's renunciation of her, that there is only a 'flux and reflux . . . in everything under the sky' (p. 448) and that no guidance is due to her from either history or nature. So Tess suffers not through any moral deficiency but because she is human in a world not devised for humans, and she is educated not when her moral lapse is clear to her but when she recognises that she lives in a Darwinian world which has neither human significance nor value. The reader is guided to this insight by the omniscient author well before it is forced on Tess herself. Our introduction to the squalid Durbeyfield family is followed by the oblique authorial comment:

> all these young souls were passengers in the Durbeyfield ship . . . If the heads of the Durbeyfield household chose to sail into difficulty, disaster, starvation, disease, degradation, death, thither were these half-dozen little captives under hatches compelled to sail with them – six helpless creatures, who had never been asked if they wished for life

> on any terms, much less if they wished for it on such hard conditions . . . Some people would like to know whence the poet . . . gets his authority for speaking of 'Nature's holy plan'. (p. 24)

This passage, of extraordinary intensity for Hardy, who nearly always preserves a tone of cool irony in his interpolations, forms an extended metaphor working on at least three levels. The microcosmic world of the Durbeyfield family with its bungling fecundity mirrors the universe of the novel as a whole. On the surface the image of the ship's passengers suggests the nexus of social and domestic mores in which the children and Tess, too, for she is a part of all this, are snared. The implicit freedom of 'passengers' does not eliminate the possibility of social betterment; but as Hardy's indignation gathers force the passengers become 'captives under hatches' – just as the children are the slaves of circumstances so is Tess herself shackled by her ancestry. At both these levels were are still in the human world. But at the bottom of the metaphor the Durbeyfield ship is an Ark, and not a secure Noah's Ark under divine protection but a frail bark only just proof against the ravening natural order; the children therein are young animals naked before nature's blank and unyielding visage, condemned to suffer by having the unwanted gift of awareness forced upon them so that birth itself is 'an ordeal of degrading personal compulsion' (p. 456). The attack on Wordsworth sets into motion a current of speculation which runs right through the novel and which continues the reversal of Wordsworthian pantheism which Hardy began sometime in the previous decade with the poem 'Nature's Questioning'. Instead of the poet himself going to receive impulses from the vernal wood, he represents bewildered creatures coming to him for enlightenment. Hardy offers in that poem a range of answers, but in *Tess* the impression generally is that there is no 'holy plan'; or if there is a plan it must be wholly alien and, far from being holy, must be positively diabolic. One charge against Hardy is that his nature rejects its theoretical neutrality and strews rather more pains about his characters' pilgrimages than blisses; and this is indeed true in *Tess*. In fact, there is no consistent attitude towards nature. The scenes at Talbothays, with their images of agricultural prosperity and burgeoning vitality, are strongly set off against the denunciation of Wordsworthian pantheism, and this has caused some readers to dismiss all such sequences as no more than extended pathetic fallacies serving as 'notations of the immediate context, pointing up and identifying, in a quite limited way, the effects being sought'.[44] But this is to short-circuit a central issue of the biological debate, and the power of the novel is diminished by insisting on an excluded-middle logic. The novel is ambivalent towards nature as malignant or benignant precisely because the new biology was itself ambivalent. What we do have is a ransacking of all the resources of Nature and Nurture for the bases of morality; and rock-bottom is reached with Tess's poignant insight 'once victim, always victim – that's the law!' (p. 423). Her fate confirms its truth. It is a truth iterated in Hardy's exploration of the human meaning of sexual selection (a subsidiary theme, as it is in Darwin a subsidiary hypothesis), starting with the May dance which is itself a stub of a fertility ritual. Here Tess's 'pedigree, ancestral skeletons, monumental record' (p. 16) fail to secure her a partner in the sexual battle, for acting without preference Angel takes the first to hand. In the consummatory scene where Angel carries the four milkmaids over the flooded road the girls, with insects caught in their gauzey dresses, are themselves 'entrapped flies and butterflies' (p. 183). Though they bovinely accept the natural dispensation ('such supplanting was to be'), their rational awareness does not release them from individual torture. Every detail is right in the following scene in the communal bedchamber. Each girl seeks escape in sleep; the semi-tropical weather adds to the languor; the dripping cheesewring marks time passing, the emotional pressure and the mechanistic relentlessness. Sexual selection *hurts*:

> the air of the sleeping-chamber seemed to palpitate with the hopeless passion of the girls. They writhed feverishly under the oppressiveness of an emotion thrust on them by cruel Nature's law – an emotion which they had neither expected nor desired . . . each was but a portion of one organism called sex. (p. 187)

The dramatic bite of the suffering here derives from the twin irresponsibilities of society and nature. Tess's own sexual guilt is, to be sure, presented as the artefact of an arbitrary moral code; but is that code any more arbitrary than nature itself? What closer interaction of the social machinery could remedy the plight of the lovesick milkmaids? Only some such abolishment of the sexual pressure altogether as Hudson's dream solution in *A Crystal Age*. The social order in *Tess* is oppressive only because human institutions mirror the universal callousness, just as man's interference can only temporarily palliate the flow of suffering towards the maximum possible for his species. As for nature, not only its moral but also its aesthetic values are suspect, for the latter only bear an accidental reference to the human condition, and frequently mock it. The horse's blood on the road shows an 'iridescence of coagulation' (p. 36) despite the horrific accident which has released it; the refracted light in the water-bottle continues to work on the 'chromatic problem' of forming a spectrum despite Tess's catastrophic disclosure to Angel (p. 291). In short, nature and society unify in a single conspiracy against the individual – an idea which is conveyed with the utmost economy and precision in the lounging figure of the postilion, unemployed and burdened with the weeping chancre on his leg – a stigma which has its origin in his social role (the bruise is caused by his manner of sitting) but its persistent refusal to heal in the imponderables of physiology.

This almost surrealistic figure also condenses into a single image the strong theme of the degenerative collapse both of the rural life generally and of personal ancestry. The disrepair and dilapidation of farm buildings; the impoverishment leading to urban drift and a painful uprooting from a native locale; the seasonal shifts from high summer to dreary late autumn and ravaging winter – all these modulate insensibly into a meditation on the decay of families and indeed of the human rural germ plasm itself. This latter idea probably owes little to Weismann (although the 'On heredity' essay does contain a clear explication of his idea of panmixia) and more to Hardy's genealogical researches into the Durberville family. Still, rural decay is a direct function of its inability to cope with change, for adaptation to a particular way of life has caused fatal rigidities in structure and habit. The city and the country are offered metaphorically as two mutually incompatible organisms competing for the same territory and nutrition. Along the railway line urbanity reaches out its 'steam feeler', a pseudopodium which touches rurality and recoils 'as if what it touched had been uncongenial' (p. 239). The greater vitality of the city is swamping the country; the fitter are prevailing (and the little detail about the watered milk marks Hardy's equivocation); the weaker are failing to adapt and exploit a new niche. Chaseborough has become a 'decayed market-town' (p. 75), Kingsbere a 'half-dead townlet' (p. 461) and the d'Urbervilles a 'worn-out family' (p. 213) for exactly the same reason that the emigrant farmworkers in Brazil die out in the hostile environment: the 'workfolk' are a 'species' trapped by their excessive specialisation – the loads on the wagons, instinctively built the same in every village, are 'as peculiar, probably, to the rural labourer as the hexagon to the bee' (p. 459) – just as the Insectivora have trapped themselves for stability's sake in an evolutionary cul-de-sac. The stirring up of these communities by the new economic conditions has disturbed 'the backbone' of the villages and has induced the mindless migrations from place to place: the frantic mobility is that of the disordered anthill. We are encouraged to read even the history of individual families under the aspect of natural history; for when the rivet-holes in the despoiled tombs of the d'Urbervilles are compared to 'martin-holes in a sand-cliff' (p. 464) it is as if families, like species and individuals, can only enjoy a certain lifespan before the edge of their reproductive capacities is turned. And so we find a succession of images related to medallions and coining, clearly to suggest the stamping of innumerable generations with a consequent blunting of the die. The antiquarian-parson Tringham on seeing the Durbeyfield profile exclaims 'that's the d'Urberville nose and chin – a little debased' (p. 4), and Alec, too, has 'full lips, badly moulded' (p. 44). By contrast The Slopes, home of the Stoke-d'Urberville parvenus, looks like 'the last coin issued from the Mint' (p. 43). And Tess herself, in whose person her vigorous ancestry finds its recrudescence, shows while milking a profile 'keen as a cameo cut from the dun background of the cow' (p. 192).

Hardy's use of Weismann's belief that the germ plasm, the material principle of heredity, is the sole biological reality leads him to impose on the reader the temporal insignificance of the individual carrier of it. In the opening conversation, in fact, the jocular contempt of Tringham towards John Durbeyfield is clear from the way he uses 'you' as though the mere idea of an extinct family has more reality than its degraded offshoots. But, then, that is indeed one of the prime assertions of the novel, as it is in the poem 'Heredity'. There we noticed the image of projection: the casting of inherited characteristics on to a screen of ever-renewed flesh. The same image is picked up in the quiet, superbly handled scene of evening milking-time at Talbothays. As the shadows of the cows are cast on the wall the declining sun copies them 'as diligently as it had copied Olympian shapes on marble *façades* long ago'. Here the biological order has a certain stability, and these individual cows seem in shadow to be as permanent as their species itself. But the illusion of one matches the illusion of the other, the shadows part and parcel with those in Plato's Cave allegory. The sole reality is mutability, the flow of the germ plasm itself; individuals are nothing; generations of cows 'have passed to an oblivion almost inconceivable in its profundity' (pp. 136–7). Hardy, like Darwin in the *Origin*, leaves us to make the final connections. Time opens up and, as though in parallel mirrors, we see receding replications of the organic world – including humanity. Hardy goes even farther than this. Though paradoxically the portrait of Tess – and, to a lesser degree, of Angel and Alec, too – is so intense because he makes her exist not merely as a fully rounded character but even, given our knowledge of her heredity, as a four-dimensional one, another part of Tess's education (and ours) consists in Hardy's showing her and us that her heredity is the thing 'inherent in the universe' which, valueless though irresistible, helps to destroy her; and this despite her dreams of 'an aged and dignified face . . . furrowed with incarnate memories' (p. 45). Hardy may well insist in a later intrusion that 'to visit the sins of the fathers upon the children . . . does not mend the matter' (p. 91), but the question remains: then what does? For the tragedy *can* be traced back to Tess's inheritance of the qualities of her ancestors; to the fact that on her father's side she is an atavism, a 'belated seedling of an effete aristocracy' (p. 297) as Angel angrily but accurately calls her, and that on her mother's side her florescent charms ('unknightly, unhistorical') are the source of her destructive sexual magnetism through which the purely animal powers of nature flow untrammelled. In his attack on Tess after the wedding-night revelations at Wellbridge, Angel alienates the reader with his gross, unjust and hypocritical anger; yet we are forced to perceive that he does penetrate to the centre of Tess's condition and that his recrimination is meaningful: 'I cannot help associating your decline as a family with this other fact – of your want of firmness. Decrepit families imply decrepit wills, decrepit conduct' (p. 297). The words are outrageous enough but, striking out the moral obloquy, we are surely meant to think that they are scarcely more outrageous than the truth. Though Tess at crucial points in her history believes she chooses her destiny, the movement of the novel is such that it hammers away ruthlessly at the tragic self-deception here; for at exactly those points where she feels most free (free to write a letter, free to repulse Alec's advances) something happens to negate that feeling: never a *deus ex machina*, however, for between them heredity and physiology provide the most reductive, the most behaviouristic, of explanations. Weismannian inheritance was deterministic. That Hardy fully accepted it as such emerges from his handling of the numerous pathetic fallacies *Tess* contains. Sometimes these collapse under the weight of the significance that Hardy tries to charge them with; but more often it is very clear that neo-Darwinism allowed Hardy to empathise with natural processes in a way not at all fallacious: that is, as a 'negative Romantic'. At Talbothays, for example, Tess in a state of anguish hears a 'crack-voiced' sparrow sing 'in a sad, machine-made tone' (p. 173). Reading that last odd adjective with post-Darwinian eyes we see that more is going on here than a simple transfer of emotion: it can be read as being, on the most commonplace level of fact, literally true. A nightingale singing with unpremeditated art is Romantic poetic licence: Tess's sparrow sings because its hormones make it sing, because a billion years of selective evolution deny even the possibility of its doing anything else. Given the persistent identification of Tess's fate with that of birds, we can hardly expect her to be allotted any more freedom.[45] But there is one important distinction – though it is not an advantage – to be found in her greater human consciousness, her more exposed nerve endings. At the crisis of the novel, when Tess is actually reduced to the status of a hunted animal, her situation both compares and contrasts with that of the wounded pheasants (ch. XLI). Hardy makes his identification by deliberately breaking down the barriers between artificial and natural breeding, showing the absurdity (in an existentialist sense) of the second in terms of the first. The pheasants are brought into being solely for the shotgun. Such breeding is in every sense unnatural, and it remains so after we recognise the paradox: nature itself can be unnatural. The sufferings of Tess, like the pheasants', are terminated in nature only by the inability to bear more. But the contrast is insistent. Their miseries are quickly relieved by a broken neck; Tess's greater potentiality for pain drives her on, but only to the same fate. The pointlessness is appalling.

A reading of *Tess* from the point of view of inheritance theory, then, must assert the novel's rigid determinism, its total pessimism, and its fidelity to one coherent scientific vision of the world. The large number of coincidences in the plot have traditionally been thought to spoil its determinism. But are those coincidences really offered to us as collocated but random events? Do not most of them stem from a sense of character as fate? When Tess goes to the Clare house and finds them at church, that is coincidence. But her subsequent retreat is fully explicable in terms of her slightly vacillating temper, and is proffered as such. Michael Millgate agrees that accidents in *Tess* 'are rarely if ever a product of the random inexplicable intrusion of fate but rather of the inevitable – or at least, credible – outcome of the immediate narrative context of their own personalities as conditioned and limited by the forces of heredity and environment'.[46] Exactly: Hardy distinguishes between random and merely chance events just as the Victorian biologists found it necessary to do. A large number of chance events may simulate purposiveness, and therefore suggest either helpful design or malevolence – for these are opposite sides of the same illusion – in the natural world. But Hardy the novelist agrees with the naturalist that to invoke chance is only to admit ignorance of the exact links in the causal chain. Neither doubts that the chain exists and is unbreakable. In general, then, the coldest possible eye is cast throughout on the human condition, though one must always remember that Hardy's adoption of neo-Darwinian inheritance theory and with it a set of dismal inferences was always a purely voluntary action, for his reading was a selection from the facts, even from the scientific facts, just as was Butler's advocacy of neo-Lamarckism. There is no doubt, however, that Hardy himself regarded his novel as an epiphany, a 'moment of vision', to use the phrase which he applied to a group of poems in which he speaks out with his own voice. One of these poems, 'The Pedigree' (dated 1916 by Hardy), strongly supports this contention. As the poet studies his family-tree late at night circumambient nature echoes the degradation he feels: the clouds are 'green-rheumed' and the moon is 'in its old age'. He perceives that the biological leverage his ancestors exert is irresistible, that 'every heave and coil and move I made/Within my brain' was 'As long forestalled by their so making it'. And so on to the blank conclusion

> 'I am merest mimicker and counterfeit! –
> Though thinking, *I am I,*
> *And what I do I do myself alone'*

which coincides exactly with Tess's refusal to learn history: 'what's the use of learning that I am one of a long row only . . . The best is not to remember that your nature and your past doings have been just like thousands' and thousands' ' (p. 162). But Tess is forced against her will to remember it. In this her tragedy is absolute and is not relieved by her undoubted nobility of character; for her struggles, her fine-spirited refusal to submit to the inevitable, merely serve to lime her the more thoroughly. And what is true of Tess, this poem suggests, is equally true of Hardy himself. The malignity of nature, which so often seems to be our active opponent, is actually a subjective impression filtered through a consciousness too refined for a Darwinian world of defect where botching and making do are the very means of advancement. What hope there is in the novel – little indeed – springs not from Tess's individual heroism but from that single vital source, the force of inheritance itself. In the very last statement of the book, as 'Liza-Lu and Angel join hands and go on in a kind of genetic parable, Tess's last hope that 'if she were to become yours it would almost seem as if death had not divided us' (p. 505) is fulfilled and the genotype is made secure. The sheer heartlessness of the conception is fitting.

Notes: Chapter 7

1 The near-contemporary biographies of both men are minutely detailed, recording even the most trivial acquaintances. Hardy does not figure in Henry Festing Jones, *Samuel Butler, Author of 'Erewhon' (1835–1902): A Memoir*; nor does Butler in Florence Emily Hardy, *The Life of Thomas Hardy, 1840–1928*. Hereafter cited as *Memoir* and *Life* respectively.

2 That *The Way of All Flesh* is defective as a novel because it fails to absorb its scaffolding is implicit throughout Malcolm Muggeridge's scathing biography of Butler, *The Earnest Atheist: A Study of Samuel Butler*. A brief but concentrated defence is made by P. N. Furbank in his *Samuel Butler*; but most recently William H. Marshall in *The World of the Victorian Novel* has boldly tried to set the novel up as a truly unified work rather than a 'non-symbolic document' by pointing out the metaphorical transformations of concepts which in fact do *not* appear in the 'evolution' books. As will be evident, the reading which follows starts from much the same basis as Muggeridge's.

3 'Before Life and After', *The Complete Poems of Thomas Hardy*, ed. James Gibson, p. 277.

4 Probably this phrase is Butler's own, but it is actually taken from Jones's summary of a letter from Butler to T. W. G. Butler of 18 February 1876: *Memoir*, Vol. 2, p. 445.

5 Samuel Butler, *Ernest Pontifex, or The Way of All Flesh* ed. Daniel F. Howard, p. 77. Howard's edition is certainly not free of its own introduced errors, but it is the only authoritative text and is used here throughout.

6 This phrase is from a letter to Charles Darwin dated 15 April 1873, outlining his plans for what turned into *The Way: Memoir*, Vol. 1, p. 189. At this time Butler was a minor acquaintance of Darwin's, though on good enough terms to visit him twice in that year. Butler also supplied him with some drawings used in the second edition of *The Expression of the Emotions in Man and Animals* (1873).

7 Letter to E. M. A. Savage of 16 June 1872: *Memoir*, Vol. 1, p. 160.

8 *Life and Habit* (IV), p. 43. Unless noted otherwise, all Butler quotations are from the Shrewsbury Edition, ed. H. F. Jones and A. R. Bartholomew. The roman numerals identify the volume number of this collected edition.

9 Letter to E. M. A. Savage, dated by Jones 15 January 1876: *Memoir*, Vol. 1, p. 233.

10 Claude Thomas Bissell, 'A study of *The Way of All Flesh*', in Herbert Davis, William C. DeVane and Robert Cecil Bald (eds), *Nineteenth Century Studies*, p. 288.

11 Letter to E. M. A. Savage of 3 December 1883: *Memoir*, Vol. 1, p. 400.

12 Appendix D, ibid., Vol. 2, pp. 468–71.

13 'The Deadlock in Darwinism' (XIX), p. 53. The interpolations have been added to make the connection between novel and theory clearer, but it is reasonable to suppose that Butler had his complete but unpublished novel in mind when he wrote this.

14 For such a reading, see for instance R. Balfour Daniels, 'Names in the fiction of Samuel Butler', *South Central Bulletin*, vol. 29 (Winter 1969), pp. 129–32, and William H. Marshall, 'The use of symbols in *The Way of All Flesh*', *Tennessee Studies in Literature*, vol. 10 (1965), pp. 109–21.

15 *Life and Habit* (IV), p. 64.

16 In his preface to *The Way of All Flesh* (XVII) Streatfeild casually remarks without a hint of qualification that the novel 'may be taken as a practical illustration of the theory of heredity embodied in [*Life and Habit*]' (p. xxxiii).

17 But, also as usual, *only* superficially. That card-playing and dicing were Butler's private concretions for the method of what he called 'Charles-Darwinian evolution' is obvious from the several times that he brings these activities and the theory together. Thus:

'according to Messrs. Darwin and Wallace, we may have evolution, but are on no account to have it as mainly due to intelligent effort . . . We are to set it down to the shuffling of cards, or the throwing of dice without the play.' See 'Deadlock in Darwinism' (XIX), pp. 9–10; the same comparison is to be found in *Luck, or Cunning?* (VIII), p. 67.

18 *Life and Habit* (IV), p. 50.

19 Nothing in the novel displays more markedly Butler's limitation as a novelist than these scenes in Coldbath Fields which were composed, we are told, after he paid a social call on the Governor (*Memoir*, Vol. 2, p. 11). Only official publicity could account for these kindly warders, so deferentially anxious for Ernest's recovery so he can start learning a useful trade.

20 This should not carry the implication that Ernest's decisions are all fully conscious ones, or that he believes himself 'free'. David Daiches, in *Some Late Victorian Attitudes*, mentions Butlerism as a determinist evolutionary theory, but as his section on the novel starts by cutting away the approach being taken here (he calls it 'not a novel about evolution or even heredity': p. 58) he does not follow through with the task of showing exactly how Ernest's path is a pre-selected one. We can concede that Butler's biological work is never perfectly lucid on this subject, but *The Way* surely is.

21 *Life and Habit* (IV), p. 202.

22 'Ramblings in Cheapside', *Collected Essays*, Vol. 2 (XIX). First published unchanged in *Universal Review* (December 1890).

23 Bissell compares Towneley with the Ancients of Shaw's *Back to Methuselah* to the detriment of the former, coming to the conclusion that Butler's vision is 'restricted and relatively crude' ('A study', p. 303). But there is surely no evidence in the text for his further claim that the portrayal of Towneley has an element of contempt in it by the end. Towneley is, and is meant to be, an object of unblinking admiration.

24 *Athenaeum*, 9 January 1892, pp. 49–50.

25 Harvey Curtis Webster, *On a Darkling Plain: The Art and Thought of Thomas Hardy*, p. 175.

26 Arnold Kettle, *An Introduction to the English Novel*, Vol. 2, pp. 50, 61.

27 John Holloway, 'Hardy's major fiction', in his *The Charted Mirror: Literary and Critical Essays*, pp. 99, 105.

28 Leo Henkin's *Darwinism in the English Novel 1860–1910: The Impact of Evolution on Victorian Fiction* restricts his treatment to two minor novels which mention Darwinian concepts in passing, and he therefore quite obscures the subtleties of which Hardy's art is capable. Elliot B. Gose, 'Psychic evolution: Darwinism and initiation in *Tess of the d'Urbervilles*', *Nineteenth Century Fiction*, vol. 18 (December 1963), pp. 261–72, concerns himself with anthropology rather than with biology. Jean R. Brooks, 'Darwinism in Thomas Hardy's major novels', unpublished MA thesis, Birkbeck College, University of London, provides an excellent overview, though her determination to find a biological concept to match each novel – 'environment' in *The Woodlanders*, 'variability' in *Jude*, etc. – makes her argumentative structure highly inflexible. Her assertion that Hardy read Weismann 'after finishing his adaptation of *Tess* for serial publication' (p. 20) is unsupported and, on the evidence of the *Life*, false. Her *Thomas Hardy: The Poetic Structure* abandons the Darwinian approach.

29 Trevor Johnson, *Thomas Hardy*, pp. 114–16. On the other hand, Raymond Williams in *The Country and the City* (London, 1973) makes another valuable point, again with reference to *Tess*, in saying that to write about spring in terms of germs and inorganic particles as well as sights and sounds is Hardy's special gift – 'the native place and experience but also the education, the conscious inquiry' (p. 206) – and this rehabilitation of Hardy as a novelist with a central and not peripheral imaginative debt to his reading in the natural sciences has continued with Bruce Johnson's ' "The perfection of species" and Hardy's Tess', in U. C. Knoepflmacher and G. B. Tennyson (eds), *Nature and the Victorian Imagination*. It is true that Hardy, at his rare best, does achieve what seemed to be impossible for most other late-Victorian writers: a fully satisfying aesthetic-scientific vision.

30 Webster, *On a Darkling Plain*, p. 132.

31 In 1911 and again in 1924. For the earlier statement, see Carl J. Weber, *Hardy of Wessex:*

His Life and Literary Career, pp. 246–7. The second, made in a letter of 21 June 1924, is printed by the recipient Ernest Brennecke at the front of his *Thomas Hardy's Universe: A Study of a Poet's Mind* (first printed 1926).

32 *Life*, p. 254. Out of eight notables whom Hardy mentions by name, no less than five were, according to the *Dictionary of National Biography*, professionally involved in the life sciences.

33 *Life*, p. 259. At the time Hardy knew him Crichton-Browne was Lord Chancellor's Visitor in Lunacy and the author of many works on mental disorders. Possibly Henry Drummond's lectures brought the topic of altruism under discussion, although his *Ascent of Man* appeared in the following year. Petr Kropotkin's articles, later collected as *Mutual Aid: A Factor of Evolution*, were at the time appearing regularly in *Nineteenth Century*. Crichton-Browne was not a specialist on heredity, of course, so this is further proof of how common were melioristic interpretations of Darwinism in the late 1880s. Roy Morrell, *Thomas Hardy: The Will and the Way*, grossly distorts Hardy's own feelings about biological altruism by quoting this note as 'his firm belief' (p. 99).

34 Lionel Stevenson, *Darwin among the Poets*, pp. 257–8, suggests that Lamarckism *did* attract Hardy, and he quotes in support the lines from the poem 'San Sebastian': 'Maybe we shape our offspring's guise/From fancy, or we know not what.' But this is to confuse Hardy's ideas with the speaker's, who is a remorseful old soldier. Another of Hardy's ex-military characters, Granfer Cantle in *The Return of the Native*, complains that 'all that soldiering and smartness in the world seems to count for nothing in forming the nature of the son'. Should we take that to be Hardy's deliberate *anti*-Lamarckian sentiment? In fact the ancient folk-belief in telegony goes back as far as Genesis and Hippocrates. In its pseudo-scientific form, the example of the naturally short-foreskinned Jew (i.e. as a result of generations of circumcision) was being peddled by H. J. Voigt at least as early as 1789. None of these has anything to do with the speeding-up of the evolutionary process, which was the cause of the late-Victorian interest in Lamarckism.

35 The censored serial ran in the *Graphic* from July to December 1891. The excised scene of the baptism of Sorrow was published first in the *Fortnightly Review*, vol. 49 (May 1891), pp. 695–701.

36 Weber, *Hardy of Wessex*, pp. 170, 174. If Weber's reconstruction is accurate, the first quotation argues strongly for some restructuring of the sixteen chapters mentioned above, and hence for a possible Weismannian influence on all parts of the novel. This may nowhere have amounted to more than a pointing-up of the imagery.

37 John T. Laird, *The Shaping of 'Tess of the d'Urbervilles'*.

38 August Wisemann, *Essays upon Heredity and Kindred Biological Problems*, trans. Edward B. Poulton, Selmar Schönland and Arthur E. Shipley (Oxford, 1889), p. x. Hardy's copy of this book – if he ever owned one – seems to have vanished. Neither Colby College nor the Dorset County Museum has it among their items from Hardy's own library. It was not one of the 309 items from this library sold in May 1938, although the editor of the reprint of the sales catalogue says that these were only a portion of Hardy's books. See J. Stephens Cox (ed.), *The Library of Thomas Hardy, OM*, p. 193. In Lennart A. Björk's two-volume *The Literary Notes of Thomas Hardy* there is only one mention of Weismann, and that is copied from a review.

39 *Complete Poems*, p. 434. It was probably written between 1914 and 1918, although an earlier date is possible. J. O. Bailey, *The Poetry of Thomas Hardy: A Handbook and Commentary*, argues that the seed of this poem is the diary entry of 19 February 1889 (*Life*, p. 217). But the poem is hardly 'the story of a face'; and the interpolation that 'this idea was to some extent carried out in . . . the poem entitled "Heredity" ' is almost certainly F. E. Hardy's. The tone of the poem to me suggests a theoretical, not an anecdotal, source.

40 George John Romanes, 'Weismann's theory of heredity', *Contemporary Review*, vol. 57 (May 1890), p. 698.

41 Although he is not mentioned in the *Life*, it is certain that Hardy also knew something of the work of Galton, for he copied a passage from Galton into his Literary Notebook

I. What is more, this is a clear source for part of the conversation between Tess and Abraham in chapter IV, and shows how Hardy wove quite recondite ideas into his dialogue – in this case, not very successfully. Brooks, *Thomas Hardy*, p. 216, notes of Tess's impulsive temperament that Galton attributed this trait to full-blood Normans in his *Hereditary Genius* (1869). However, there is no evidence that Hardy read this before 1889–90. There may have been some influence, but the Galton-Hardy link is too tenuous to warrant further consideration.

42 Thomas Hardy, *Tess of the d'Urbervilles*, the Wessex Edition (London, 1920), p. 114. All further references in the text are from this edition.

43 Maurice Maeterlinck, 'The tragical in daily life', *The Treasure of the Humble*, trans. Alfred Sutro, p. 108.

44 Michael Millgate, *Thomas Hardy: His Career as a Novelist*, p. 271. F. B. Pinion agrees: 'Hardy's argumentative intrusions had placed him in an illogical position. On the one hand Nature is cruel . . . on the other, the "Innocence" of the natural world is made the basis for attacking the moral code' (*A Hardy Companion*, p. 173).

45 This identification is noticed, for instance, in Philip Mahone Griffith, 'The image of the trapped animal in Hardy's *Tess of the d'Urbervilles*', *Tulane Studies in English*, vol. 13 (1963), pp. 85–94. What is most notable in this essay is that, although Griffith sees that this major strand of imagery illuminates Tess's plight, he does not see its central purpose as being to embed her the more firmly in natural process, but instead says it represents 'the malicious and destructive force of societal, as opposed to natural, laws' (p. 86). But Hardy sets up no such dichotomy: in this novel the inadequacies of society mirror those of nature. And, if any qualities characterise Hardy's Darwinian nature, they are exactly those of malice and destructiveness. Consider the black-humorous treatment of John Durbeyfield's fat round the heart: fat which is closing inexorably like a trap. The medical complaint is, of course, devoid of value in itself. But notice how in the *human* world Durbeyfield is at last put to death at the very worst moment. The malevolence is passive, not active; but it is still very real.

46 Millgate, *Thomas Hardy*, p. 272.

Conclusion

> Conceding for a moment that there is any analogy between a bee and a man in a shirt and pantaloons (which I deny), and that it is settled that the man is to learn from the bee (which I also deny), the question still remains, what is he to learn? To imitate? Or to avoid? When your friends the bees worry themselves to that highly fluttered extent about their sovereign, and become perfectly distracted touching the slightest monarchical movement, are we men to learn the greatness of Tuft-hunting, or the littleness of the Court Circular? I am not clear, Mr Boffin, but that the hive may be satirical.
>
> CHARLES DICKENS (1864)

Aldous Huxley once attempted the difficult task of rating the various pure sciences on an imaginative index – that is to say, on a scale which would show the relative intrinsic appeal of each science's revelations for the literary artist. Huxley placed biology very high on this scale, and in giving his reasons for doing so he made no reference to that science's state of development at different times. His assessment is, or claims to be, a universal one.

> Biology, it is obvious, is more immediately relevant to human experience than are the exacter sciences of physics and chemistry. Hence, for all writers, its special importance. The sciences of life can confirm the intuitions of the artist, can deepen his insights and extend the range of his vision.[1]

We need to bear in mind that novelists, especially those interested in an analytic way in the aesthetics of the novel, tend to *ex cathedra* pronouncements which do little more than justify their own practice. Without doubt Huxley is here inflating his own predilections into an expansive generalisation: one recalls that in *Point Counter Point* the character Philip Quarles, himself a novelist and an obvious Huxley persona, reads up assiduously on ichthyology before writing a scene full of fishy symbolism to be set in front of the tanks of an aquarium. Discounting the special pleading, then, is it really so 'obvious' that biology does indeed have that 'special importance' which Huxley claims for it? Is it universally true of all the life sciences from virology to ethology and at every stage of their development from the anecdotal to the rigidly statistical? Certainly if we draw the limits extensively enough; if, say, we define biology as the science of living things – the definition which Huxley's grandfather helped to publicise – then the biologist can legitimately concern himself with the entire span of human experience and activity. It reaches down and becomes entangled with chemical and physical reductionism; it reaches upwards into history, psychology, anthropology and maybe even parapsychology. Every student of the human world is a biologist, if a sufficiently catholic view be taken; and we might even say that the literary artist is not merely free to draw on biology for inspiration but that in exercising his craft he is himself a biologist. When pushed to its limit, therefore, Huxley's remark is nothing more than a banality; and short of that limit it is hard to see how it can be true. Could *any* imaginative writer, no matter how inclusive his genius, make, say, the matter of population genetics grist to his mill? One ventures to doubt it.

But, if in this sense the universality of Huxley's claim appears somewhat ridiculous, in another – his insistence that the human relevance of biology is a permanent feature of that science – is worth dwelling on. Though the foregoing chapters have stressed the unusual literary utility of late-Victorian biology, no treatment of the relations between science and letters in that period should allow any impression to linger that those relations ceased with the final triumph of the neo-Darwinian 'modern synthesis'. There are in fact many continuities which have lasted up to the present day and which therefore testify to the permanent validity of Huxley's assessment. In the Introduction an opinion was quoted that the melting of the critical Darwinian issues into the cultural background has meant an equivalent ebbing of literary interest in biological themes, the sole survivors being by 1910 only cheap romances or 'comic strips and radio programmes exploiting the far-flung future and man's caveman past'.[2] As we have argued throughout, such a judgement will only hold up if one takes a somewhat myopic view of the main drive of biological investigation (a drive which far transcended the issue of Darwinism as such) up to and beyond the end of the century. The truth is, rather, that after the establishment of Mendelism both speculative and purely literary interest in biological theory continued unabated. Certainly we must concur that, to a degree, the speculative writer has been forced in recent times to become much more cautious as some of the accretions which clung about the technical problems of a century ago have been shorn away by the patient labours of unimaginative specialists. No writer today, we must suppose, could gain a hearing for long gloomy prognostications about the future of European civilisation based on information about the evolutionary development of the liver-fluke. Pulpit science, 'bio-optimism' of the brand peddled by Drummond, has quite gone out of fashion. The catastrophes of modern history have left us little room or inclination to worry, as Greg and Hudson and Wells in their day worried, about the risk of permanent dull social harmony inducing racial senescence. We can now see pretty clearly that particular apocalypse as the vision of a society essentially at peace with itself, for good or bad. Nevertheless, the impulse to read out the future by inspecting the bowels of a science remains as strong as it ever was, and this is still specially so in that no-man's land between severely technical and frankly literary prose. The line whose last representative in our period was Samuel Butler continued fertile in the first part of the present century and, in a somewhat more professional form, has lasted to this day. One might mention in illustration the treatise by the South African statesman Jan Christiaan Smuts (1870–1950), *Holism and Evolution* (1926, but in part dating from 1910); several works by the political philosopher and pioneer sociologist Leonard Trelawney Hobhouse (1864–1929) such as *Development and Purpose* (1913); and the 1916–18 Gifford Lectures by his friend Samuel Alexander (1859–1938), later published as *Space, Time and Deity* (1920). These particular works originated in the confusion of the first decades of the century and ride a number of hobby-horses in presenting what are actually different brands of Emergent Evolution. But it would be very premature indeed to predict that our present neo-Darwinian biology has at long last stifled for good the evolutionary philosopher. The old debate, which might finally be summed up as the question 'Does biology tell us anything that might justify our extrapolation from minds to Mind?', continues apace. Conjectures in their own way as bold as anything in Winwood Reade are to be found in the pages of Alister Hardy's influential *The Living Stream* (1965). Certainly the tradition of 'biophilosophy' has responded to the shift of interest, with the rise of genetics and biochemistry, into much more carefully defined biological problems. For the interpretation of 'chance' as meaning, effectively, human ignorance of causes (one of the interpretations, we recall, authorised by Darwin himself) Mendelism substituted the much more suggestive demonstration of pure randomness at the microscopic level of gene interchange – a process able to be described by probability theory. Randomness so observed is a far more perplexing element to incorporate into any evolutionary philosophy – into any form of evolutionism – since it by definition excludes any further introduction of teleology. Something of the spirit if not the message of Henry Drummond lived on, despite the conceptual restrictions, in Teilhard de Chardin's *Phenomenon of Man* (1959). The prose-poetry of evolutionism may be dead, but Chardin has sung of the emergence and the rise of the spirit of man out of the meaningless ebb and flow of energy transformations. Even in his terminology ('zoosphere', 'noosphere') he is but reworking the ideas behind Huxley's parable of the Unweeded Garden.

Other channels of expression – the novel of scientific versus religious controversy, the conversational set-pieces on Darwinism, the polemic pamphlets – have not survived. Naturalism, which forced many very different writers to seek the springs of character and motive in abstractions like Hardy's Immanent Will or in the genetic lottery, has proved a passing phase. Utopian fiction, prose epics and, above all, mainstream science fiction: these have proven to be the most hospitable genres for writers with biological interests, and many outstanding examples date from after the

evolution at length sweep away even the most long-lived and successful varieties of humankind. Stapledon's fiction (novel? philosophical treatise? – it is hard to classify it) is notable for its handling of eugenics. Man remaking himself is not the endpoint of evolution that Wells thought it might be: Stapledon was a writer of the succeeding generation, one disillusioned about such easy promises. Stapledon's scope is so great that, though he can indeed show eugenic changes, he can also show unaided natural selection transforming man many times over. He deals very harshly with his predecessors' dreams of evolutionary control, feeling that it expresses nothing more than enormous hubris. It is in just this sense that Stapledon, clearly in the grip of the Weismannism of his youth, makes his biological fantasia genuinely a *Darwinian* fantasia. *Last and First Men* has rightly been called 'by far the most ambitious and sustained attempt to create an "evolutionary bible" ',[5] but while it is certainly part of Stapledon's intention to create an eschatology convincing to post-Darwinian man he cannot be said to grant much significance to the individual. His salvation is of no moment; in Stapledon's bleak universe he is at best just a very transient note in a great symphony. Stapledon's influence is perceptible, too, in some other attempts at evolutionary prose epics (for instance in Vardis Fisher's twelve-novel cycle *The Testament of Man*, 1943–60), but most writers have been content to investigate the less grandiose possibilities of contemporary biology and create on a much smaller scale. Much of this work fits more and more consciously into mainstream science fiction. This genre at first made use of only the crudest biology – like the living fossils in Conan Doyle's *The Lost World* (1912) – but in the last half-century sophistication has reached the point that the technical discovery and the story exploiting it have become almost simultaneous events.

Continuities such as the ones we have mentioned here help to vindicate Huxley's claim for the special imaginative import of biology. But, as Huxley himself is careful to note, writers are not normally interested in sciences as coherent structures of thought, and will simply ignore what is of no use to them. They will pick and choose: and may choose, as we have seen, the exotic, the absurd, or simply the inaccurate. Biology, therefore, as Huxley cautiously and properly implies, *grants* nothing: at most it can deepen an intuition – or confirm a prejudice. We cannot reasonably expect writers to distinguish between good data and bad; nor can we expect them to engage with issues which the biologists themselves were incapable of confronting. But all the interested parties will share areas of concern, and certain phrases, certain parables, certain metaphors will express that sharing. All the themes we have traced, for instance, have one common bond at least: each of them may be said to display one version of a uniform concern over the transmission and elaboration of human culture; and each theme poses the question in a different form of how far the natural biological mechanisms can be trusted for this task. One striking omission is any real appreciation in our period that there might be other modes of transmitting information down the generations and these other modes might be of much greater human significance than the genetic ones. Despite the cry from all the evolutionistic optimists that the time was right for man to remake himself in his own image, and despite the insistence from some others that he must cut himself adrift from natural process, there was no firm awareness that the onset of urban civilisation had itself inaugurated a new phase of evolution. There were, to be sure, some imprecise formulations of what might be termed 'cultural evolutionism' even in J. S. Mill's pre-Darwinian essay 'Nature': he speaks there of a 'self-culture' that would be impossible 'without aid from the general sentiment of mankind delivered through books, and from the contemplation of exalted characters real or ideal', calling this 'the only nature which it is ever commendable to follow'.[6] Mill's concept of 'self-culture' reaches beyond more personal education. He is aware that the permanent records of society supply an impersonal inheritance, an 'artificially perfected nature' where truly human values may be forged and stored in a repository not subject to the defects of the flesh. But on this matter the four main interpreters of early Darwinian biology – Spencer, Darwin, Wallace and Huxley – were curiously silent. Spencer has a few comments about 'superorganic' evolution in his *Principles of Biology* of 1864–7, but nothing more. Darwin's handling in the *Descent* of the part played by man's cultural, as opposed to biological, inheritance is grotesquely inadequate and in fact constitutes his most considerable omission. In accounting for the aesthetic sensibilities, for instance, which to us are largely or wholly products of cultural conditioning, Darwin appeals at one point to genetic endowment and at another to personal training. Some of the more refined tastes, he argues (without defining them), are never found among barbarous or uneducated people in any society. He gives very insufficient weight to the supreme power of long tradition, and barely mentions the inventions which allow the invested capital (both physical and mental) to increase cumulatively despite the brief generations. Darwin's imperception is shared by Wallace who, despite his optimism about educational advancement, has nothing but a casual aside in his essay 'Human progress' to the effect that 'each generation benefits by the trials and failures of the preceding generation'[7] – an oddly insipid remark. As for Huxley, he was aware that it is language, letters and the printing press which in modern times have permitted people of genius to stand on one another's shoulders. And yet he never advanced beyond the point, expressed so vigorously in his last serious philosophical analysis *Evolution and Ethics*, that all human refinement can only be temporary ground wrested forcibly from a recalcitrant nature, and that the ethical struggle is now and forever in deadly conflict with the cosmic. Possibly Huxley was aware of the dubious logic of opposing one part of the evolutionary process against another. His simile of two hands driven by one brain in opposite directions to snap a piece of string (in the 'Prolegomena') is a nod in this direction. At any rate, he never anywhere gives cultural evolution its rightful due, and never uses the concept to overcome any of his fundamental theoretical perplexities.

It was not until the middle 1890s at the earliest that the novel suspicion began to assert itself that evolutionary biology as then understood had only the most limited contribution to make to an understanding of the tiny span of all recorded history and prehistory. The suspicion began to be articulated only as the controversy mounted over the claim, touched on in Chapter 3, that only supernatural interventions could account for the advanced qualities of the human intellect. Men began to look about them for alternative explanations, and soon they settled on what Romanes called 'the inestimable advantage' of a 'transmission of the effects of culture from generation to generation'.[8] This 'advantage' had not in Romanes' eyes reduced the ferocity of the evolutionary struggle one whit, but it had taken that struggle on to a new level where language and other symbolic systems were the new (and potent) instruments of aggression. Biologically speaking, the human race was now seen to be mentally, morally and physically much what it has been since the Palaeolithic. To account for any other changes there is after all no need to invoke Wallace's angelic legions: anthropology was by that time beginning to show how a vital precondition of growth into full humanity is submersion in the nourishing medium of human culture. This realisation was brought home by the 'wolf-boy' studies popular at the time, from which evidence was garnered about the amount of socialisation – more or less zero – achieved by infants raised by animals outside the pale of any human society. Such new awareness, and the consequent reduction in the significance which could reasonably be placed on the strictly biological and physiological determinants of behaviour, began to be characteristic of some of the more speculatively daring writing at the extreme end of this period. Wells, in his *The Island of Dr Moreau* (1896), introduces beasts transformed surgically into travesties of men; but what humanity these tortured monsters display flows more from their rudimentary culture, the legal and theological system (the 'Law') in which they are enmeshed, than from anatomical changes. As individuals they are freaks of a Faustian science, pathetic, contemptible; but as social units they have some dignity and acquire a cohesiveness which for a while challenges their creators. Whatever Wells's intentions were in constructing it, his fable impresses more as an anthropological than as a biological one.

What might the literary consequences have been, had these imaginative possibilities gathered about the metaphor of cultural evolutionism been drawn out earlier? It is hard to believe that the wide-ranging concern over evolutionary optimism or pessimism, over avenues of inheritance, over the ability of natural selection to bring about social reform, would have taken quite the shape displayed in previous chapters. Perhaps eugenics, and the social and political attitudes which commingle with that creed, would have been replaced by a greater interest in euthenics – the improvement of the race by the adjustment of the environment – among the biological popularisers. As it was, all the involvements which have been documented earlier took place within a frame of reference which barely admitted socio-cultural factors at all. This is not to say (of course) that these theoretical inadequacies had any dissolutive effect on the literary art based on them. The supersession of Weismannism did not bring down *Tess of the d'Urbervilles*! Nor does the failure of the 'proofs' for evolutionism spoil the optimistic rhetoric of *The Martyrdom of Man* – though that failure reduces Reade's impassioned tract for the times *to* rhetoric. What the writer who uses current science has most to fear from the flux of time is not being proven 'wrong' but being rendered incomprehensible. In three hundred years' time Hardy's 'bio-pessimism' in *Tess* may require as much elaborate glossing as Donne's natural philosophy in 'The Extasie' does today. And yet we read Donne despite his utterly obsolete physiology.

These points are well elaborated in Aldous Huxley's *Literature and Science*, and all confirm what he has to say about the universal literary potential of biology. In one way only, as this survey reaches its close, must we insist for the last time on the historical distinctiveness of biology throughout the final four decades of the last century. The tenor of the argument has been that at certain stages in its development – and not necessarily at its emergence – a science will be relatively accessible to the lay mind, while at other points it may be effectively sealed off from all but severely technical scrutiny. The sciences of chemistry and electrochemistry had a great deal of power in them for the writers of several decades before 1859, at a time when it was supposed in some quarters that the origin and processes of life are essentially electrical. Mary Shelley's Frankenstein is much more of a physicist and chemist than he is a biologist. The dread secret he uncovers is a way of animating with the vital spark a collection of dead organs stolen from charnel houses rather than a probing into physiology as such. Later on in the century, despite the very far-reaching nature of its discoveries and their practical applications, chemistry lost much of its imaginative appeal: it had become at once too recondite and too mundane. Other sciences' insights came to fill the gap – Tennyson used palaeontology in *The Princess* (1847); Ruskin meteorology in *The Storm Cloud* (1884); and Meredith astronomy in 'Meditation under Stars' (1883) – and of these biology was the most amenable to imaginative treatment. The degree to which a science may become popularly accessible is inversely related to the degree to which its findings have become quantifiable and unequivocal, and for just this reason literary interest in Victorian biology centred on the issue of teleology in evolution. Repeatedly in the previous chapters we have seen the teleologically inclined writer leaning on some supposedly empirical support for his belief that willing has a compelling power all of its own, just as real as mechanistic cause and effect. Such support was not found wanting. Cultural evolutionism made no headway because much optimistic energy flowed into the ultimately futile controversy on the inheritance of acquired characters. Even at the scientific and philosophical levels this was a strongly emotional issue. The shocking determinism of classical Darwinism revolted some idealistic minds – minds which attacked it, not because it was too revolutionary for the times but because on the contrary it was a reactionary force apparently putting a brake on brave hopes for the future. And in compensation grew up the profound wish to believe that skills acquired by individual sweat do not vanish with each generation but are to some small degree transmissible to posterity. Many social theorists and educationalists, encouraged by Darwin's own flirtation with 'soft' inheritance and perhaps also encouraged by his theory of pangenesis which gave some respectability to neo-Lamarckism, reacted with fear and distaste to the revival of Darwinism in the 1890s and the contributions of Weismann. After all, over the previous years men like Spencer had made an enormous intellectual and emotional investment in Lamarckian explanation. In his Synthetic Philosophy use-inheritance played a key role: it is proposed as the source of all inherited knowledge, as the basis of the conscience, and as the final basis of certainty for the transcendental axioms. Spencer had attempted, in his metaphysics, to marry the Kantian categorical imperative with modern biology, finding a justification for the first in the supposedly demonstrable facts of the second. It is not surprising, then, that he attacked neo-Darwinism with such ferocity, for he had stretched Lamarckism to the widest limits it ever attained. Among less totally committed thinkers, too, Weismannism threatened to dissolve the freshly laid foundations of social psychology into, in Grant Allen's words, 'a slough of pre-Darwinian and pre-Spencerian chaos'.[9] When considered coolly, of course, it is obvious that the corollary of the 'law' that educational improvements are genetically transmissible must inevitably be that evil and anti-social behaviour also has a persistency it lacks in the neo-Darwinian account. Indeed, the latter theory has this advantage: even if the valuable effects of education are not passed on, it is also true that originality or genius when once thrown up by the random permutation of the genes cannot be weakened by the direct influence of a mediocre environment in successive generations. Mendelian genetics and the particulate theory asserted that no characteristic can readily be blended away.

There are therefore massive difficulties in placing Lamarckian inheritance at the base of any optimistic social philosophy, but unfortunately the difficulty of gathering really definitive technical evidence tended to obscure this fact. It is hardly credible that as late as 1936 Robson and Richards in their important *The Variation of Animals in Nature* were saying dismissively of natural selection that

> we do not believe that Natural Selection can be disregarded as a possible factor in evolution. Nevertheless, there is so little positive evidence in its favour, so much that appears to tell against it, and so much that is as yet inconclusive, that we have no right to assign to it the main causative role in evolution.[10]

They were thus passing the same judgement in words only slightly different from Romanes' forty years earlier. One of the most influential experiments which sustained the controversy was carried out over a period of several years by the psychologist William McDougall, who taught a skill to many successive generations of rats in the expectation that it would eventually show itself to be innate. He frankly admitted to a strong personal bias to Lamarckism, and he did so in words which might have come straight from a Butler essay of the 1890s: 'on my own part, there was a feeling that a clear-cut positive result would go far to render tenable a theory of organic evolution, while a negative result would leave us in the Cimmerian darkness in which Neo-Darwinism finds itself'.[11] The Miltonic epithet intriguingly reinforces the intellectual fervour behind it, but unfortunately the daybreak when McDougall's results were positive proved to be a false dawn. His positive results were a statistical error; but these phenomena are so elusive that the experiments which decisively rejected McDougall's work consumed fifteen more years and many more generations of test animals. Slowly the belief has hardened that to inquire whether habits or skills may be inherited is, in philosophical terms, a pseudo-question because it violates Popper's principle of potential disprovability. No breeding strain of organism, it has been argued more recently, can ever be fully controlled or described; and therefore it is impossible to rule out genetic drift and hence natural selection in favour of the characteristic being looked for. If this is true, the Lamarckian controversy reduces to little more than a battle over the use of terms. Even what looked to be a clear division in nature to the Victorians has become muddied with Conrad Waddington's concept of genetic assimilation – a complex process which can, when the breeding conditions in a population are just right, simulate Lamarckian effects even though the only processes at work are Mendelian ones.[12] Gradually, then, a thicker carapace of professionalism has grown over the animosities of rival theories, and the parties no longer thrash out their differences in those long monologues fostered by the medium of the great periodicals. The splenetic hatred of a George Romanes for a Samuel Butler is no longer quite as public, though it persisted well into the present century. No Victorian biologist, after all, went so far as to blow his brains out over the Lamarckian issue, as did Paul Kammerer in 1926 when his specimens allegedly displaying induced inheritable characteristics were denounced as fakes. Just how high tempers rose before that last act, and the vindictive part played by William Bateson out of professional envy, has been well described by Arthur Koestler in *The Case of the Midwife Toad* (1971). The story is an ugly one, but it does help to bring to mind the full human reality behind what are at first sight abstractions. Then, when we know something of the passions it could arouse, the spectacle of Bernard Shaw as late in the day as 1921 cursing the Darwinians for sucking the meaning out of the universe no longer appears entirely ridiculous. By that time the best that Shaw could muster were dubious anecdotes about *idiots savants*, but he still had influential friends in the biologists' camp as did Butler forty years earlier; even if (again like Butler) he did not choose to acknowledge them. It is Shaw's rhetoric rather than his sentiments which is notable when placed against, for example, Kammerer's writings on education in the late 1920s. By teaching children properly, claims Kammerer, 'we give them more than short benefits for their own lifetime – because an extract of it will penetrate their substance which is the truly immortal part of man'.[13] If only it were true! Yet as we have seen repeatedly, and especially in Chapter 7, while disproof is lacking an erroneous scientific belief may show great vitality if it quenches the thirst of the creative imagination for human significance. So it was with neo-Lamarckism, which, simply because of the deficiencies in experimental research, is commonly and falsely understood to have been the aberration of another society more stupid and gullible than our own.

The ambiguity of the evidence about biological teleology had its impact, too, on those writers who were temperamentally unable to avert their eyes from the determinism of Darwinian biology. Even here the available options were wide indeed, ranging from a soft Arcadianism tempered with a little applied eugenics, to a harsh worship of the fittest, the 'sole humane', as Meredith termed these evolutionary successes. To the very pessimistic mind biology could even suggest that all human capacities and achievements are the outcome of blind physical forces, and could thereby encourage and endorse a materialistic personal and social psychology. By placing no weight on the value of co-operative enterprise, of mutual aid, and by emphasising that even the minutest variation has to be tested against the iron principle of survival, such a reading could inspire a tragedy of truly Sophoclean gloom, or a system of ethics of unparalleled absolutism. Such men could seek what gloomy comfort they could in the fact that the metaphysical tendency of the life sciences towards some variety of materialism has a respectable pedigree. Chaucer's Doctour of Phisik whose 'studie was but litel on the Bible' is the prototype of a line which runs, apart from the occasional *Religio Medici*, through the biologists of life and literature alike: the surgeon Julien de la Mettrie with his *L'Homme-Machine* (1747); Frankenstein; the original thought of Lamarck; the doctor Franz Gall and the phrenologists; the German physiologists of the *Darwinismus*, of whose work Wilde recorded that it gave 'curious pleasure' to Dorian Gray, as he traced 'the thoughts and passions of men to some pearly cell in the brain, or some white nerve in the body'. For a few less resilient minds (fewer, on the whole, than one would have anticipated) man's whole relationship with nature was permanently corrupted and a fountainhead of morality dried up. The command in 'The Tables Turned' at the beginning of the nineteenth century had been to 'let Nature be your teacher'. But, whereas those attending the Wordsworthian academy had majored in Pantheism, Natural Theology and Moral Earnestness, some of the new students found the curriculum changed for the compulsory disciplines of Wastefulness, Cruelty and Disharmony. But only some. To other minds the greatest challenge the century could offer intellectually was to make some facet of the evolutionary theory – and, yes, even Darwinism itself – serve as the foundation to a cheerful new theory of progress, of history, of religion, of economics, or a new technique for personal development. In Darwin's own autobiography of 1876 there are deeper and more violent oscillations of belief from one standpoint to the other than are to be found in any comparable document. Because Darwin is only too aware of taking his categorical imperatives from bloody nature, his tone at some points is positively agonised: 'what advantage', he asks, 'can there be in the sufferings of millions of the lower animals throughout almost endless time'? Here we have a man forced into misery by his fear and distaste for a natural agency of torture. Yet the determination to make what he can of an optimistic interpretation of this theory was no less strong in Darwin than it was in many other writers whose versions we have surveyed. For in the autobiography there is also a strand of thought designed to persuade himself (and his immediate family, for whom this meditation was alone intended) that 'happiness decidedly prevails', for if it did not, 'if all the individuals of any species were habitually to suffer to an extreme degree they would neglect to propagate their kind; but we have no reason to believe that this has ever or at least often occurred'.[14] Whatever we wish to make of this argument (and there is nothing much more to it than a restatement of Reade's evolutionistic tenet that past suffering is justified by present and future happiness), at least we see Darwin here abandoning his pose of aloofness: like many others, he clearly *wants* deterministic selection to work for the greater good of all. Even when Victorian materialism was expressed in its purest, most elegant and most reductive form it was not devoid of a certain romanticism. The physicist John Tyndall made the claim that 'all our philosophy, all our poetry, all our science, all our art . . . are potential in the fires of the sun',[15] which in one sense puts them in their place; but there is also the sense, which Tyndall surely wishes to touch upon, that the scientist who understands such things may gain access to colossal if shadowy powers; may be the inheritor of the mantle of both the Sibyl and Prometheus. In biology, certainly, the reductionist impulse has continued, with Mendel's statistical treatment becoming the norm and the final links being forged between probability theory and cytology. The processes of life have come to be explained more thoroughly than any Victorian could have dared – or feared – to imagine in terms of chemistry and finally physics; that is to say, in terms of organic matter. But what, after all, is matter? No scientific assumption of the period up to 1900 now seems more question-begging, after the work of Roentgen, FitzGerald, Dirac and Feynman, than that materialism is the simplest and most coherent philosophical position. After three-quarters of a century a clear victory has gone neither to the Haeckelian materialists nor to the Wallacean spiritualists. All the difference of emphasis between nineteenth- and twentieth-century biological reductionism stands revealed in the well-known apophthegm of the modern geneticist Sir Ronald Fisher that 'natural selection is a mechanism for generating improbabilities'.

We may conclude therefore that Victorian biology had something for everyone who took an interest in it, and needed to do violence to no one's deepest persuasions. Bearing no unambiguous truth, it had in itself neither a constructive nor a corrosive effect on literary activity. It was merely available for half a century or so as a repository and mine for all sorts of extra-scientific cultural preoccupations. For a short time the most flexible and suggestive of the sciences, its concepts were malleable, ready to plasticise under pressure and ready to fill every cranny of whatever mould had been prepared to receive them.

Notes: Conclusion

1 Aldous Huxley, *Literature and Science*, p. 67.
2 Leo Henkin, *Darwinism in the English Novel, 1860–1910: The Impact of Evolution on Victorian Fiction*, p. 267.
3 According to Gerald Heard, Wells privately called *Brave New World* a 'blasphemy against science' and therefore presumably took the point. Huxley himself denied that Wells was his specific target in the novel as it was finally written. See *Letters of Aldous Huxley*, ed. Grover Smith, pp. 358–9.
4 Winwood Reade, *The Martyrdom of Man* (London, 1925), p. 328.
5 By Richard Gerber in *Utopian Fantasy: A Study of English Utopian Fiction since the End of the Nineteenth Century*, p. 24.
6 'Nature', in *The Collected Works of John Stuart Mill*, ed. F. E. L. Priestley, Vol. 10, pp. 396–7.
7 Alfred Russel Wallace, 'Human progress: past and future', *Arena*, vol. 5 (January 1892), p. 145.
8 George John Romanes, *Darwin, and after Darwin . . .*, Vol. 2, *Post-Darwinian Questions: Heredity and Utility*, pp. 32–3.
9 Grant Allen, '*Essays upon Heredity*', *Academy*, vol. 37, (1 February 1890), p. 84. The linkage Allen makes here between Darwin's and Spencer's theories of inheritance looks more incongruous today than it would have in the 1890s.
10 G. C. Robson and O. W. Richards, *The Variation of Animals in Nature*, p. 316.
11 William McDougall, 'An Experiment for the testing of the hypothesis of Lamarck', *British Journal of Psychology*, vol. 17, (1927), p. 303.
12 There is a simplified account of this phenomenon in Conrad Waddington's *Nature of Life*, pp. 91–8. Even true Lamarckian inheritance at the level of the virus has recently gained an able supporter in E. J. Steele, *Somatic Selection and Adaptive Evolution: On the Inheritance of Acquired Characters*.
13 From Paul Kammerer's lecture 'The significance of the inheritance of acquired characters', as quoted in translation by Arthur Koestler, *The Case of the Midwife Toad*, p. 28.
14 Charles Darwin, *The Autobiography of Charles Darwin*, ed. Nora Barlow, pp. 90, 88.
15 John Tyndall, 'The scientific uses of the imagination', in his *Fragments of Science for Unscientific People: A Series of Detached Essays, Addresses and Reviews*, 6th edn, Vol. 2, p. 132. A paper first published in 1870.

Bibliography

The literature of Darwinism, especially the technical literature, is enormous. This bibliography is very selective indeed, being restricted for the most part to those works which in some way deal explicitly with the relations between Darwinism and the humanities in the late nineteenth century. It opens with a briefly annotated list of works which deal entirely or very substantially with these relations.

Major Sources

Appleman, Philip Dean (1955), 'Darwin and the literary critics', unpublished PhD thesis, Northwestern University. Deals with the biological data available to, and used by, Pater, Stephen, J. A. Symonds and others.

Baldwin, James Mark (1910), *Darwin and the Humanities*, 2nd edn (London: Swann Sonnenschein).

Conner, Frederick William (1973), *Cosmic Optimism: A Study of the Interpretation of Evolution by American Poets from Emerson to Robinson* (New York: Octagon Books). Contains a great deal of general material on the poetic response to evolutionary theory.

Culler, A. Dwight (1967), 'The Darwinian revolution and literary form'; see Levine and Madden, 1968. This lively essay asserts that the real Darwinian influence in Victorian literature was to invert values – a process which he traces from Butler to Carroll and on to Wilde, Pater and the literary decadence.

Henkin, Leo Justin (1963), *Darwinism in the English Novel, 1860–1910: The Impact of Evolution on Victorian Fiction* (New York: Russell & Russell). First published in 1940, this is the standard treatment, especially for minor fictional works. Major authors like Hardy are treated very sketchily.

Irvine, William (1959), 'The influence of Darwin on literature', *Proceedings of the American Philosophical Society*, vol. 103 (October), pp. 616–28.

Jones, Arthur Edward (1950), *Darwinism and Its Relationship to Realism and Naturalism in American Fiction, 1860 to 1900* (Madison, Wis.: Drew University Press).

Manly, John Matthews (1907), 'Literary forms and the new theory of the origin of species', *Modern Philology*, vol. 4 (April), pp. 577–95. Applies Vriesian mutation theory to problems of literary history, particularly the rise of Elizabethan drama.

Peckham, Morse (1959), 'Darwinism and Darwinisticism', *Victorian Studies*, vol. 3 (September), pp. 3–40. An excellent and thought-provoking discussion which deals with problems of definition and terminology.

Roppen, George (1956), *Evolution and Poetic Belief: A Study in Some Victorian and Modern Writers* (Oslo: Oslo University Press). A massive study which deals not only with Tennyson, Browning, Swinburne and Meredith but also with Butler, Shaw and Wells. However, it contains little close inspection of Victorian biological ideas.

Routh, Harold Victor (1937), *Towards the Twentieth Century: Essays in the Spiritual History of the Nineteenth* (Cambridge: Cambridge University Press). Devotes three

chapters to an examination of the part Darwinism played in the development of the English Naturalistic and Realistic schools.
Stevenson, Lionel (1932), *Darwin among the Poets* (Chicago, Ill.: University of Chicago Press). Covers the major Victorian poets, and concentrates on Tennyson, Browning and Arnold. Also deals extensively with the pre-Darwinian period.
Stevenson, Lionel (1960), 'Darwin and the novel', *Nineteenth-Century Fiction*, vol. 15 (June), pp. 29–38.
Turner, John Mills (1944), 'The response of major American writers to Darwinism, 1859–1910', unpublished PhD thesis, Harvard University.
Zirkle, Conway (1959), *Evolution, Marxian Biology, and the Social Scene* (Philadelphia, Pa: University of Pennsylvania Press). Contains a chapter on neo-Lamarckism and its influence on literature of the nineteenth and twentieth centuries.

Other Studies

'1,000,000 AD' (1893), *Punch*, vol. 105, 25 November, p. 250.
Adams, Henry (1946), *The Education of Henry Adams* (New York: Random House).
Alexander, Edward (1965), *Matthew Arnold and John Stuart Mill* (London: Routledge & Kegan Paul).
Allen, Charles Grant Blairfindie (1877), *Physiological Aesthetics* (London: King).
Allen, Charles Grant Blairfindie (1879a), '*Evolution, Old and New* . . .', *Academy*, vol. 15, 17 May, pp. 426–7.
Allen, Charles Grant Blairfindie (1879b), *The Colour Sense: Its Origin and Development. An Essay in Comparative Psychology* (London: Trübner).
Allen, Charles Grant Blairfindie (1881a), *The Evolutionist at Large* (London: Chatto & Windus).
Allen, Charles Grant Blairfindie (1881b), *Vignettes from Nature* (London: Chatto & Windus).
[Allen, Charles Grant Blairfindie] (1884), 'The Child of the Phalanstery', *Belgravia*, vol. 54 (August), pp. 163–76. Published under the pseudonym 'J. Arbuthnot Wilson', and reprinted in the same author's *Twelve Tales* (London: Grant Richards, 1899), pp. 45–64.
Allen, Charles Grant Blairfindie (1885), *Charles Darwin* (London: Longmans, Green).
Allen, Charles Grant Blairfindie (1890a), '*Essays upon Heredity*', *Academy*, vol. 37, 1 February, pp. 83–4.
Allen, Charles Grant Blairfindie (1890b), 'The girl of the future', *Universal Review*, vol. 7 (May), pp. 49–64.
Allen, Charles Grant Blairfindie (1894), *The Lower Slopes: Reminiscences of Excursions round the Base of Helicon* . . . (London: Matthews & Lane).
Allen, Charles Grant Blairfindie (1895), *The Woman Who Did* (London: John Lane).
Allen, Charles Grant Blairfindie (1897), *The Evolution of the Idea of God: An Inquiry into the Origins of Religions* (London: Grant Richards).
Allen, Garland E. (1969), 'Hugo de Vries and the reception of the "mutation theory" ', *Journal of the History of Biology*, vol. 2 (Spring), pp. 55–87.
Altick, Richard Daniel (1957), *The English Common Reader: A Social History of the Mass Reading Public, 1800–1900* (Chicago, Ill.: University of Chicago Press).
Altick, Richard Daniel (1974), *Victorian People and Ideas* (London: Dent).
Annan, Noel (1959), 'Science, religion, and the critical mind: introduction'; see Appleman, Madden and Wolff.

Antioch Review (1959), vol. 19 (Spring).
Appleman, Philip Dean (1959a), 'Darwin, Pater, and a crisis in criticism'; see Appleman, Madden and Wolff.
A[ppleman], P[hilip Dean] (1959b), 'The logic of evolution: some reconsiderations', *Victorian Studies*, vol. 3 (September), pp. 115–25.
Appleman, Philip Dean, Madden, William, and Wolff, Michael (eds) (1959), *1859: Entering an Age of Crisis* (Bloomington, Ind.: Indiana University Press).
Argyll, 8th Duke of, see Campbell, George Douglas.
Armytage, Walter Harry Green (1968), *Yesterday's Tomorrows: A Historical Survey of Future Societies* (London: Routledge & Kegan Paul).
Arnold, Matthew (1895), *Letters of Matthew Arnold, 1848–1888 . . .*, ed. George W. E. Russell, 2 vols (London: Macmillan).
Arnold, Matthew (1935), *Culture and Anarchy: An Essay in Political and Social Criticism*, ed. J. Dover Wilson (Cambridge: Cambridge University Press).
Arnold, Matthew (1960-), *The Complete Prose Works of Matthew Arnold*, ed. R. H. Super, 11 vols to date (Ann Arbor, Mich.: University of Michigan Press).
Athenaeum (1868), anonymous review of *The Principles of Geology*, 28 March, p. 455.
Athenaeum (1879), anonymous review of *Evolution, Old and New . . .*, 26 July, pp. 115–17.
Athenaeum (1889), anonymous review of *The Ascent of Man*, 20 July, pp. 87–8.
Athenaeum (1892), anonymous review of *Tess of the d'Urbervilles*, 9 January, pp. 49–50.
Badash, Lawrence (1968), 'Rutherford, Boltwood, and the age of the earth: the origin of radioactive dating techniques', *Proceedings of the American Philosophical Society*, vol. 112 (June), pp. 157–69.
Bagehot, Walter (1872), *Physics and Politics, or Thoughts on the Application of the Principles of 'Natural Selection' and 'Inheritance' to Political Society* (London: King).
Bailey, James Osler (1945), 'Hardy's "imbedded fossil" ', *Studies in Philology*, vol. 42 (July), pp. 663–74.
Bailey, James Osler (1947), *Pilgrims through Space and Time: Trends and Patterns in Scientific and Utopian Fiction* (New York: Argus Books).
Bailey, James Osler (1956), *Thomas Hardy and the Cosmic Mind: A New Reading of 'The Dynasts'* (Chapel Hill, NC: University of North Carolina Press).
Bailey, James Osler (1963), 'Evolutionary meliorism in the poetry of Thomas Hardy', *Studies in Philology*, vol. 60 (July), pp. 569–87.
Bailey, James Osler (1970), *The Poetry of Thomas Hardy: A Handbook and Commentary* (Chapel Hill, NC: University of North Carolina Press).
Baillie, John (1950), *The Belief in Progress* (London: Oxford University Press).
Barnett, Samuel Anthony (ed.) (1962), *A Century of Darwin* (London: Mercury Books).
Barton, Michael Parker (ed.) (1961), *Darwinism and the Study of Society: A Centenary Symposium* (London: Tavistock).
Barzun, Jacques Martin (1958), *Darwin, Marx, Wagner: Critique of a Heritage*, revised 2nd edn (Garden City, NY: Doubleday).
Bateson, C. Beatrice, see Bateson, William, 1928.
Bateson, William (1894), *Materials for the Study of Variation, Treated with Especial Regard for Discontinuity in the Origin of Species* (London: Macmillan).
Bateson, William (1902), *Mendel's Principles of Heredity: A Defence* (Cambridge: Cambridge University Press).

Bateson, William (*c.*1908), 'Heredity and variation in modern lights'; see Seward, 1909.

Bateson, William (1928), *William Bateson, FRS, Naturalist. His Essays and Addresses, together with a Short Account of His Life by Beatrice Bateson* (Cambridge: Cambridge University Press).

Baumer, Franklin L. (ed.) (1965), *Intellectual Movements in Modern European History* (London: Macmillan).

Beach, Joseph Warren (1936), *The Concept of Nature in Nineteenth-Century English Poetry* (London: Macmillan).

Bell, Peter Robert (ed.) (1959), *Darwin's Biological Work: Some Aspects Reconsidered* (Cambridge: Cambridge University Press).

Bellamy, Edward (1888), *Looking Backward: 2000–1887* (Boston, Mass.: Tick-nor).

Bellamy, William (1971), *The Novels of Wells, Bennett and Galsworthy: 1890–1910* (London: Routledge & Kegan Paul).

Benn, Alfred William (1962), *The History of English Rationalism in the Nineteenth Century*, 2 vols (New York: Russell & Russell).

Bentley, Eric Russell (1957), *Bernard Shaw, 1856–1950* (Norfolk, Conn.: New Directions).

Beresford, John Davys, and Tyson, Esmé Wynn ([1944]), *The Riddle of the Tower* (London: Hutchinson).

Bergonzi, Bernard (1960a), '*The Time Machine:* an ironic myth', *Critical Quarterly*, vol. 2 (Winter), pp. 293–305.

Bergonzi, Bernard (1960b), 'The publication of *The Time Machine*, 1894–5', *Review of English Studies*, vol. 11, pp. 42–51.

Bergonzi, Bernard (1961), *The Early H. G. Wells: A Study of the Scientific Romances* (Manchester: Manchester University Press).

Berry, Minta Sue (1963), 'Creation's groan: late nineteenth-century thought as reflected in the works of Thomas Hardy', unpublished PhD thesis, Vanderbilt University.

Besant, Walter (1889), *The Eulogy of Richard Jefferies*, 2nd edn (London: Chatto & Windus).

Bevington, Merle Mowbray (1966), *The Saturday Review, 1855–1868: Representative Educated Opinion in Victorian England* (New York: AMS Press).

Bibby, Cyril (1959), 'Huxley and the reception of the *Origin*', *Victorian Studies*, vol. 3 (September), pp. 76–86.

Bibby, Cyril (1972), *Scientist Extraordinary: The Life and Scientific Work of Thomas Henry Huxley, 1825–1895* (Oxford: Pergamon Press).

Bissell, Claude Thomas (1940), 'A study of *The Way of All Flesh*'; see Davis, DeVane and Bald, 1940.

Bissell, Claude Thomas (1961), 'The Butlerian inheritance of G. B. Shaw', *Dalhousie Review*, vol. 41, pp. 159–73.

Björk, Lennart A. (ed.) (1974), *The Literary Notes of Thomas Hardy* (Göteborg: Gothenburg Studies in English, 29).

Blacker, Carlos Paton (1952), *Eugenics: Galton and After* (London: Duckworth).

Bleiler, Everett Franklin (1948), *The Checklist of Fantastic Literature: A Bibliography of Fantasy, Weird and Science Fiction Books, Published in the English Language* (Chicago, Ill.: Shasta).

Blind, Mathilde (1886), *Shelley's View of Nature Contrasted with Darwin's* (printed for private distribution).

Blind, Mathilde (1899), *The Ascent of Man*, with an introduction by A. R. Wallace (London: T. Fisher Unwin).

Blinderman, Charles S. (1961), 'T. H. Huxley and Kingsley', *Victorian Newsletter*, vol. 20, pp. 25–8.

Blinderman, Charles S. (1962), 'Semantic aspects of T. H. Huxley's literary style', *Journal of Communication*, vol. 22, pp. 171–8.

Boll, T. E. M. (1966), 'Tess as an animal in nature', *English Literature in Transition*, vol. 9, pp. 210–11.

Bowman, Sylvia Edmonia (1958), *The Year 2000: A Critical Biography of Edward Bellamy* (New York: Bookman Associates).

Bowman, Sylvia Edmonia, and others (1962), *Edward Bellamy Abroad: An American Prophet's Influence* (New York: Twane).

Brennecke, Ernest (1966), *Thomas Hardy's Universe: A Study of the Poet's Mind* (New York: Folcroft).

Brett, Raymond L. (1960), 'The influence of Darwin upon his contemporaries', *South Atlantic Quarterly*, vol. 59 (Winter), pp. 69–81.

Breuer, Hans-Peter (1970), 'A critical and annotated edition of Samuel Butler's *Erewhon*', unpublished PhD thesis, Stanford University.

Bridges, J. H. (1877), 'Evolution and Positivism', *Fortnightly Review*, vol. 22 (June), pp. 853–74, and (July), pp. 89–114.

Brogan, Howard O. (1957), 'Science and narrative structure in Austen, Hardy, and Woolf', *Nineteenth-Century Fiction*, vol. 11 (March), pp. 276–87.

Brome, Vincent (1951), *H. G. Wells: A Biography* (London: Longmans, Green).

Brooks, Jean R. (1961), 'Darwinism in Thomas Hardy's major novels', unpublished MA thesis, Birkbeck College, University of London.

Brooks, Jean R. (1971), *Thomas Hardy: The Poetic Structure* (Ithaca, NY: Cornell University Press).

Bryan, Daniel Vance (1953), 'Samuel Butler: creative evolution in literature', unpublished PhD thesis, University of Minnesota.

Buchmann, Ralf (1941), *Martin F. Tupper and the Victorian Middle Class Mind* (Bern: Francke).

Bucke, Richard Maurice (1948), *Cosmic Consciousness: A Study in the Evolution of the Human Mind* (New York: Dutton).

Buckle, George Earle (1871), 'Natural selection insufficient to the development of man', *Popular Science Review*, vol. 10, pp. 14–24.

Buckley, Jerome Hamilton (1945), *William Ernest Henley: A Study in the 'Counter-Decadence' of the 'Nineties* (Princeton, NJ: Princeton University Press).

Buckley, Jerome Hamilton (1966a), *The Triumph of Time: A Study of the Victorian Concepts of Time, History, Progress and Decadence* (Cambridge, Mass.: Harvard University Press).

Buckley, Jerome Hamilton (1966b), *The Victorian Temper: A Study in Literary Culture* (London: Cass).

Bulwer-Lytton, Edward (1895), *The Novels and Romances of Edward Bulwer-Lytton*, 40 vols (Boston, Mass.: Little, Brown).

Burchfield, Joe Donald (1975), *Lord Kelvin and the Age of the Earth* (London: Macmillan).

Burnaby, John (1959), *Darwin and the Human Situation* (Cambridge: Heffer).

Burrow, John Wyan (1963), 'Evolution and anthropology in the 1860s: the Anthropological Society of London, 1863–71', *Victorian Studies*, vol. 7 (December), pp. 137–54.

Burrow, John Wyan (1966), *Evolution and Society: A Study in Victorian Social Theory* (Cambridge: Cambridge University Press).
Bury, John Bagnall (1920), *The Idea of Progress: An Inquiry into Its Origin and Growth* (London: Macmillan).
Bush, John Nash Douglas (1950), *Science and English Poetry: A Historical Sketch, 1590–1950* (London: Oxford University Press).
Butler, Samuel (1962), *The Correspondence of Samuel Butler with His Sister May*. . ., ed. Daniel F. Howard (Berkeley, Calif.: University of California Press).
Butler, Samuel (1965), *Ernest Pontifex, or The Way of All Flesh: A Story of English Domestic Life* . . ., ed. Daniel F. Howard (London: Methuen).
Butler, Samuel (1968), *The Shrewsbury Edition of the Works of Samuel Butler*, ed. Henry Festing Jones and A. T. Bartholomew, 20 vols (New York: AMS Press).
Campbell, George Douglas, 8th Duke of Argyll (1867), *The Reign of Law* (London: Strahan).
Campbell, George Douglas, 8th Duke of Argyll (1884), *The Unity of Nature* (London: Strahan).
Campbell, George Douglas, 8th Duke of Argyll (1894), *The Burdens of Belief and Other Poems* (London: Murray).
Campbell, George Douglas, 8th Duke of Argyll (1906), *Autobiography and Memoirs* . . ., ed. Duchess of Argyll, 2 vols (London: Murray).
Campbell, John Angus (1970), 'Darwin and the origin of species: the rhetorical ancestry of an idea', *Speech Monographs*, vol. 37 (March), pp. 1–14.
Cannon, Herbert Graham (1959), *Lamarck and Modern Genetics* (Manchester: Manchester University press).
Cannon, Walter F. (1961), 'The bases of Darwin's achievement: a revaluation', *Victorian Studies*, vol. 5 (December), pp. 109–34.
Cannon, Walter F. (1967), 'Darwin's vision in *On the Origin of Species*'; see Levine and Madden, 1968.
Carey, G. O. (1964), 'Butler's theory of evolution: a summary', *English Literature in Transition*, vol. 7, pp. 230–3.
Carpenter, Edward (1889), *Civilisation: Its Cause and Cure, and Other Essays* (London: Swann Sonnenschein).
Carter, George Stuart (1957), *A Hundred Years of Evolution* (London: Sidgwick & Jackson).
Cazamian, Madeleine L. (1923), *Le Roman et les ideés en Angleterre: l'influence de la science (1860–1890)*, Publications de la Faculté des Lettres de l'Université de Strasbourg, Fascicule 15 (London: Oxford University Press).
'Cecil, Hugh Mortimer', see Newman, Ernest.
[Chambers, Robert] (1844), *Vestiges of the Natural History of Creation* (London: Churchill).
Cherry, Douglas (1966), 'The two cultures of Matthew Arnold and T. H. Huxley', *Wascana Review*, vol. 1, pp. 53–61.
Chesterton, Gilbert Keith (1937), *Autobiography* (London: Hutchinson).
Chesterton, Gilbert Keith (1966), *The Victorian Age in Literature* (London: Oxford University Press).
Clagett, Marshall (ed.) (1959), *Critical Problems in the History of Science* (Madison, Wis.: University of Wisconsin Press).
Clarke, Ignatius Frederick (1958), 'The nineteenth-century Utopia', *Quarterly Review*, vol. 296, pp. 80–91.

Clarke, Ignatius Frederick (1961), *The Tale of the Future: From the Beginning to the Present Day. A Checklist . . . between 1644 and 1960* (London: Library Association).

Clarke, Ignatius Frederick (1966), *Voices Prophesying War, 1763–1984* (London: Oxford University Press).

Clarke, Ignatius Frederick (1979), *The Pattern of Expectation, 1644–2001* (London: Cape).

Clodd, Edward (1900), *Grant Allen: A Memoir, with a Bibliography* (London: Grant Richards).

Clodd, Edward (1916), *Memories* (London: Chapman & Hall).

Cole, George Douglas Howard (1947), *Samuel Butler and 'The Way of All Flesh'* (London: Home & Van Thal).

Cole, Susan Ablon (1971), 'The Utopian plays of George Bernard Shaw: a study of the plays and their relationship to the fictional Utopias of the period from the early 1870s to the early 1920s', unpublished PhD thesis, Brandeis University.

Coleman, William (1971), *Biology in the Nineteenth Century: Problem of Form, Function and Transformation* (New York: Wiley).

Coleridge, Samuel Taylor (1836), *Specimens of the Table Talk of Samuel Taylor Coleridge*, 2nd edn (London: Murray).

Collins, James (1959), 'Darwin's impact on philosophy', *Thought*, vol. 34, pp. 185–248.

Comfort, Alexander (1961), *Darwin and the Naked Lady: Discursive Essays on Biology and Art* (London: Routledge & Kegan Paul).

Cowan, Ruth Leah Schwartz (1968 [published 1971]), 'Sir Francis Galton and the continuity of the germ-plasm: a biological idea with political roots', *Actes du XIle Congrès International d'Histoire des Sciences*, vol. 8, pp. 181–6.

Cowan, Ruth Leah Schwartz (1969), 'Sir Francis Galton and the study of heredity in the nineteenth century', unpublished PhD thesis, Johns Hopkins University.

Cowles, Thomas (1937), 'Malthus, Darwin, and Bagehot: a study in the transference of a concept', *Isis*, vol. 26 (March), pp. 341–8.

Cox, J. Stevens (ed.) (1969), *The Library of Thomas Hardy, OM* (St Peter Port: Toucan Press).

Croll, James (1877), 'Evolution by force impossible: a new argument against materialism', *British Quarterly Review*, vol. 77, pp. 35–71.

Crum, Ralph B. (1966), *Scientific Thought in Poetry* (New York: AMS Press).

Cruse, Amy (n.d.), *The Victorians and Their Reading* (Boston, Mass.: Houghton Mifflin).

Cunliffe, John W. (1963), *Leaders of the Victorian Revolution* (New York: Russell & Russell).

Daiches, David (1969), *Some Late Victorian Attitudes* (London: Deutsch).

Daniels, R. Balfour (1969), 'Names in the fiction of Samuel Butler', *South Central Bulletin*, vol. 29 (Winter), pp. 129–32.

Darlington, Cyril Dean (1959a), *Darwin's Place in History* (Oxford: Blackwell).

Darlington, Cyril Dean (1959b), 'The origin of Darwinism', *Scientific American*, vol. 200 (May), pp. 60–6.

Darwin, Charles Robert (1862), *On the Various Contrivances by which British and Foreign Orchids are Fertilised by Insects . . .* (London: Murray).

Darwin, Charles Robert (1870), 'Notes on the habits of the Pampas Woodpecker (*Colaptes campestris*)', *Proceedings of the Zoological Society of London*, 1 November, pp. 705–6.

Darwin, Charles Robert (1872), *The Expression of the Emotions in Man and Animals* (London: Murray).

Darwin, Charles Robert (1903), *More Letters of Charles Darwin: A Record of His Work in a Series of Hitherto Unpublished Letters . . .*, ed. Francis Darwin and A. C. Seward, 2 vols (London: Murray).

Darwin, Charles Robert (1905), *The Variation of Animals and Plants under Domestication*, 2nd edn revised, 2 vols (London: Murray).

Darwin, Charles Robert (1909), *The Descent of Man, and Selection in Relation to Sex*, 2nd edn (London: Murray).

Darwin, Charles Robert (1945), *The Formation of Vegetable Mould, through the Action of Worms, with Observations on Their Habits . . .*, with an introduction by Sir Albert Howard (London: Faber, 1945).

Darwin, Charles Robert (1958), *The Autobiography of Charles Darwin, 1809–1882, with Original Omissions Restored . . .*, ed. Nora Barlow (London: Collins).

Darwin, Charles Robert (1959), *The Origin of Species by Charles Darwin: A Variorum Text*, ed. Morse Peckham (Philadelphia, Pa: University of Pennsylvania Press).

Darwin, Francis (1887), *The Life and Letters of Charles Darwin, Including an Autobiographical Chapter . . .*, 3 vols (London: Murray).

[Davis, Ellis James] (1875), *Pyrna: A Commune, or Under the Ice* (London: Bickers).

Davis, Herbert, DeVane, William C., and Bald, Robert Cecil (eds) (1940), *Nineteenth Century Studies* (Ithaca, NY: Cornell University Press).

Dawes, Ben (1952), *A Hundred Years of Biology* (London: Duckworth).

De Beer, Gavin Rylands (1961), 'The origins of Darwin's ideas on evolution and natural selection', *Proceedings of the Royal Society of London, Series B, Biological Sciences*, vol. 155 (November), pp. 321–8.

De Beer, Gavin Rylands (1963), *Charles Darwin: Evolution by Natural Selection* (London: Nelson).

Delage, Yves (1903), *L'Hérédité et les grands problèmes de la biologie générale*, 2nd edn (Paris: Schleicher Frères).

DeMaria, Robert (1959), 'From Bulwer-Lytton to George Orwell: the Utopian novel in England, 1870–1950', unpublished PhD thesis, Columbia University.

Derham, William (1713), *Physico-Theology, or A Demonstration of the Being and Attributes of God, from His Works of Creation* (London).

Descartes, René (1931), *The Philosophical Works of René Descartes*, trans. E. S. Haldane and G. R. T. Ross, 2 vols (Cambridge: Cambridge University Press).

Dewey, John (1951), *The Influence of Darwin on Philosophy and Other Essays in Contemporary Thought* (New York: Peter Smith).

Dickson, Lovat (1972), *H. G. Wells: His Turbulent Life and Times* (Harmondsworth: Penguin). First published 1969.

Dobzhansky. Theodosius (1959), 'Genetics and the destiny of Man', *Antioch Review*, vol. 19 (Spring), pp. 57–68.

Dowden, Edward (1878), *Studies in Literature, 1789–1877* (London: Kegan Paul).

Drachman, Julian Moses (1931), *Studies in the Literature of Natural Science* (New York: Macmillan).

'Drew, D'Arcy', see Mivart, St George Jackson.

Drummond, Henry (1885), *Natural Law in the Spiritual World*, 16th edn (London: Hodder & Stoughton).

Drummond, Henry (1894), *The Lowell Lectures on the Ascent of Man* (London: Hodder & Stoughton).

Dudley, Fred A. (1939), 'Matthew Arnold and science', unpublished PhD thesis, State University of Iowa.

Duffin, Henry Charles (1937), *Thomas Hardy: A Study of the Wessex Novels*, 3rd edn (Manchester: Manchester University Press).

Duncan, David (1908), *The Life and Letters of Herbert Spencer* (London: Methuen).

Eastman, Max Forrester (1931), *The Literary Mind: Its Place in an Age of Science* (New York: Schribner's).

Eiseley, Loren Corey (1956), 'Charles Darwin', *Scientific American*, vol. 194 (February), pp. 62–72.

Eiseley, Loren Corey (1959a), 'Charles Darwin, Edward Blyth, and the theory of natural selection', *Proceedings of the American Philosophical Society*, vol. 103 (February), pp. 94–158.

Eiseley, Loren Corey (1959b), 'Alfred Russel Wallace', *Scientific American*, vol. 200 (February), pp. 70–84.

Eiseley, Loren Corey (1961), *Darwin's Century: Evolution and the Men Who Discovered It* (New York: Anchor).

Eiseley, Loren Corey (1965), 'Darwin, Coleridge, and the theory of unconscious creation', *Daedalus*, vol. 94 (Summer), pp. 588–602.

Ellegård, Henrik Alvar (1957), *The Readership of the Periodical Press in Mid-Victorian Britain* (Göteborg: Göteborg University Press).

Ellegård, Henrik Alvar (1958), *Darwin and the General Reader: The Reception of Darwin's Theory of Evolution in the British Periodical Press, 1859–1872* (Göteborg: Gothenburg Studies in English).

Ellegård, Henrik Alvar (1969), 'The Darwinian theory and the argument from design'; see Hardin, 1969.

Elliot, Robert C. (1970), *The Shape of Utopia: Studies in a Literary Genre* (Chicago, Ill.: University of Chicago Press).

Ellis, William (1866), *Thoughts on the Future of the Human Race* (London: Smith, Elder).

Essays and Reviews (1861), 8th edn (London: Longman, Green, Longman & Roberts).

Etymonia (1875) (London: Tinsley).

Evans, Benjamin Ifor (1954), *Literature and Science* (London: Allen & Unwin).

Everett, Edwin Mallard (1971), *The Party of Humanity: The 'Fortnightly Review' and Its Contributors, 1865–1874* (New York: Russell & Russell).

Firor, Ruth A. (1931), *Folkways in Thomas Hardy* (New York: Barnes).

Fisher, Ronald Aylmer (1930), *The Genetical Theory of Natural Selection* (Oxford: Clarendon Press).

Fleming, Donald (1959), 'The centenary of the *Origin of Species*', *Journal of the History of Ideas*, vol. 20 (June), pp. 437–46.

Fleming, Donald (1961), 'Charles Darwin, the anaesthetic man', *Victorian Studies*, vol. 4 (March), pp. 219–36.

Forrest, Derek William (1975), *Francis Galton: The Life and Work of a Victorian Genius* (London: Elek).

Forrester, J. M. (1971), 'George Eliot and physiology', *Proceedings of the Royal Society of Medicine*, vol. 64 (July), pp. 724–6.

Foskett, Douglas John (1953), 'Wilberforce and Huxley on evolution', *Nature*, vol. 172, 14 November, p. 920.

Foskett, Douglas John (1955), 'T. H. Huxley and his presentation of scientific material', unpublished MA thesis, University of London.

Fothergill, Phillip (1952), *Historical Aspects of Organic Evolution* (London: Hollis & Carter).
Frazer, James George (1968), *Creation and Evolution in Primitive Cosmogonies and Other Pieces* (London: Dawsons).
Frederick, John T. (1972), *William Henry Hudson* (New York: Twayne).
[Froude, James Anthony] (1847), *Shadows of the Clouds* (London: Ollivier).
Fuller, Watson (ed.) (1971), *The Social Impact of Modern Biology* (London: Routledge & Kegan Paul).
Furbank, Philip Nicholas (1948), *Samuel Butler* (Cambridge: Cambridge University Press).
Gale, Barry G. (1972), 'Darwin and the concept of a struggle for existence: a study in the extra-scientific origins of scientific ideas', *Isis*, vol. 63 (September), pp. 321–44.
Galton, Francis (1865), 'Hereditary talent and character', *Macmillan's Magazine*, vol. 12 (June), pp. 157–66, and (August), pp. 318–27.
Galton, Francis (1908), *Memories of My Life* (London: Methuen).
Galton, Francis (1909), *Essays in Eugenics* (London: Eugenics Education Society).
Galton, Francis (1911), *Inquiries into Human Faculty and Its Development*, 2nd edn (London: Dent).
Galton, Francis (1962), *Hereditary Genius: An Inquiry into Its Laws and Consequences*, with an introduction by C. D. Darlington, 2nd edn (London: Collins).
Gantz, Kenneth Franklin (1939), 'The beginnings of Darwinian ethics, 1859–1871', *University of Texas Studies in English*, vol. 8 (July), pp. 180–209.
Gardner, Joseph H. (1970), 'A Huxley essay as "poem" ', *Victorian Studies*, vol. 14 (December), pp. 177–91.
Garnett, Edward (1917), 'W. H. Hudson', *Dial*, vol. 62, 8 February, pp. 83–7.
Garrett, John Charles (1968), *Utopias in Literature since the Romantic Period* (Christchurch: University of Canterbury).
Gasking, Elizabeth B. (1959), 'Why was Mendel's work ignored?', *Journal of the History of Ideas*, vol. 20 (January), pp. 60–84.
Geison, Gerald L. (1969), 'Darwin and heredity: the evolution of his hypothesis of pangenesis', *Journal of the History of Medicine and Allied Sciences*, vol. 24 (October), pp. 375–411.
George, Wilma (1964), *Biologist-Philosopher: A Study of the Life and Writings of Alfred Russel Wallace* (London: Abelard-Schuman).
Gerber, Richard (1955), *Utopian Fantasy: A Study of English Utopian Fiction since the End of the Nineteenth Century* (London: Routledge & Kegan Paul).
Ghiselin, Michael T. (1969), *The Triumph of the Darwinian Method* (Berkeley, Calif.: University of California Press).
Ghiselin, Michael T. (1971), 'The individual in the Darwinian revolution', *New Literary History*, vol. 3 (Autumn), pp. 113–34.
Gibbons, Tom (1973), *Rooms in the Darwin Hotel: Studies in English Literary Criticism and Ideas, 1880–1920* (Nedlands: University of Western Australia Press).
Gibson, Walker (1958), 'Behind the veil: a distinction between poetic and scientific language in Tennyson, Lyell, and Darwin', *Victorian Studies*, vol. 2 (September), pp. 60–8.
Gillispie, Charles Coulston (1960), *The Edge of Objectivity: An Essay in the History of Scientific Ideas* (London: Oxford University Press).
Glass, Hiram Bentley, and others (1959), *Forerunners of Darwin, 1745–1859* (Baltimore, Md: Johns Hopkins Press).

Glicksberg, Charles I. (1952), 'Hardy's scientific pessimism', *Western Humanities Review*, vol. 6, pp. 273–83.

Good, Rankine (1954), 'The origin of *The Origin*: a psychological approach', *Biology and Human Affairs*, vol. 20 (October), pp. 10–16.

Goodwin, Carolyn Ann (1971), 'Charles Kingsley and the vision of childhood: a study of *The Water-Babies* as a response to Darwinism', unpublished MA thesis, Queens University, Ontario.

Gose, Elliot B. (1963), 'Psychic evolution: Darwinism and initiation in *Tess of the d'Urbervilles*', *Nineteenth-Century Fiction*, vol. 18 (December), pp. 261–72.

Gosse, Philip Henry (1857), *Omphalos: An Attempt to Untie the Geological Knot* . . . (London).

[Grant, Charles William] (1872), *Our Blood Relations, or The Darwinian Theory* (London/Bath: Simkin Marshall/R. E. Peach). Attribution in Halkett and Laing, Vol. 4, p. 277.

Gray, Asa (1963), *Darwiniana: Essays and Reviews Pertaining to Darwinism*, ed. A. Hunter Dupree (Cambridge, Mass.: Harvard University Press).

Green, Martin (1964), *Science and the Shabby Curate of Poetry: Essays about the Two Cultures* (London: Longmans).

'Green, Nunsowe' (1882), *A Thousand Years Hence: Being Personal Experiences as Narrated by Nunsowe Green* (London: Sampson Low, Marston, Searle & Rivington).

Greene, John C. (1959), 'Biology and social theory in the nineteenth century: Auguste Comte and Herbert Spencer'; see Clagett, 1959.

Greene, John C. (1961), *Darwin and the Modern World View* (Baton Rouge, La: Louisiana State University Press).

[Greg, William Rathbone] (1868), 'On the failure of "natural selection" in the case of man', *Fraser's Magazine*, vol. 78 (September), pp. 353–62.

Greg, William Rathbone (1872), *Enigmas of Life* (Edinburgh).

Greg, William Rathbone (1874), *Rocks Ahead, or The Warnings of Cassandra* (London).

Griffen, Ernest C. (1961–2), 'Hardy and the growing consciousness of the Immanent Will: a study in the relationship of philosophy to literary form', *Cairo Studies in English*, pp. 121–33.

Griffith, Philip Mahone (1963), 'The image of the trapped animal in Hardy's *Tess of the d'Urbervilles*', *Tulane Studies in English*, vol. 13, pp. 85–94.

Grinnell, George James (1969), 'The Darwin Case: a computer analysis of scientific creativity', unpublished PhD thesis, University of California at Berkeley.

Grove, Richard Stanley (1969), 'A re-examination of Darwin's argument in *On the Origin of Species*', unpublished PhD thesis, University of Missouri.

Gruber, Howard E. (1974), *Darwin on Man: A Psychological Study of Scientific Creativity . . . together with Darwin's Early and Unpublished Notebooks Transcribed and Annotated by Paul H. Barrett* (London: Wildwood House).

Gruber, Jacob W. (1960), *A Conscience in Conflict: The Life of St George Jackson Mivart* (New York: Columbia University Press).

Haight, Gordon S. (1957), 'H. G. Wells's "The man of the year million" ', *Nineteenth-Century Fiction*, vol. 12 (March), pp. 323–6.

Haldane, John Burdon Sanderson (1932), *The Causes of Evolution* (London: Longmans, Green).

Haldane, John Burdon Sanderson (1959), 'Natural selection'; see Bell, 1959.

Haller, Mark Hughlin (1963), *Eugenics: Hereditarian Attitudes in American Thought* (New Brunswick, NJ: Rutgers University Press).

Halliday, R. J. (1971), 'Social Darwinism: a definition', *Victorian Studies*, vol. 14 (June), pp. 389–405.
Hamilton, Robert (1946), *W. H. Hudson: The Vision of Earth* (London: Dent).
Hanawalt, Mary Wheat (1937), 'Charles Kingsley and science', *Studies in Philology*, vol. 34 (October), pp. 589–611.
Hardin, Garrett (1961), *Nature and Man's Fate* (New York: Mentor).
Hardin, Garrett (ed.) (1969), *Population, Evolution, and Birth Control: A Collage of Controversial Ideas* (San Francisco, Calif.: Freeman).
Hardy, Alister (1965), *The Living Stream* (London: Collins).
Hardy, Florence Emily (1962), *The Life of Thomas Hardy, 1840–1928* (London: Macmillan).
Hardy, Thomas (1891a), 'The science of fiction III', *New Review*, vol. 4 (April), pp. 315–19.
Hardy, Thomas (1891b), 'The midnight baptism: a study in Christianity', *Fortnightly Review*, vol. 49 (May), pp. 695–701.
Hardy, Thomas (1912–31), *The Wessex Edition of the Works of Thomas Hardy in Prose and Verse*, 21 vols (London: Macmillan).
Hardy, Thomas (1976), *The Complete Poems of Thomas Hardy*, ed. James Gibson (London: Macmillan).
Hartog, Marcus (1893), 'The Spencer–Weismann controversy II', *Contemporary Review*, vol. 64 (July), pp. 54–9.
Hartog, Marcus (1914), 'Samuel Butler and recent mnemic biological theories', *Scientia*, vol. 15, pp. 38–52.
Hawkins, Desmond (1950), *Thomas Hardy* (London: Arthur Barker).
Haymaker, Richard Eugene (1954), *From Pampas to Hedgerows and Downs: A Study of W. H. Hudson* (New York: Bookman Associates).
Helfand, Michael S. (1977), 'T. H. Huxley's "Evolution and ethics": the politics of evolution and the evolution of politics', *Victorian Studies*, vol. 20 (Winter), pp. 159–77.
Henderson, Archibald (1956), *George Bernard Shaw: Man of the Century* (New York: Appleton-Century-Crofts).
Henderson, Philip (1953), *Samuel Butler: The Incarnate Bachelor* (London: Cohen & West).
Hillegas, Mark Robert (1967), *The Future as Nightmare: H. G. Wells and the Anti-Utopians* (London/New York: Oxford University Press).
Himmelfarb, Gertrude (1959), *Darwin and the Darwinian Revolution* (New York: Doubleday).
Himmelfarb, Gertrude (1968), *Victorian Minds* (London: Weidenfeld & Nicolson).
Hofstadter, Richard (1964), *Social Darwinism in American Thought* (Boston, Mass.: Beacon Press).
Holloway, John (1960), *The Charted Mirror: Literary and Critical Essays* (London: Routledge & Kegan Paul).
Holt, Lee E. (1964), *Samuel Butler* (New York: Twayne).
Hopkins, William (1860), 'Physical theories of the phenomena of life', *Fraser's Magazine*, vol. 61 (June), pp. 739–52.
Hough, Graham (1947), 'The natural theology of *In Memoriam*', *Review of English Studies*, vol. 23 (July), pp. 244–56.
Houghton, Walter Edwards (1948–9), 'The rhetoric of T. H. Huxley', *University of Toronto Quarterly*, vol. 18, pp. 159–75.
Houghton, Walter Edwards (1959), *The Victorian Frame of Mind, 1830–1870* (New Haven, Conn.: Yale University Press).

Hubenka, Lloyd John (1966), 'The religious philosophy of Bernard Shaw', unpublished PhD thesis, University of Nebraska.

Hudson, Derek (1949), *Martin Tupper: His Rise and Fall* (London: Constable).

Hudson, William Henry (1870), [Third letter on the ornithology of Buenos Ayres], *Proceedings of the Zoological Society of London*, 24 March, pp. 158–60. Dated 28 January, according to the manuscript version printed in Hudson, 1951.

Hudson, William Henry (1887), *A Crystal Age* (London: T. Fisher Unwin).

Hudson, William Henry (1904), *Green Mansions: A Romance of the Tropical Forest* (London: Duckworth).

Hudson, William Henry (1918), *Far Away and Long Ago: A History of My Early Life* (London: Dent).

Hudson, William Henry (1919), *A Crystal Age*, 2nd edn, 4th impression (London: Duckworth).

Hudson, William Henry (1922–3), *The Uniform Edition of the Works of William Henry Hudson*, 24 vols (London: Dent).

Hudson, William Henry (1924), *Birds and Man*, new edn (London: Duckworth).

Hudson, William Henry (1925), *Letters from W. H. Hudson to Edward Garnett*, new edn (London: Dent).

Hudson, William Henry (1928), *Men, Books and Birds: with Notes, Some Letters, and an Introduction by Morley Roberts* (London: Cape).

Hudson, William Henry (1939), *The Book of a Naturalist*, 2nd edn (London: Hodder & Stoughton).

Hudson, William Henry (1951), *Letters on the Ornithology of Buenos Ayres . . .*, ed. David R. Dewar (Ithaca, NY: Cornell University Press).

Hudson, William Henry (1951), *Nature in Downland* (London: Dent).

Hughes, H. Stuart (1967), *Consciousness and Society: The Reorientation of European Social Thought, 1890–1930* (London: MacGibbon & Kee).

Hull, David L. (1973), *Darwin and His Critics* (London: Oxford University Press).

Huxley, Aldous Leonard (1932), *Brave New World* (London: Chatto & Windus).

Huxley, Aldous Leonard (1947), *The Olive Tree and Other Essays* (London: Chatto & Windus).

Huxley, Aldous Leonard (1962), *Island* (London: Chatto & Windus).

Huxley, Aldous Leonard (1963), *Literature and Science* (London: Chatto & Windus).

Huxley, Aldous Leonard (1969), *Letters of Aldous Huxley*, ed. Grover Smith (London: Chatto & Windus).

Huxley, Julian Sorell (1927), *Religion without Revelation* (London: Benn).

Huxley, Julian Sorell (1942), *Evolution: The Modern Synthesis* (London: Allen & Unwin).

Huxley, Julian Sorell (1958a), 'The emergence of Darwinism', *Journal of the Linnean Society of London*, vol. 44 (July), pp. 1–14.

Huxley, Julian Sorell (1958b), Introduction to: Charles Darwin, *The Origin of Species* (New York: Mentor).

Huxley, Leonard (1900), *Life and Letters of Thomas Henry Huxley*, 2 vols (London: Macmillan).

Huxley, Thomas Henry (1971), 'Mr Darwin's critics', *Contemporary Review*, vol. 18 (November), pp. 471–9.

Huxley, Thomas Henry ([1887]), 'On the reception of the *Origin of Species*'. Undated, but before 1887; see Darwin, Francis, 1887.

Huxley, Thomas Henry (1892), 'Westminster Abbey: October 12, 1892', *Nineteenth*

Century, vol. 32 (November), pp. 831–2. This poem appeared in Huxley, 1913, with some minor changes.
Huxley, Thomas Henry (1894–5), *Collected Essays*, 9 vols (London: Macmillan).
Huxley, Thomas Henry (1913), [Three poems], in *Poems of Henrietta A. Huxley with Three of Thomas Henry Huxley* (London: Duckworth).
Hyman, Stanley Edgar (1959a), 'The *Origin* as scripture', *Virginia Quarterly Review*, vol. 35 (Autumn), pp. 540–52.
Hyman, Stanley Edgar (1959b), 'The whole round world', *New Mexico Quarterly*, vol. 29 (Autumn), pp. 277–91.
Hyman, Stanley Edgar (1966), *The Tangled Bank: Darwin, Marx, Frazer, Freud as Imaginative Writers* (New York: Universal Library).
Irvine, William (1956), *Apes, Angels and Victorians: A Joint Biography of Darwin and Huxley* (London: Weidenfeld & Nicolson).
Isaacs, Leonard (1977), *Darwin to Double Helix: The Biological Theme in Science Fiction* (London: Butterworth).
Jefferies, John Richard (1885), *After London, or Wild England* . . . (London: Cassell).
Jefferies, John Richard (1898), *The Story of My Heart: My Autobiography* (London: Longmans).
Jefferies, John Richard (1948), *The Nature Diaries and Note-Books of Richard Jefferies* . . ., ed. Samuel J. Looker (London: Grey Walls Press).
[Jenkin, Henry Charles Fleeming] (1867), 'The origin of species', *North British Review*, vol. 46, pp. 277–318.
Jenkin, Henry Charles Fleeming (1887), *Papers Literary, Scientific, etc.* . . ., ed. Sidney Colvin and J. A. Ewing, with a memoir by R. L. Stevenson, 2 vols (London: Longmans).
Jensen, John Vernon (1959), 'The rhetoric of T. H. Huxley and Robert G. Ingersoll in relation to the conflict between science and theology', unpublished Phd thesis, University of Minnesota.
Jepsen, Glenn L. (1949), 'Selection, "orthogenesis", and the fossil record', *Proceedings of the American Philosophical Society*, vol. 93 (December), pp. 479–500.
Joad, Cyril Edwin Mitchinson (1924), *Samuel Butler, 1835–1902* (London: Parsons).
Johnson, Bruce (1977), ' "The perfection of species" and Hardy's Tess'; see Knoepflmacher and Tennyson, 1977.
Johnson, Trevor (1968), *Thomas Hardy* (London: Evans).
Johnson, Arthur (1959), '*The Water-Babies*: Kingsley's debt to Darwin', *English*, vol. 12 (Autumn), pp. 215–19.
Jones, Ernest (1959), *Free Associations: Memories of a Psycho-Analyst* (London: Hogarth Press).
Jones, Frederick L. (1952), 'Macaulay's theory of poetry in *Milton*', *Modern Language Quarterly*, vol. 13 (December), pp. 356–62.
Jones, Henry Festing (1919), *Samuel Butler, Author of 'Erewhon' (1835–1902): A Memoir*, 2 vols (London: Macmillan).
Jones, Joseph Jay (1960), *The Cradle of Erewhon: Samuel Butler in New Zealand* (Melbourne: Melbourne University Press).
Judd, John Wesley (1910), *The Coming of Evolution: The Story of a Great Revolution in Science*, Cambridge Manuals of Science and Literature (Cambridge: Cambridge University Press).
Kaleb, George (1963), *Utopia and Its Enemies* (Glencoe, Ill.: Free Press of Glencoe).
Keith, Arthur (1955), *Darwin Revalued* (London: Watts).

Keith, William John (1965), *Richard Jefferies: A Critical Study* (Toronto: Toronto University Press).
Kellog, Vernon Lyman (1907), *Darwinism Today: A Discussion of Present-Day Scientific Criticism of the Darwinian Selection Theories, together with a Brief Account of the Principal Other Proposed Auxiliary and Alternative Theories of Species-Forming* (London: Bell).
Kendall, Guy (1946), *Charles Kingsley and His Ideas* (London: Hutchinson).
Kettle, Arnold (1953), *An Introduction to the English Novel*, 2 vols (London: Hutchinson University Library).
Kidd, Benjamin (1885a), 'The habits and intelligence of bees', *Longman's Magazine*, vol. 6 (June), pp. 173–85.
Kidd, Benjamin (1885b), 'Humble bees', *Longman's Magazine*, vol. 7 (December), pp. 196–210.
Kidd, Benjamin (1895), *Social Evolution*, 2nd edn (London: Macmillan).
Kidd, Benjamin (1921), *A Philosopher with Nature* (London: Methuen).
Kingsley, Charles (1855), *Glaucus, or The Wonders of the Shore* (Cambridge: Macmillan).
Kingsley, Charles (1870), 'Strange noises heard at sea off Grey Town', *Nature*, vol. 2, 19 May, p. 46. Letter to the editor.
Kingsley, Charles (1880), *Scientific Lectures and Essays* (London: Macmillan).
Kingsley, Charles (1889), *Madam How and Lady Why, or First Lessons in Earth Lore for Children* (London: Macmillan).
Kingsley, Charles (1901–3), *The Life and Works of Charles Kingsley*, 19 vols (London: Macmillan).
[Kingsley, Frances E.] (1877), *Charles Kingsley: His Letters and Memories of His Life* . . ., 2 vols (London: King).
Kingsmill, Hugh (1929), *After Puritanism, 1850–1900* (London: Duckworth).
Knoepflmacher, U. C., and Tennyson, G. B. (eds) (1977), *Nature and the Victorian Imagination* (Berkeley, Calif.: University of California Press).
Koestler, Arthur (1971), *The Case of the Midwife Toad* (London: Hutchinson).
Kogan, B. R. (ed.) (1960), *Darwin and His Critics: The Darwinian Revolution* (San Francisco, Calif.: Wadsworth).
Kopp, Karl Caton (1963), 'The origin and characteristics of "decadence" in British literature of the 1890s', unpublished PhD thesis, University of California at Berkeley.
Kottler, Malcolm J. (1974), 'Alfred Russel Wallace, the origin of man, and spiritualism', *Isis* vol. 65, pp. 145–92.
Kropotkin, Petr (1972), *Mutual Aid: A Factor of Evolution* (London: Allen Lane). First published as a series of articles in *Nineteenth Century* between 1890 and 1896.
Laird, John T. (1966), 'The manuscript of Hardy's *Tess of the d'Urbervilles* and what it tells us', *AUMLA*, vol. 25 (May), pp. 68–82.
Laird, John T. (1975), *The Shaping of 'Tess of the d'Urbervilles'* (Oxford: Clarendon Press).
Lankester, Edwin Ray (1880), *Degeneration: A Chapter in Darwinism* (London: Macmillan).
Lankester, Edwin Ray (1890), *The Advancement of Science: Occasional Essays and Addresses* (London: Macmillan).
Lautz, Richard Eugene (1967), 'Bulwer-Lytton as novelist', unpublished PhD thesis, University of Pennsylvania.
Lawrence, William (1819), *Lectures on Physiology, Zoology, and the Natural History of Man* . . . (London).

Layard, George Somes (1901), *Mrs Lynn Linton: Her Life, Letters, and Opinions* (London: Methuen).

Le Gallienne, Richard (1910), *Attitudes and Avowals: with Some Retrospective Reviews* (London: John Lane).

Leach, Chester Raymond (1966), 'A critical edition of Tennyson's *In Memoriam*', unpublished PhD thesis, University of Michigan.

Lennox, Cuthbert (1901), *Henry Drummond: A Biographical Sketch* (London: Melrose).

Lerner, I. Michael (1959), 'The concept of natural selection: a centennial view', *Proceedings of the American Philosophical Society*, vol. 103, pp. 173–82.

Lester, John Ashby (1968), *Journey through Despair, 1880–1914: Transformations in British Literary Culture* (Princeton, NJ: Princeton University Press).

Levine, George Lewis, and Madden, William Anthony (eds) (1968), *The Art of Victorian Prose* (London: Oxford University Press).

Levine, Richard A. (ed.) (1967), *Backgrounds to Victorian Literature* (San Francisco, Calif.: Chandler).

Lewes, George Henry (1857), *The Biographical History of Philosophy, from Its Origin in Greece down to the Present Day*. . ., Library Edition (London: Parker).

Lewes, George Henry (1859–60), *The Physiology of Common Life*, 2 vols (London/Edinburgh).

Limoges, Camille (1970), *La Sélection naturelle: étude sur la première constitution d'un concept (1837–1859)* (Paris: Presses Universitaires de France).

Linton, Eliza Lynn (1894), 'Professor Henry Drummond's discovery', *Fortnightly Review*, n.s., vol. 56 (September), pp. 448–57.

Loewenberg, Bert James (1959a), *Darwin, Wallace and the Theory of Natural Selection, Including the Linnean Society Papers* (Cambridge, Mass.: Arlington Books).

Loewenberg, Bert James (1959b), 'The mosaic of Darwinian thought', *Victorian Studies*, vol. 3 (September), pp. 3–18.

Loewnberg, Bert James (1962), 'Darwin and the tragic vision', *American Quarterley*, vol. 14, pp. 618–22.

Loewenberg, Bert James (1965), 'Darwin and Darwin studies, 1959–63', *History of Science*, vol. 4, pp. 15–54.

Looker, Samuel J., and Porteus, Crichton (1965), *Richard Jefferies: Man of the Fields* (London: John Baker).

Lovejoy, Arthur Oncken (1936), *The Great Chain of Being: A Study of the History of an Idea* (Cambridge, Mass.: Harvard University Press).

Lovejoy, Arthur Oncken (1965), 'Some aspects of Darwin's influence upon modern thought'; see Baumer, 1965.

Lunn, Arnold Henry Moore (1931), *The Flight from Reason: A Study of the Victorian Heresy* (London: Eyre & Spottiswoode).

Lyell, Charles (1830–3), *Principles of Geology* (London).

Lyell, Charles (1863), *The Geological Evidences of the Antiquity of Man, with Remarks on Theories of the Origin of Species by Variation* (London).

Macaulay, Thomas Babington (1897), *The Life and Works of Thomas Babington Macaulay*, 10 vols (London: Longmans, Green).

Macbeth, Norman (1972), *Darwin Retried: An Appeal to Reason* (Boston, Mass.: Gambit Press).

Mackenzie, Norman and Jeanne (1973), *The Time Traveller: The Life of H. G. Wells* (London: Weidenfeld & Nicolson).

Mackintosh, Robert (1899), *From Comte to Benjamin Kidd: The Appeal to Biology, or Evolution for Human Guidance* (London: Macmillan).
Maddox, John (1973), 'Scientific journalism', *The Listener*, vol. 90, 6 December, pp. 761–2.
Maeterlinck, Maurice (1908), *The Treasure of the Humble*, trans. Alfred Sutro (London: Allen & Unwin).
Maitland, Edward (1873), *By and By: An Historical Romance of the Future* (London).
Malthus, Thomas Robert (1872), *An Essay on the Principle of Population . . .*, 6th edn (London: Reeves & Turner).
Manuel, Frank Edward (ed.) (1965a), *Utopias and Utopian Thought* (Boston, Mass.: Houghton Mifflin).
Manuel, Frank Edward (1965b), 'Toward a psychological history of Utopias'; see Manuel, 1965a.
Marchand, Leslie A. (1971), *The Athenaeum: A Mirror of Victorian Culture* (New York: Octagon Books).
Marchant, James (1916), *Alfred Russel Wallace: Letters and Reminiscences* (New York/London: Harper).
Marshall, William H. (1965), 'The use of symbols in *The Way of All Flesh*', *Tennessee Studies in Literature*, vol. 10, pp. 109–21.
Marshall, William H. (1967), *The World of the Victorian Novel* (New York: Yoseloff).
Martin, Robert Bernard (1959), *The Dust of Combat: A Life of Charles Kingsley* (London: Faber).
Mazzeo, Joseph Antony (1967), *The Design of Life: Major Themes in the Development of Biological Thought* (London: Macdonald).
McConnaughey, Gloria (1950), 'Darwin and social Darwinism', *Osiris*, vol. 9, pp. 397–412.
McDougall, William (1927), 'An experiment for the testing of the hypothesis of Lamarck', *British Journal of Psychology*, vol. 17, pp. 268–304.
McKinney, H. Lewis (1972), *Wallace and Natural Selection* (New Haven, Conn.: Yale University Press).
Meadows, A. J. (1969), *The High Firmament: A Survey of Astronomy in English Literature* (Leicester: Leicester University Press).
Meeker, Joseph W. (1974), *The Comedy of Survival: Studies in Literary Ecology* (New York: Scribner's).
Mill, John Stuart (1969), *The Collected Works of John Stuart Mill*, ed. F. E. L. Priestley (Toronto: University of Toronto Press).
Millgate, Michael (1971), *Thomas Hardy: His Career as a Novelist* (London: Bodley Head).
Millhauser, Milton (1956), 'The literary impact of *Vestiges of Creation*', *Modern Language Quarterly*, vol. 17 (September), pp. 213–26.
Millhauser, Milton (1959), *Just before Darwin: Robert Chambers and 'Vestiges'* (Middletown, Conn.: Wesleyan University Press).
Millhauser, Milton (1969), 'Tennyson, *Vestiges*, and the dark side of science', *Victorian Newsletter*, vol. 20, pp. 22–5.
Mills, Carl Henry (1965), 'The intellectual and literary background of George Bernard Shaw's *Man and Superman*', unpublished PhD thesis, University of Nebraska.
Milner, Philip Alan (1972), 'The moral vision of Thomas H. Huxley', unpublished PhD thesis, University of Notre Dame.
Mivart, St George Jackson (1871), *On the Genesis of Species* (London).

Mivart, St George Jackson (1876), *Lessons from Nature as Manifested in Mind and Matter* (London: Murray).
[Mivart, St George Jackson] (1894), *Henry Standon, or Love's Debt to Duty. By D'Arcy Drew*, 3 vols (London).
Montagu, M. F. Ashley (1942), 'Further comment [on Social Darwinism]', *Science and Society*, vol. 6 (Winter), pp. 74–5.
Montagu, M. F. Ashley (ed.) (1946), *Studies and Essays in the History of Science and Learning* . . . (New York: Schuman).
Morgan, Thomas Hunt (1903), *Evolution and Adaptation* (London: Macmillan).
Morley, John (1908), *Miscellanies. Fourth Series* (London: Macmillan).
Morley, John (1911), *Science and Literature* (Oxford: privately printed).
Morrell, Roy (1962), 'Hardy in the Tropics: some implications of Hardy's attitude towards nature', *Review of English Literature*, vol. 3 (January), pp. 7–21.
Morrell, Roy (1965), *Thomas Hardy: The Will and the Way* (Kuala Lumpur: University of Malaya Press).
Morris, William (1970), *News from Nowhere, or An Epoch of Rest: Being Some Chapters from a Utopian Romance* . . ., ed. James Redmond (London: Routledge & Kegan Paul).
Morton, Arthur Leslie (1952), *The English Utopia* (London: Lawrence & Wishart).
Mudford, Peter G. (1966), 'The impact of the theory of evolution on the late nineteenth century, with special reference to Thomas Henry Huxley and Samuel Butler', unpublished BLitt thesis, Oxford University.
Muggeridge, Malcolm (1936), *The Earnest Atheist: A Study of Samuel Butler* (London: Eyre & Spottiswoode).
Muller, Herbert J. (1943), *Science and Criticism: The Humanistic Tradition in Contemporary Thought* (New Haven, Conn.: Yale University Press).
Muncy, Raymond Lee (1973), *Sex and Marriage in Utopian Communities* (Bloomington, Ind.: Indiana University Press).
Myers, Frederic William Henry (1893), *Science and a Future Life; with Other Essays* (London: Macmillan).
Myers, Walter Lawrence (1927), *The Later Realism: A Study of Characterisation in the British Novel* (Chicago, Ill.: University of Chicago Press).
Negley, Glen Robert, and Patrick, J. Max (1952), *The Quest for Utopia: An Anthology of Imaginary Societies* (New York: Schuman).
Neill, Thomas Patrick (1959), *1859 in Review* (Westminster, Md: Newman Press).
Nelson, James G. (1966), *William Watson* (New York: Twayne).
Newman, Ernest (1897), *Pseudo-Philosophy at the End of the Nineteenth Century* . . ., Vol. 1, *An Irrationalist Trio: Kidd – Drummond – Balfour* (London: London University Press). No further volumes published.
Newman, John Henry (1964), *The Idea of a University, Defined and Illustrated in Nine Discourses* . . ., ed. Martin J. Svaglic (New York: Holt, Reinhart & Winston).
Newton, K. M. (1974), 'George Eliot, George Henry Lewes and Darwinism', *Durham University Journal*, vol. 66, pp. 278–93.
Newton, William (1951a), 'Chance as employed by Hardy and the naturalists', *Philological Quarterly*, vol. 30 (April), pp. 154–75.
Newton, William (1951b), 'Hardy and the naturalists: their use of physiology', *Modern Philology*, vol. 49 (August), pp. 28–41.
Nicholson, Norman (1950), *H. G. Wells* (London: Arthur Barker).
Nietzsche, Friedrich (1964), *The Complete Works of Friedrich Nietzsche*, ed. Oscar Levy, 18 vols (New York: Russell & Russell).

Nordau, Max Sloman (1968), *Degeneration* (New York: Howard Fertig). A translation from the second German edition.
'Notes on development (no. 1)' (1863), *Punch*, vol. 44, 21 March, p. 122.
Nursall, J. R. (1963), 'The consequences of Darwinism', *Dalhousie Review*, vol. 42, pp. 472–80.
Olmsted, James Montrose Duncan (1946), *Charles Edouard Brown-Séquard: A Nineteenth Century Neurologist and Endocrinologist* (Baltimore, Md: Johns Hopkins University Press).
Ong, Walter J. (1966), 'Evolution, myth, and poetic vision', *Comparative Literature Studies*, vol. 3, pp. 1–20.
Orwell, George (1961), *Collected Essays* (London: Secker & Warburg).
Osborn, Henry Fairfield (1894), *From the Greeks to Darwin: An Outline of the Development of the Evolution Idea* (New York: Columbia University Press).
Paley, William (1822), *Natural Theology, or Evidences of the Existence and Attributes of the Deity, Collected from the Appearances of Nature* (Edinburgh).
Parrinder, Patrick (ed.) (1972), *H. G. Wells: The Critical Heritage* (London: Routledge & Kegan Paul).
Patrick, J. Max (1956), 'The portrait of Huxley in Mallock's *New Republic*', *Nineteenth-Century Fiction*, vol. 11 (June), pp. 61–9.
Pearson, Karl (1897), *The Chances of Death and Other Studies in Evolution*, 2 vols (London: Arnold).
Pearson, Karl (1905), *National Life from the Standpoint of Science*, 2nd edn (Cambridge: Cambridge University Press).
Pearson, Karl (1914–30), *The Life, Letters and Labours of Francis Galton*, 4 vols (Cambridge: Cambridge University Press).
Peckham, Morse (1965), *Man's Rage for Chaos: Biology, Behavior and the Arts* (New York: Chilton).
Penrose, Lionel Sharples (1971), 'Ethics and eugenics'; see Fuller, 1971.
Persons, Stow (ed.) (1968), *Evolutionary Thought in America* (New York: Archon Books).
Pfeifer, Edward Justin (1957), 'The reception of Darwinism in the United States, 1859–1880, unpublished PhD thesis, Brown University.
Philmus, Robert Michael (1969) '*The Time Machine*; or, the fourth dimension as prophecy', *PMLA*, vol. 84 (May), pp. 530–5.
Philmus, Robert Michael (1970), *Into the Unknown: the Evolution of Science Fiction from Francis Godwin to H. G. Wells* (Berkeley/Los Angeles, Calif.: University of California Press).
Philmus, Robert Micheal, and Hughes, David Y. (eds) (1975), *H. G. Wells: Early Writings in Science and Science Fiction* (Berkeley, Calif.: University of California Press).
'Physicus', see Romanes, George John.
Pickens, Donald K. (1968), *Eugenics and the Progressives* (Nashville, Tenn.: Vanderbilt University Press).
Pinion, Francis Bertram (1968), *A Hardy Companion: A Guide to the Works of Thomas Hardy and Their Background* (London: Macmillan).
Plato (1955), *The Republic*, trans. H. D. P. Lee (Harmondsworth: Penguin).
Platzner, R. L. (1969), 'Wells's Jungle Book: the influence of Kipling on *The Island of Dr Moreau*', *Victorian Newsletter*, vol. 36, pp. 19–22.
Polak, Frederick Lodewijk (1961), *The Image of the Future: Enlightening the Past,*

close of our period. Of the dozens of eutopias and dystopias written in the century since 1870 very few do not begin with evolutionary presuppositions or reckon with possible developments in the biological sciences. Huxley's own *Brave New World* (1932), for instance, is at once a gross realisation and a shocking condemnation of post-Galtonian eugenics. It makes final mock of his grandfather's tenuous hope for eugenics and of Wells's bumptious optimism in *A Modern Utopia*; and it expands into harsh and gripping satire all of Wells's ambivalent feelings about that hand sticking out of the Selenite breeding-bottle.[3] Not many of the more recent Utopias have been so antagonistic to biology. Even Aldous Huxley's quarrel with it later proved to be with the applications rather than with the science itself; for in his last novel, the geographical eutopia *Island* (1962), he celebrates in detail never before attempted the applied biological sciences – pharmacology, neurophysiology, agronomy – as potent means of individual liberation.

As for the evolutionary prose epic, there were flamboyant calls from the secular wing right from the beginning for a panegyric on the ascent of man. The call came loudest from the proud atheists and positivists eager to begin the writing of the Gospel of Man and eager, too, to have an epic standing in the same relation to the *Origin* as does *Paradise Lost* to the Bible. Winwood Reade, as might be expected, bellowed his affirmation that the biological history of man is 'a splendid narrative, the materials of which it is for science to discover, the glories of which it is for poets to portray'.[4] No one in the nineteenth century rose to Reade's challenge – no one had quite his sanguinary verve – except for the very minor exception of Mathilde Blind, a poet who did in fact compose an *Ascent of Man* in 1888 – an epic by turns lugubrious and exalted, but always feeble – and later persuaded Alfred Russel Wallace to write an introduction to it. Victorian epics, when written at all, were written within the boundaries of limited models like Icelandic saga. The concepts of evolutionary biology were simply too expansive, too disconnected and too transitional to supply a stable metaphysical frame to any epic; nor was there an audience sufficiently homogenous in its beliefs to find the imaginative transformation convincing. In more recent years some epics of this kind have found more of an audience as the neo-Darwinian orthodoxy has grown more familiar and lost some of its initial and more unreasonable terrors. But such epics have been written as prose fiction, and directed towards the future rather than the past. Thomson's and Wells's vision of solar decay has been replaced by more immediate catastrophe; or, as in the novels of Olaf Stapledon (1886–1950), by great cycles of decay and rebirth. Stapledon's *Last and First Men* (1930) purports to tell nothing less than the whole future history of the human species, covering a timespan of a hundred million years. Stapledon took to heart Wells's strictures about static Utopias. On his huge canvas he has room enough for long periods of technological triumph when social mores seem fixed immutably, but the slow-acting pressures of

Orientating the Present, Forecasting the Future, trans. Elise Boulding, 2 vols (Leyden/New York: Oceana).

Pope-Hennessy, Una Constance (1948), *Canon Charles Kingsley: A Biography* (London: Chatto & Windus).

Potter, George Reuben (1937), 'Tennyson and the biological theory of mutability in species', *Philological Quarterly*, vol. 16 (October), pp. 321–43.

Potter, George Reuben (1959), *Macaulay* (London: Longmans, Green).

Poulton, Edward Bagnall (1908), *Essays on Evolution, 1889–1907* (Oxford: Clarendon Press).

Proctor, Richard Anthony (1868), 'Lands and seas of another world', *Fraser's Magazine*, vol. 78 (August), pp. 249–56.

Purdy, Richard Little (1954), *Thomas Hardy: A Bibliographical Study* (London: Oxford University Press).

Radl, Emanuel (1930), *The History of Biological Theories* (London: Oxford University Press). Translated by E. J. Hatfield from the first German edition of 1909.

Raknem, Ingvald (1962), *H. G. Wells and His Critics* (Bergen: Universitetsforlaget).

Reade, William Winwood (1925), *The Martyrdom of Man*, with an introduction by F. Legge (London: Rationalist Press). Reprinted from the original edition of 1872, published by Watts.

Reade, William Winwood (1933), *The Outcast* (London: Watts).

Richards, Ivor Armstrong (1970), *Poetries and Sciences: A Reissue of 'Science and Poetry' (1926, 1935), with Commentary* (New York: Norton).

Roberts, Morley (1924), *W. H. Hudson: A Portrait* (London: Eveleigh Nash & Grayson).

Roberts, Morley (1930), 'W. H. Hudson', *Virginia Quarterley Review*, vol. 6 (October), pp. 507–21.

Robson, Guy Coburn, and Richards, Owain Westmacott (1936), *The Variation of Animals in Nature* (London: Longmans, Green).

Rogers, James A. (1972), 'Darwinism and Social Darwinism', *Journal of the History of Ideas*, vol. 33, pp. 265–80.

Rogers, Robert Emmons (ed.) (1921), *The Voice of Science in Nineteenth-Century Literature: Representative Prose and Verse . . .* (Boston, Mass.: Atlantic Monthly Press).

Romanes, Ethel (1896), *The Life and Letters of George John Romanes, Written and Edited by His Wife*, new edn (London: Longmans).

[Romanes, George John] (1878), *A Candid Examination of Theism, by Physicus* (London).

Romanes, George John (1881), 'Unconscious memory . . .', *Nature*, vol. 23, 27 January, pp. 285–7.

R[omanes], G[eorge] J[ohn] (1889), *Poems, 1879–1889* (London: Harrison).

Romanes, George John (1890), 'Weismann's theory of heredity', *Contemporary Review*, vol. 57 (May), pp. 686–99.

Romanes, George John (1892–7), *Darwin, and after Darwin: An Exposition of the Darwinian Theory and a Discussion of Post-Darwinian Questions*, 3 vols (London: Longmans, Green). The volumes were published separately and successively as: *The Darwinian Theory*, 1892; *Post-Darwinian Questions: Heredity and Utility*, 1895; *Post-Darwinian Questions: Isolation and Physiological Selection*, 1897. The whole treatise is based on two courses of lectures delivered between 1888 and 1890, though much expanded.

Romanes, George John (1893a), *An Examination of Weismannism* (London: Longmans, Green).
Romanes, George John (1893b), 'Mr Herbert Spencer on "natural selection" ', *Contemporary Review*, vol. 63 (April), pp. 499–517.
Romanes, George John (1893c), 'The Spencer–Weismann controversy I (with a note by Herbert Spencer)', *Contemporary Review*, vol. 64 (July), pp. 50–4.
Romanes, George John (1896), *A Selection from the Poems. . .*, with an introduction by T. H. Warren (London: Longman).
Rosenberg, John David (1962), *The Darkening Glass: A Portrait of Ruskin's Genius* (New York: Columbia University Press).
Ruskin, John (1884), *The Storm Cloud of the Nineteeth Century: Two Lectures Delivered at the London Institution, February 4th and 11th, 1884* (Orpington: George Allen).
Ruskin, John (1903–12), *The Works of John Ruskin*, ed. E. T. Cook and Alexander Wedderburn, 39 vols (London: George Allen).
Russell, Bertrand (1955), 'John Stuart Mill', *Proceedings of the British Academy*, vol. 41, pp. 43–59.
Rutland, William R. (1940), 'Tennyson and the theory of evolution', *Essays and Studies by Members of the English Association*, vol. 26, pp. 7–29.
Rutland, William R. (1962), *Thomas Hardy: A Study of His Writings and Their Background* (New York: Russell & Russell).
Salt, Henry Stephens (1892), *Animals' Rights Considered in Relation to Social Progress* (London: Bell).
Salt, Henry Stephens (1905), *Richard Jefferies: His Life and His Ideals* (London: Fifield).
Sandow, Alexander (1938), 'Social factors in the origin of Darwinism', *Quarterly Review of Biology*, vol. 13 (September), pp. 315–26.
Sarton, George (1937), 'Darwin's conception of the theory of natural selection', *Isis*, vol. 26 (March), pp. 336–40.
Schurman, Jacob Gould (1887), *The Ethical Import of Darwinism* (New York: Scribner's).
Schneider, Herbert W. (1945), 'The influence of Darwin and Spencer on American philosophical theology', *Journal of the History of Ideas*, vol. 6 (January), pp. 3–18.
Sears, Paul B. (1950), *Charles Darwin: The Naturalist as Cultural Force* (New York: Scribner's).
Seeber, Hans U. (1971), 'Bulwer-Lytton's underworld', *Moreana*, vol. 30, pp. 39–40.
Seward, Albert Charles (1909), *Darwin and Modern Science: Essays in Commemoration of the Centenary of the Birth of Charles Darwin and of the Fiftieth Anniversary of the Publication of 'The Origin of Species'. . .* (Cambridge: Cambridge University Press).
Shaw, George Bernard (1932), *Prefaces* (London: Constable).
Shaw, George Bernard (1965–72), *The Collected Letters of George Bernard Shaw*, ed. Dan H. Laurence, 2 vols (London: Max Reinhardt).
Shryock, Richard Harrison (1946), 'The strange case of Wells's theory of natural selection (1813)'; see Montagu, 1946.
Simpson, George Gaylord (1959), foreword to Francis Darwin, *The Life and Letters of Charles Darwin . . .*, 2 vols (New York: Basic Books).
Simpson, George Gaylord (1961), 'Lamarck, Darwin and Butler: three approaches to evolution', *American Scholar*, vol. 30 (Spring), pp. 238–49.
Singer, Charles (1962), *A History of Biology to about the Year 1900: A General Introduction to the Study of Living Things*, 3rd edn (London: Abelard-Schuman).

Smiles, Samuel (1859), *Self-Help; with Illustrations of Character and Conduct* (London: Murray).

Smith, G. Elliot (1932), 'Conversion in science', *Huxley Memorial Lectures, 1925–1932* (London: Macmillan), pp. 31–8.

Smith, George Adam (1900), *The Life of Henry Drummond*, 4th edn (London: Hodder & Stoughton).

Smith, John Maynard (1965), 'Eugenics and Utopia', *Daedalus*, vol. 94 (Spring), pp. 487–505.

Southerington, Frank Rodney (1971), *Hardy's Vision of Man* (London: Chatto & Windus).

Southworth, James Granville (1947), *The Poetry of Thomas Hardy* (New York: Columbia University Press).

Spencer, Herbert (1857), 'Progress: its law and cause', *Westminster Review*, n.s., vol. 11 (April), pp. 445–85. Nominally a review of A. von Humbolt, *Cosmos*; C. Lyell, *Principles of Geology*, 9th edn; W. B. Carpenter, *Principles of Comparative Physiology*, 4th edn.

Spencer, Herbert (1864–7), *The Principles of Biology*, 2 vols (London: Williams & Norgate).

Spencer, Herbert (1886), 'The factors of organic evolution', *Nineteenth Century*, vol. 19 (April), pp. 570–89, and (May), pp. 749–70.

Spencer, Herbert (1890), 'The inheritance of acquired characters', *Nature*, vol. 41, 6 March, pp. 414–15.

Spencer, Herbert (1893a), *Education: Intellectual, Moral, and Physical*, cheap edn (London: Williams & Norgate).

Spencer, Herbert (1893b), 'The inadequacy of "natural selection" ', *Contemporary Review*, vol. 63 (February), pp. 153–66, and (March), pp. 439–56.

Spencer, Herbert (1893c), 'Professor Weismann's theories', *Contemporary Review*, vol. 63 (May), pp. 743–60.

Spencer, Herbert (1893d), 'A rejoinder to Professor Weismann', *Contemporary Review*, vol. 64 (December), pp. 893–913.

Spencer, Herbert (1894), 'Weismannism once more', *Contemporary Review*, vol. 66 (October), pp. 592–608.

Spencer, Herbert (1898–9), *The Principles of Biology*, revised and enlarged edn, 2 vols (London: Williams & Norgate).

Spencer, Herbert (1899), 'The filiation of ideas; see Duncan, 1908.

Stanley, Oma (1957), 'T. H. Huxley's treatment of "nature" ', *Journal of the History of Ideas*, vol. 18 (January), pp. 120–7.

Stapledon, William Olaf (1930), *Last and First Men: A Story of the Near and Far Future* (London: Methuen).

Starkey, Margaret M. (1952), 'The history of ideas and literary studies', *Modern Language Quarterly*, vol. 13 (September), pp. 264–7.

Stasny, John F. (1961), 'W. Winwood Reade's *The Martyrdom of Man*: a Darwinian history', *West Virginia University Bulletin: Philological Papers*, vol. 13 (December), pp. 37–49.

Steck, James Sperow (1968), 'The fiction of W. H. Hudson', unpublished PhD thesis, Pennsylvania State University.

Steele, Edward John (1979), *Somatic Selection and Adaptive Evolution: On the Inheritance of Acquired Characters* (Toronto: Williams & Wallace).

Steiner, George (1971), *In Bluebeard's Castle* (London: Faber).

S[tephen], L[eslie] (1872), 'Darwinism and divinity', *Fraser's Magazine*, vol. 5 (April), pp. 409–21.

Stephen, Leslie (1882), *The Science of Ethics* (London: Smith, Elder).

Stevenson, Robert Louis (1895), *New Arabian Nights*, new edn (London: Chatto & Windus).

Stevenson, Robert Louis (1910), *Talk and Talkers* (London: Chatto & Windus).

Stevenson, Robert Louis (1912), *Memoir of Fleeming Jenkin* (London: Longmans). See also Jenkin, 1887.

Stillman, Clara Gruening (1932), *Samuel Butler: A Mid-Victorian Modern* (London: Secker).

Street, Brian Vincent (1975), *The Savage in Literature: Representations of 'Primitive' Society in English Fiction, 1858–1920* (London: Routledge & Kegan Paul).

Sullivan, Thomas Richard (1968), 'A "way to a better": Hardy's two views of evolution', unpublished PhD thesis, University of Iowa.

Sully, James (1881), 'Scientific optimism', *Nineteenth Century*, vol. 10 (October), pp. 573–87.

Sussman, Herbert L. (1968), *The Victorians and the Machine: The Literary Response to Technology* (Cambridge, Mass.: Harvard University Press).

Suvin, Darko (1973), '*The Time Machine* versus *Utopia* as a structural model for science fiction', *Comparative Literature Studies*, vol. 10 (December), pp. 334–49.

Symonds, John Addington (1890), *Essays, Speculative and Suggestive*, 2 vols (London: Chapman & Hall).

Tax, Sol (ed.) (1960), *Evolution after Darwin: The University of Chicago Centennial*, 3 vols (Chicago, Ill.: University of Chicago Press).

Taylor, Gordon Rattray (1958), *The Angel-Makers: A Study in the Psychological Origins of Historical Change, 1750–1850* (London: Heinemann).

Teggart, Frederick John (1929), *The Idea of Progress: A Collection of Readings . . .* (Berkeley, Calif.: University of California Press).

Tennyson, Alfred, Lord (1969), *The Works of Alfred, Lord Tennyson*, ed. Christopher Ricks (London: Longmans).

Thomson, William (1862), 'On the age of the sun's heat', *Macmillan's Magazine*, vol. 5 (March), pp. 388–93.

Thomson, William (1891–4), *Popular Lectures and Addresses*, 2nd edn, 3 vols (London: Macmillan).

Thomson, William (1899), 'The age of the earth as an abode fitted for life', *Science*, vol. 9, 12 May, pp. 665–74.

Tietze, Frederick Inglebrit (1952), 'Tennyson, science and the poetic sensibility', unpublished PhD thesis, University of Wisconsin.

Tomalin, Ruth (1954), *W. H. Hudson* (London: Witherby).

Tupper, Martin Farquahar (1876), *Proverbial Philosophy: A Book of Thoughts and Arguments, Originally Treated, in Four Series Complete* (London).

Turner, Frank Miller (1974), *Between Science and Religion: The Reaction to Scientific Naturalism in Late Victorian England* (New Haven, Conn.: Yale University Press).

Tyndall, John (1879), *Fragments of Science for Unscientific People: A Series of Detached Essays, Lectures and Reviews*, 6th edn, 2 vols (London: Longmans).

Vorzimmer, Peter J. (1969), 'Darwin, Malthus, and the theory of natural selection', *Journal of the History of Ideas*, vol. 30 (December), pp. 527–42.

Vorzimmer, Peter J. (1970), *Charles Darwin: The Years of Controversy. 'The Origin of Species' and Its Critics, 1859–1882* (Philadelphia, Pa: Temple University Press).

Waddington, Conrad Hal (1961), *The Nature of Life* (London: Allen and Unwin).

Wagar, W. Warren (1971), *H. G. Wells and the World State* (New Haven, Conn.: Yale University Press).

Wagar, W. Warren (1972), *Good Tidings: The Belief in Progress from Darwin to Marcuse* (Bloomington, Ind.: Indiana University Press).

Wagner, Geoffrey (1965), 'A forgotten satire: Bulwer-Lytton's *The Coming Race*', *Nineteenth-Century Fiction*, vol. 19 (March), pp. 379–85.

Wallace, Alfred Russel (1864), 'The origin of the human races and the antiquity of Man deduced from the theory of natural selection', *Journal of the Anthropological Society of London*, vol. 2, pp. clviii-clxx.

Wallace, Alfred Russel (1867), 'Creation by law', *Quarterly Journal of Science*, vol. 4 (October), pp. 471–88.

[Wallace, Alfred Russel] (1869), 'Sir Charles Lyell on geological climates and the origin of species', *Quarterly Review*, vol. 126, (April), pp. 359–94.

Wallace, Alfred Russel (1870), *Contributions to the Theory of Natural Selection: A Series of Essays* (London: Macmillan).

Wallace, Alfred Russel (1879), 'Organisation and intelligence', *Nature*, vol. 19, 27 March, pp. 477–80.

Wallace, Alfred Russel (1889), *Darwinism: An Exposition of the Theory of Natural Selection, with Some of Its Applications* (London: Macmillan).

Wallace, Alfred Russel (1890), 'Human selection', *Fortnightly Review*, n.s., vol. 48 (September), pp. 325–37.

Wallace, Alfred Russel (1892), 'Human progress: past and future', *Arena*, vol. 5 (January), pp. 143–59.

Wallace, Alfred Russel (1895), *Natural Selection and Tropical Nature: Essays on Descriptive and Theoretical Biology*, new edn (London: Macmillan).

Wallace, Alfred Russel (1905), *My Life: A Record of Events and Opinions*, 2 vols (London: Chapman & Hall).

Wallace, Alfred Russel (1907), *Is Mars Habitable? A Critical Examination . . .* (London: Macmillan).

Wallace, Alfred Russel (1910), *The World of Life: A Manifestation of Creative Power, Directive Mind and Ultimate Purpose* (London: Bell).

Walsh, Chad (1962), *From Utopia to Nightmare* (London: Bles).

Watt, Ian (ed.) (1971), *The Victorian Novel: Modern Essays in Criticism* (London: Oxford University Press).

Webb, Beatrice (1926), *My Apprenticeship* (London: Longmans, Green).

Weber, Carl Jefferson (1965), *Hardy of Wessex: His Life and Literary Career*, revised edn (New York: Columbia University Press).

Webster, Harvey Curtis (1964), *On a Darkling Plain: The Art and Thought of Thomas Hardy* (London: Cass).

Weismann, Friedrich Leopold August (1885), *Die Continuität des Keimplasma's als Grundlage einer Theorie der Vererbung* (Jena: Naumburg a/s).

Weismann, Friedrich Leopold August (1891–2), *Essays upon Heredity and Kindred Biological Problems*, ed. Edward B. Poulton, Selmar Schönland and Arthur E. Shipley, 2nd edn, 2 vols (Oxford: Clarenden Press). The contents of the first edition (1889) are reprinted in this edition as the first volume. Volume 2 contains four extra essays.

Weismann, Friedrich Leopold August (1893), 'The all-sufficiency of natural selection. I. A reply to Herbert Spencer' and 'II. Conclusion', *Contemporary Review*, vol. 64 (September), pp. 309–38, and (October), pp. 596–610.

Weismann, Friedrich Leopold August (1904), *The Evolution Theory . . .*, trans. J.

Arthur Thomson and Margaret R. Thomson, 2 vols (London: Arnold). First German edition 1902.

Wellek, René, and Warren, Austin (1963), *Theory of Literature*, 3rd edn (Harmondsworth: Penguin).

Wells, Geoffrey Harry, see 'West, Geoffrey'.

Wells, Herbert George (1891), 'Zoological retrogression', *Gentleman's Magazine*, vol. 271 (September), pp. 246–53.

Wells, Herbert George (1893a), 'The man of the year million: a scientific forecast', *Pall Mall Budget*, November, pp. 1796–7.

Wells, Herbert George (1893b), *Text-Book of Biology*, with an introduction by G. B. Howes, 2 vols (London: Clive).

Wells, Herbert George (1894), 'The extinction of man: some speculative suggestions', *Pall Mall Gazette*, 25 September, p. 3.

Wells, Herbert George (1896a), 'Intelligence on Mars', *Saturday Review*, vol. 81, 4 April, pp. 345–6.

Wells, Herbert George (1896b), 'Human evolution, an artificial process', *Fortnightly Review*, n.s. vol. 60 (October), 590–5.

Wells, Herbert George (1896c), *The Island of Dr Moreau* (London: Heinemann).

Wells, Herbert George (1898), *Certain Personal Matters: A Collection of Material, Mainly Autobiographical* (London: Lawrence & Bullen).

Wells, Herbert George (1902), 'The discovery of the future: a discourse delivered at the Royal Institution on Friday, January 24 by Mr H. G. Wells', *Nature*, vol. 65, 6 February, pp. 326–31.

Wells, Herbert George (1905), *A Modern Utopia* (London: Chapman & Hall).

Wells, Herbert George (1914), *Anticipations of the Reaction of Mechanical and Scientific Progress upon Human Life and Thought*, new edn (London: Chapman & Hall).

Wells, Herbert George (1934), *Experiment in Autobiography: Discoveries and Conclusions of a Very Ordinary Brain – since 1866*, 2 vols (London: Gollancz).

Wells, Herbert George (1948), *The Short Stories of H. G. Wells*, (London: Benn).

Wells, Herbert George (1951), *The War of the Worlds* (London: Heinemann).

Wells, Herbert George (1969), *The Time Machine: An Invention* (London: Heinemann).

Wells, Herbert George (1973), *The First Men in the Moon* (London: Heinemann).

Wells, Herbert George, and Gregory, Richard Arman (1893), *Honours Physiography* (London: Hughes).

Wells, Herbert George, Huxley, Julian, and Wells, G. P. (1931), *The Science of Life* (London: Cassell).

Wells, Kentwood D. (1971), 'Sir William Lawrence (1783–1867): a study of pre-Darwinian ideas on heredity and variation', *Journal of the History of Ideas*, vol. 6 (Fall), pp. 361–91.

West, Anthony (1957), 'H. G. Wells', *Encounter*, vol. 8 (February), pp. 52–9.

'West, Geoffrey' [Geoffrey Harry Wells] (1930), *H. G. Wells: A Sketch for a Portrait* (London: Howe).

'West, Geoffrey' [Geoffrey Harry Wells] (1937), *Charles Darwin: The Fragmentary Man* (London: Routledge).

Whitehead, Alfred North (1925), *Science and the Modern World . . . Lowell Lectures, 1925* (Cambridge: Cambridge University Press).

[Wilberforce, Samuel] (1860), 'On the origin of species . . .', *Quarterly Review*, vol. 108 (July), pp. 225–64.

Wilde, Oscar (1948), *The Works of Oscar Wilde*, ed. G. F. Maine (London: Collins).
Willey, Basil (1960), *Darwin and Butler: Two Versions of Evolution* (London: Chatto & Windus).
Willey, Basil (1961), 'Darwin's place in the history of thought'; see Barton, 1961.
Willey Basil (1966), *More Nineteenth-Century Studies: A Group of Honest Doubters* (New York: Harper).
Williams, George C. (1966), *Adaptation and Natural Selection: A Critique of Some Current Evolutionary Thought* (Princeton, NJ: Princeton University Press).
Williams, Raymond (1972), 'Social Darwinism', *The Listener*, 23 November, pp. 696–700.
Williams, Raymond (1973), *The Country and the City* (London: Chatto & Windus).
Williams-Ellis, Amabel (1966), *Darwin's Moon: A Biography of Alfred Russel Wallace* (London: Blackie).
Williamson, John Stewart (1964), 'H. G. Wells, critic of progress: a study of the early fiction', unpublished PhD thesis, University of Colorado.
Wilson, Andrew (1881), 'Degeneration', *Gentleman's Magazine*, vol. 250 (April), pp. 470–94.
Wiltshire, Bruce (ed.) (1968), *Romanticism and Evolution: The Nineteenth Century* (New York: Putnam).
Wingfield-Stratford, Esmé (1932), *The Victorian Sunset* (London: Routledge).
Winsor, William Tillinghast (1963), 'W. H. Hudson, naturalist and master of prose: a study of the man, his work, and his interpretation of nature', unpublished PhD thesis, Columbia University.
Winsten, Stephen (1951), *Salt and His Circle* (London: Hutchinson).
Woods, Thomas (1961), *Poetry and Philosophy: A Study in the Thought of John Stuart Mill* (London: Hutchinson).
Wordsworth, William (1950), *The Poetical Works of William Wordsworth*, ed. Thomas Hutchinson, new edn revised by Ernest de Selincourt (London: Oxford University Press).
Wordsworth, William (1960), *The Prelude, or The Growth of a Poet's Mind (Text of 1805)*, ed. Ernest de Selincourt, revised impression by Helen Darbishire (London: Oxford University Press).
Worth, G. J. (1961), ' "Popular culture" and the seminal books of 1859', *Victorian Newsletter*, vol. 19, pp. 24–7.
Wreden, William P. (1938), *A Selection of Books from the Library of Thomas Hardy, OM*, Catalogue No. 11 (Burlinghame, Calif.).
Young, George Malcolm (1960), *Victorian England: Portrait of an Age* (London: Oxford University Press).
Young, Robert M. (1974), 'The role of psychology in the nineteenth-century evolutionary debate', unpublished manuscript.
Zabel, Morton Dauwen ([1957]), 'Samuel Butler: the Victorian insolvency'; see Watt, 1971.
'Zeta', see Froude, James Anthony.
Zirkle, Conway (1941), 'Natural selection before the *Origin*', *Proceedings of the American Philosophical Society*, vol. 84 (April), pp. 71–123.
Zirkle, Conway (1946), 'The early history of the idea of the inheritance of acquired characters and of pangenesis', *Transactions of the American Philosophical Society*, vol. 35 (March), pp. 91–151.

INDEX

For Product Safety Concerns and Information please contact our EU representative GPSR@taylorandfrancis.com
Taylor & Francis Verlag GmbH, Kaufingerstraße 24, 80331 München, Germany

www.ingramcontent.com/pod-product-compliance
Lightning Source LLC
LaVergne TN
LVHW020626300926
849161LV00004B/622

* 9 7 8 1 1 3 8 7 9 9 2 4 0 *